Springer Theses

Recognizing Outstanding Ph.D. Research

Aims and Scope

The series “Springer Theses” brings together a selection of the very best Ph.D. theses from around the world and across the physical sciences. Nominated and endorsed by two recognized specialists, each published volume has been selected for its scientific excellence and the high impact of its contents for the pertinent field of research. For greater accessibility to non-specialists, the published versions include an extended introduction, as well as a foreword by the student's supervisor explaining the special relevance of the work for the field. As a whole, the series will provide a valuable resource both for newcomers to the research fields described, and for other scientists seeking detailed background information on special questions. Finally, it provides an accredited documentation of the valuable contributions made by today’s younger generation of scientists.

Theses are accepted into the series by invited nomination only and must fulfill all of the following criteria

- They must be written in good English.
- The topic should fall within the confines of Chemistry, Physics, Earth Sciences, Engineering and related interdisciplinary fields such as Materials, Nanoscience, Chemical Engineering, Complex Systems and Biophysics.
- The work reported in the thesis must represent a significant scientific advance.
- If the thesis includes previously published material, permission to reproduce this must be gained from the respective copyright holder.
- They must have been examined and passed during the 12 months prior to nomination.
- Each thesis should include a foreword by the supervisor outlining the significance of its content.
- The theses should have a clearly defined structure including an introduction accessible to scientists not expert in that particular field.

More information about this series at http://www.springer.com/series/8790

Alaa Abdulhady Jaber

Design of an Intelligent Embedded System for Condition Monitoring of an Industrial Robot

Doctoral Thesis accepted by
Newcastle University, UK

Springer

Author
Dr. Alaa Abdulhady Jaber
School of Mechanical and Systems Engineering
Newcastle University
Newcastle upon Tyne
UK

and

Mechanical Engineering Department
University of Technology
Baghdad
Iraq

Supervisor
Prof. Robert Bicker
School of Mechanical and Systems Engineering
Newcastle University
Newcastle upon Tyne
UK

ISSN 2190-5053 ISSN 2190-5061 (electronic)
Springer Theses
ISBN 978-3-319-44931-9 ISBN 978-3-319-44932-6 (eBook)
DOI 10.1007/978-3-319-44932-6

Library of Congress Control Number: 2016949598

Printed on acid-free paper

This Springer imprint is published by Springer Nature
The registered company is Springer International Publishing AG
The registered company address is: Gewerbestrasse 11, 6330 Cham, Switzerland

This thesis is dedicated to my beloved father, the martyr Abdulhady Jaber, who was given the honour of martyrdom on 29 December 1986, and also to all the martyrs of Iraq, who sacrificed themselves in order to keep Iraq and its people secure and safe. All the love, appreciation and respect are due to them.

Supervisor's Foreword

The University of Newcastle upon Tyne has enthusiastically nominated this Ph.D. thesis for a 2016 Springer Theses series because it was considered to be an excellent example of independent and well-conducted research. Dr. Alaa Abdulhady Jaber joined Newcastle University as a postgraduate student in January 2012 to study for Ph.D. and quickly defined the area of his research topic in the initial months, and having undertaken a critical review of the research literature in the area of condition monitoring, he established that very little published research was available on its application to industrial robots. Having revealed a significant research opportunity, Dr. Jaber then committed himself wholeheartedly to his studies, tackling the challenging objectives in a very professional and systematic way, ensuring the work was accomplished in a timely manner.

Dr. Jaber has successfully designed and commissioned an intelligent health monitoring system, specifically for use on an industrial robot, which is able to predict the onset of mechanical wear and failure in the joints of the geared transmissions. The design of the two-level embedded wireless condition monitoring system utilizes a 3-axis accelerometer to capture the operational arm vibration in real time which it then analyses against healthy signature data to establish the emergence of a potential fault. The developed system is capable of monitoring a number of robots simultaneously and also lends itself very well for application on any power transmission equipment in which the loads and speeds are not constant, particularly where access is restricted. As such, this provides significant scope for future research.

The work presented in this thesis represents three significant achievements. First is the development of a condition monitoring algorithm based on vibration analysis of an industrial robot for fault detection and diagnosis. The combined use of a statistical control chart with time-domain signal analysis for detecting a mechanical fault via an arm-mounted wireless processor system represents the first stage of fault detection. Second, the design and development of a sophisticated embedded microprocessor base station (the second stage) for the online processing of the intelligent condition monitoring algorithm, and third, the implementation of a discrete wavelet transform, using an artificial neural network, with statistical feature extraction for the robot fault diagnosis in which the vibration signals are decomposed into eight levels of wavelet coefficients. The strategy for detecting and analysing potential faults in the robot used in the case study was considered both practical and inexpensive and as such provides a significant contribution towards solving complex mechanical faults in a wide range of power transmission systems and opens up the opportunity for taking this research forward to address real-world applications.

At this time, Dr. Jaber has published the results of his work at six International Conferences and already has six Journal publications to his name, with a further four submitted and under review, and three more in preparation. He has also received a number of awards and certificates of merit from the Iraqi Ministry of Higher Education for his achievements, in testimony to his efforts to successfully promote his work.

I supervised over 25 Ph.D. students over the years at Newcastle and am proud to say that I consider Dr. Jaber as being one of the most dedicated and committed students which it was my pleasure to supervise. His continued enthusiasm and hard work enabled him to overcome the many technical challenges posed, and in summary, I feel this is a sufficient proof that this work is worthy of a nomination for the Springer Theses and take this opportunity to wish him every success for the future.

Newcastle upon Tyne, UK
May 2016

Prof. Robert Bicker

Abstract

Industrial robots have long been used in production systems in order to improve productivity, quality and safety in automated manufacturing processes. There are significant implications for operator safety in the event of a robot malfunction or failure, and an unforeseen robot stoppage, due to different reasons, has the potential to cause an interruption in the entire production line, resulting in economic and production losses. Condition monitoring (CM) is a type of maintenance inspection technique by which an operational asset is monitored and the data obtained are analysed to detect signs of degradation, diagnose the causes of faults and thus reduce maintenance costs. So, the main focus of this research is to design and develop an online, intelligent CM system based on wireless embedded technology to detect and diagnose the most common faults in the transmission systems (gears and bearings) of the industrial robot joints using vibration signal analysis.

To this end an old, but operational, PUMA 560 robot was utilized to synthesize a number of different transmission faults in one of the joints (3—elbow), such as backlash between the gear pair, gear tooth and bearing faults. A two-stage condition monitoring algorithm is proposed for robot health assessment, incorporating fault detection and fault diagnosis. Signal processing techniques play a significant role in building any condition monitoring system, in order to determine fault–symptom relationships, and detect abnormalities in robot health. Fault detection stage is based on time-domain signal analysis and a statistical control chart (SCC) technique. For accurate fault diagnosis in the second stage, a novel implementation of a time-frequency signal analysis technique based on the discrete wavelet transform (DWT) is adopted. In this technique, vibration signals are decomposed into eight levels of wavelet coefficients and statistical features, such as standard deviation, kurtosis and skewness, are obtained at each level and analysed to extract the most salient feature related to faults; the artificial neural network (ANN) is then used for fault classification. A data acquisition system based on National Instruments (NI) software and hardware was initially developed for preliminary robot vibration analysis and feature extraction. The transmission faults induced in the robot can change the captured vibration spectra, and the robot's natural frequencies were established using experimental modal analysis, and also the fundamental fault frequencies for the gear transmission and bearings were obtained and utilized for preliminary robot condition monitoring.

In addition to simulation of different levels of backlash fault, gear tooth and bearing faults which have not been previously investigated in industrial robots, with several levels of severity, were successfully simulated and detected in the robot's joint transmission. The vibration features extracted, which are related to the robot healthy state and different fault types, using the data acquisition system were subsequently used in building the SCC and ANN, which were trained using part of the measured data set that represents the robot's operating range. Another set of data, not used within the training stage, was then utilized for validation. The results indicate the successful detection and diagnosis of faults using the key extracted parameters. A wireless embedded system based on the ZigBee communication protocol was designed for the application of the proposed CM algorithm in real time, using an Arduino DUE as the core of the wireless sensor unit attached on the robot arm. A Texas Instruments digital signal processor (TMS320C6713 DSK board) was used as the base station of the wireless system on which the robot's fault diagnosis algorithm is run. To implement the two stages of the proposed CM algorithm on the designed embedded system, software based on the C programming language has been developed. To demonstrate the reliability of the designed wireless CM system, experimental validations were performed, and high reliability was shown in the detection and diagnosis of several seeded faults in the robot.

Optimistically, the established wireless embedded system could be envisaged for fault detection and diagnostics on any type of rotating machine, with the monitoring system realized using vibration signal analysis. Furthermore, with some modifications to the system's hardware and software, different CM techniques, such as acoustic emission (AE) analysis or motor current signature analysis (MCSA), can be applied.

Keywords Condition monitoring · Industrial robot fault detection and diagnosis · Experimental modal analysis · Embedded system · Wireless · Vibration signal analysis · Wavelet transform · Statistical control chart · Artificial neural network

Publications and Awards

Journal Papers

1. Alaa Abdulhady Jaber and Robert Bicker, '**Fault Diagnosis of Industrial Robot Gears Based on Discrete Wavelet Transform and Artificial Neural network**', Insight—Non-Destructive Testing and Condition Monitoring, Vol. 58, No. 4, 2016.
2. Alaa Abdulhady Jaber and Robert Bicker, '**Design of a Wireless Sensor Node for Vibration Monitoring of Industrial Machinery**', International Journal of Electrical and Computer Engineering-IJECE, Vol. 6, No. 1, 2016.
3. Alaa Abdulhady Jaber and Robert Bicker, '**Industrial Robot Fault Detection Based on Statistical Control Chart**', American Journal of Engineering and Applied Sciences-AJEAS, Vol. 9, Issue. 1, 2016.
4. Alaa Abdulhady Jaber and Robert Bicker, '**Industrial Robot Backlash Fault Diagnosis Based on Discrete Wavelet Transform and Artificial Neural Network**', American Journal of Mechanical Engineering-AJME, Vol. 4, No. 1, pp. 21–31, 2016.
5. Alaa Abdulhady Jaber and Robert Bicker, '**Real-Time Wavelet Analysis of a Vibration Signal Based on Arduino-UNO and LabVIEW**', International Journal of Material Science and Engineering-IJMSE, Vol. 3, No. 1, pp. 66–20, 2015.
6. Alaa Abdulhady Jaber and Robert Bicker, '**The State of the Art in Research into the Condition Monitoring of Industrial Machinery**', International Journal of Current Engineering and Technology-IJCET, Vol. 4, No. 3, pp. 1986–2001, 2014.
7. Alaa Abdulhady Jaber and Robert Bicker, '**Fault Diagnosis of Industrial Robot Bearings Using Combined DWT and FFT**', submitted to International Journal of Condition Monitoring-IJCM.
8. Alaa Abdulhady Jaber and Robert Bicker, '**Fault Diagnosis of Industrial Robot Bearings Based on Discrete Wavelet Transform and Artificial Neural network**', submitted to International Journal of Prognostics and Health-IJPHM.
9. Alaa Abdulhady Jaber and Robert Bicker, '**Wireless Fault Detection System for an Industrial Robot Based on Statistical Control Chart**', submitted to Journal of Failure Analysis and Prevention.

In the following are papers planned to be submitted to international journals:

10. Alaa Abdulhady Jaber and Robert Bicker, '**Design of an Intelligent System for an Industrial Robot Condition Monitoring**', to be submitted to Journal of Intelligent and Robotic Systems.

11. Alaa Abdulhady Jaber and Robert Bicker, '**Design Methodology of an Intelligent Embedded System for an Industrial Robot Condition Monitoring**', to be submitted to Sensors Journal, IEEE.
12. Alaa Abdulhady Jaber and Robert Bicker, '**Design and Implementation of an Intelligent Embedded System for an Industrial Robot Condition Monitoring**', to be submitted to Sensors Journal, IEEE.

Conference Papers

13. Alaa Abdulhady Jaber and Robert Bicker, '**Wavelet–Fourier Transforms for an Industrial Robot Fault Detection**', presented at the 12th International Conference on Condition Monitoring and Machinery Failure Prevention Technologies, **Oxford/UK**, 2015.
14. Alaa Abdulhady Jaber and Robert Bicker, '**The Optimum Selection of Wavelet Transform Parameters for the Purpose of Fault Detection in an Industrial Robot**', presented at the 4th IEEE International Conference on Control System, Computing and Engineering, **Penang/Malaysia**, 2014.
15. Alaa Abdulhady Jaber and Robert Bicker, '**A Simulation of Non-stationary Signal Analysis Using Wavelet Transform Based on LabVIEW and Matlab**', presented at the 8th European Modelling Symposium. **Pisa/Italy**, 2014.
16. Alaa Abdulhady Jaber and Robert Bicker, '**Industrial Robot Fault Detection Based on Wavelet Transform and LabVIEW**', presented at the first IEEE International Conference on Systems Informatics, Modelling and Simulation, **Sheffield/UK**, 2014.
17. Alaa Abdulhady Jaber and Robert Bicker, '**A Systematic Strategy to Find the Natural Frequencies of an Industrial Robot**', presented at the International Conference on Advances in Mechanical and Robotics Engineering, **Zurich/Switzerland**, 2013.

Posters

18. Alaa Abdulhady Jaber and Robert Bicker, '**Design of an Intelligent Embedded System for Condition Monitoring of an Industrial Robot**', presented at Post Graduate Conference 2013, **Newcastle University/UK**, 2014.

Awards

- Award and certificate of merit from **the Iraqi Minister of Higher Education and Scientific Research, Dr Hussein Ibrahim Al-Shahristani**, for showing very good performance during my PhD study, London/UK, 2015.
- Award and certificate of achievement from the **School of Mechanical and Systems Engineering/ Newcastle University** for getting the second place

poster presentation in the postgraduate conference, Newcastle upon Tyne/UK, 2014.

- Award and certificate of merit from the **Iraqi Cultural Attaché in London** for publishing papers in different international journals and conferences, Newcastle upon Tyne/UK, 2014.

Acknowledgements

First and foremost, all praise and deep thanks are due to Allah (the Creator), who helped and guided me through the challenges of my study. Glory is to Allah who has given me the strength, patience and knowledge to continue and finish my Ph.D. journey. It would not have been possible to write this thesis without the help and support of many kind people around me. Due to space constraints, it is impossible to acknowledge all of them here; a few, however, would be mentioned.

First, I want to express my profound gratitude to my principal supervisor, Dr. R. Bicker, who introduced me to different fascinating scientific fields such as mechatronics, robotics and artificial intelligence, and for valuable guidance, support, encouragement and advice throughout my research journey. Also, I would like to thank my second supervisor, Dr. J. Hedley, for his help and support. Besides my supervisors, I would like to thank all the technicians and staff at the School of Mechanical and Systems Engineering at Newcastle University, in particular Mr. Ken Madden, Mr. Michael Foster, Mr. Stephen Charlton and Mr. Brian Stoker.

It would not have been possible to conduct this research project without the kind sponsorship of the Iraqi government, represented by the Ministry of Higher Education and Scientific Research, the Iraqi Cultural Attaché in London, and the University of Technology in Baghdad, and so I am grateful to extend sincere thanks and appreciation to them.

In a special way, I acknowledge the inspirational and moral support to continue studying and face the challenges received from my martyr father (1960–1986), Abdulhady Jaber. May God bless his soul. I owe the deepest gratitude to my mother and grandmother who have overwhelmed me with their prayers and best wishes throughout my years of study and in life generally. Special and heartfelt thanks go to my wife and children for their understanding, persistent endurance and moral encouragement and support throughout my four years of studies. I offer my appreciation to my brother, sister, uncles and aunts for their continuous prayers, support and encouragement.

Sincere thanks must also go to all of my friends and well-wishers for their prayer and support ever since I embarked upon the Ph.D. journey. Lastly, I would like to thank my colleagues, the Ph.D. students in the MEMS and Sensors Research Group for their valued companionship during my time at Newcastle University. I wish them all everything I would wish for myself.

Contents

Abbreviations

ADC	Analog-to-digital converter
AE	Acoustic emission
AI	Artificial intelligence
ANN	Artificial neural network
ASP	Analog signal processing
BPFI	Ball pass frequency of inner race
BPFO	Ball pass frequency of outer race
BSF	Ball spin frequency
CCS	Code composer studio
CF	Crest factor
CISC	Complex instruction set computer
CL	Centre line
CM	Condition monitoring
CPLD	Complex programmable logic device
CPU	Central processing unit
CUSUM	Cumulative sum control chart
CWT	Continuous wavelet transform
DAD	Data acquisition device
db	Daubechies wavelet family
DFT	Discrete Fourier transform
DIP	Dual in-line package switches
DOF	Degree of freedom
DSK	Development starter kit
DSP	Digital signal processor
DWT	Discrete wavelet transform
EDM	Electrical discharge machining
EDMA	Enhanced direct memory access controller
EMA	Experimental modal analysis
EMD	Empirical mode decomposition
EWMA	Exponentially weighted moving average chart

FDI	Fault detection and isolation
FFT	Fast Fourier transform
FLS	Fuzzy logic system
FPGA	Field programmable gate array
FRF	Frequency response function
FT	Fourier transform
FTC	Fault tolerant control
FTF	Fundamental train (or cage) frequency
GA	Genetic algorithm
GF	Genetic fuzzy system
HP	High-pass filter
HPI	Host port interface
iCBM	Intelligent condition-based monitoring
ICP	Integrated circuit piezoelectric
IDE	Integrated development environment
ISM	Industrial, scientific and medical radio bands
Ku	Kurtosis
LCL	Lower control limit
LED	Light-emitting diode
LP	Low-pass filter
LWL	Lower warning limit
MCSA	Motor current signature analysis
MCU	Micro-controller unit
MEMS	Micro-electro-mechanical sensors
MFLOPS	Millions of floating-point operations per second
MIPS	Millions of instructions per second
MLPN	Multilayer perceptron neural network
MSE	Mean square error
NI	National Instrument company
PAN	Personal area network
PCA	Principle component analysis
PDF	Probability density function
PK	Peak value
PUMA	Programmable universal machine for assembly
RAM	Random access memory
RCFA	Root cause failure analysis
RIA	Robot Institute of America
RISC	Reduced Instruction Set Computer
RMS	Root mean square
RNN	Recurrent neural network
RSSI	Received signal strength indicator
RTDX	Real-time data exchange
SCC	Statistical control chart
Sk	Skewness
SSE	Summed squared error

STD	Standard deviation
STFT	Short-time Fourier transform
SVM	Support vector machine
sym	Symlet wavelet families
TE	Transmission error
TI	Texas Instrument company
UART	Universal asynchronous receiver/transmitter
UCL	Upper control limit
UWL	Upper warning limit
VLIW	Very long instruction word
Wi-Fi	Wireless fidelity
WLAN	Wireless local area network
WSN	Wireless sensor network
WT	Wavelet transform
X-CTU	XBee configuration and testing utility

List of Figures

List of Tables

Chapter 1
Introduction

Abstract This chapter provides the background to the research presented in this thesis and outlines the motivation for and the rationale behind the study. The main purposes of the different maintenance strategies available are described and condition monitoring technique is discussed in detail and its main steps are explained. Finally, the research aims and objectives are stated, and the structure of the thesis is briefly described.

1.1 Background to Industrial Robotics

The robot institute of America (RIA) has defined the industrial robot as a reprogrammable multifunctional manipulator designed to move material, parts, tools, or specialized devices through variable programmed motions for the performance of a variety of tasks (Spong et al. 2005). The mechanical structure of a standard industrial robot is composed of number of links and joints. Links are the main bodies that make the mechanism and are connected to each other by joints. The kinematic chain of the robot is defined by the configuration of links and joints. The number of joints identifies the number of manipulated degrees of freedom (DOF) of a robot. The most common configuration of industrial robots is the six DOF with serial kinematics and revolute joints, denoting that links are connected in series through joints allowing for rotational movements.

Industrial robots are now commonly used in production systems in order to improve productivity, quality and safety in automated manufacturing processes. Many functions can be carried out by industrial robots and they represent the basic building blocks of the production sector. Recent developments involve using robots cooperatively with production line operatives, and they are now routinely used in healthcare, nuclear plants, and other hazardous environments. Regardless of application, there are significant implications for operator safety in the event of a robot malfunction or failure, and the consequent downtime has a significant impact

A.A. Jaber, *Design of an Intelligent Embedded System for Condition Monitoring of an Industrial Robot*, Springer Theses,
DOI 10.1007/978-3-319-44932-6_1

Robots doing a pick and place job | Robots in the car manufacturing factory

Fig. 1.1 Examples of industrial robot applications (http://www.gizmag.com; http://singularityhub.com)

on productivity in manufacturing. As demonstrated by the applications shown in Fig. 1.1, an unforeseen robot stoppage due to different reasons has the potential to cause an interruption in the entire production line, resulting in economic and production losses. Availability and maintainability, which can be defined as the probability of a system operating satisfactorily in any time period and its capability of being repaired, are therefore critical for industrial robots. Automated supervision of the robot system is desirable as this can increase robot availability and maintainability and reduce operator effort. Currently, there are few commercially available solutions that allow for the automated monitoring of the mechanical components of the robot, thus the ability to continuously monitor the status and condition of robots has become an important research topic in recent years and is now receiving considerable attention. This is the main focus of the present study. Additionally, advances in the fields of electronics and micro-electromechanical technologies, where miniaturized and high-precision sensors and embedded processors are readily available at affordable prices, has led to rapid developments in the areas of mechanical systems and process monitoring. Not surprisingly, the attention of the research community has shifted to focus on building remote and automated condition monitoring (CM) solutions with lower number of sensors.

1.2 Industrial Robot Faults

The term fault is generally understood to mean an unpermitted variation of one or more characteristic features of a system away from the normal, reasonable, and standard behaviour (Isermann 2005). There are two types of fault that may occur in machine plant systems. These can be categorized in terms of their behaviour over time, and by their way they affect the system (Isermann 2005). The first type is classified as an abrupt fault, which is a fault that occurs very suddenly in the system. This is difficult to predict and affects the operation of the system immediately, as, for example, with a power supply drop. The second type is intermittent faults. These are faults which affect the system at certain time intervals, and include incipient faults which occur gradually over time. This fault type can be considered to be an additive fault which is effectively added to the plant's inputs, and changes its parameters. These faults can be diagnosed before critical degradation takes place, therefore, the interest in this thesis is limited to monitoring the second category of faults.

Joints in industrial robot are commonly actuated by electrical motors. The permanent magnet servomotors are a popular choice to produce the driving force to move the robot joints because of their easy operation and high power density and performance (Halme 2006). In general, servomotors are electromechanical components, whose faults can be originated from electrical, mechanical and from other external reasons. Mechanical faults sources are bearing failures, movement in winding, and rotor eccentricity, among others. For the electrical faults, overheating, overloading, or short circuit will increase the resistance of components or break the wires. These lead to decrease of rotor output power and extra power loss.

Also, in order to transform the motor power to the robot joints, mechanical reduction gears in the transmission system are used. The power is then transmitted from the input to the output shaft through the gear contacts and mesh. However, the main fault mechanisms that may appear in the joint gearbox are basically the same as that arising in the other types of gearboxes. The most common gear tooth failure types are scuffing, cracking, macro and micro pitting, wear, bending fatigue, fracture due to overload as well as backlash between mating teeth. Typically gear faults initiate at a local point on the tooth surface and then progress until they lead to gear wheel fracture. Moreover, since gears are normally supported on rolling element bearings, faults in these bearings, such as wear in the inner or outer race, represent another typical type of faults in the gear transmission. Due to the importance of the moving parts in the robot, the focus in this thesis will be on monitoring faults in gears and bearings of a robot's gear transmission.

1.3 Overview of Maintenance Strategies

Maintenance can be defined as the combination of all technical and administrative actions, including supervisory actions, intended to retain an item in, or restore it to, a state in which it can perform a required function (ISO 14224:2004). It is performed to keep equipment and systems running efficiently for at least the design life of the components. Maintenance strategies can be divided into four different types (Girdhar 2004) as shown in Fig. 1.2, as described below:

- Breakdown or run-to-failure maintenance: This is the oldest form of maintenance, and its basic idea is to allow the machine to run until it fails, and then conduct repairs. This is a good approach for machines when downtime is not

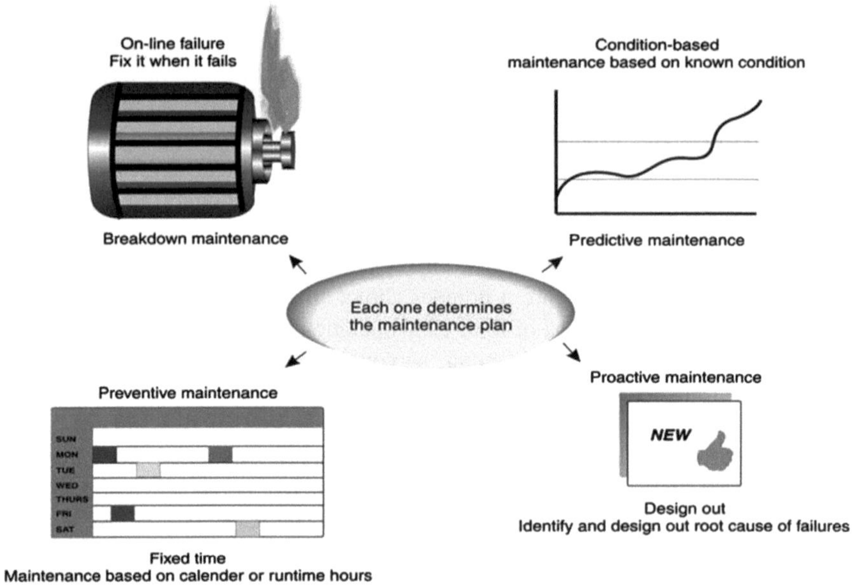

Fig. 1.2 Maintenance strategies (Girdhar 2004)

important and does not have any significant effect on the production process. This method can be associated with a high risk of catastrophic failure, unexpected shutdown, and high labour and maintenance costs. Additionally, it increases health, safety, and environmental risks.

- Preventive or time-based maintenance: This is a set of scheduled maintenance activities that are performed on machines at predetermined time intervals regardless of their condition in order to prevent any deterioration in their operating conditions. This technique is suitable for machines which do not run continuously. However, time-based maintenance always requires parts in good condition to be replaced along with those in a deteriorated condition. This may triple the cost involved: the cost of the parts themselves, the maintenance cost, and the downtime needed to replace those parts.
- Predictive or condition-based maintenance: This is a set of activities which are performed only when a functional failure is identified. This approach depends on continuous monitoring of changes in the physical condition of equipment (signs of failure) in order to maximize its service life without increasing the risk of failure.
- Proactive or prevention maintenance: This maintenance philosophy relies on connecting all failures to their root causes. It is also called root cause failure analysis (RCFA) as it uncovers the latent causes of any undesirable event in the cvvmachine. Therefore, additional efforts are required to thoroughly

investigate the cause of the failure and to determine appropriate approaches to improve the reliability of the machine and prevent potential problems from occurring repeatedly. The main disadvantage of proactive maintenance is that knowledgeable and well-trained personnel are required along with the procurement of specialized equipment.

1.4 Background to Condition-Based Monitoring (Maintenance)

Condition monitoring (CM) was first developed for the American nuclear industry in the 1960s (Wild 1994). It is termed online CM when it runs concurrently during the normal operation of the system. If, however, the system needs to be run in a particular manner, it is called offline CM. On- or off-line CM can be defined as a type of maintenance inspection where an operational asset is monitored and the data obtained is then analysed to detect signs of degradation, diagnose the causes of faults, and also predict how long it can be safely or economically run (Beebe 2004). The basis of CM is illustrated in Fig. 1.3 below.

The philosophy reflected by the above figure is that suitable parameters need to be chosen which can be used as metrics to provide good and reliable indicators of the internal condition of the system and its key components. Periodic measurements are taken at appropriate time intervals to establish whether or not these parameters remain within the repeatability band. The values of parameters will tend to lie outside the repeatability band as system performance becomes degraded, eventually leading to a fault occurring, and thus the state of the system can be established.

However, various monitoring techniques can be used to establish the status of a machine, including vibration, acoustic emission, wear particles, and thermal monitoring. One or more of these methods can be applied depending on the machine's

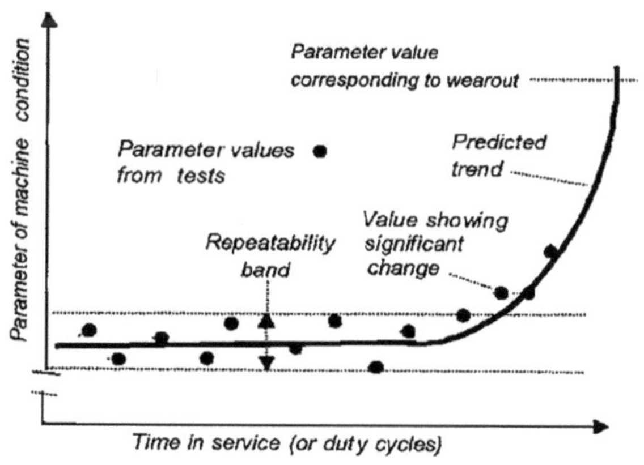

Fig. 1.3 The principle of condition monitoring (Beebe 2004)

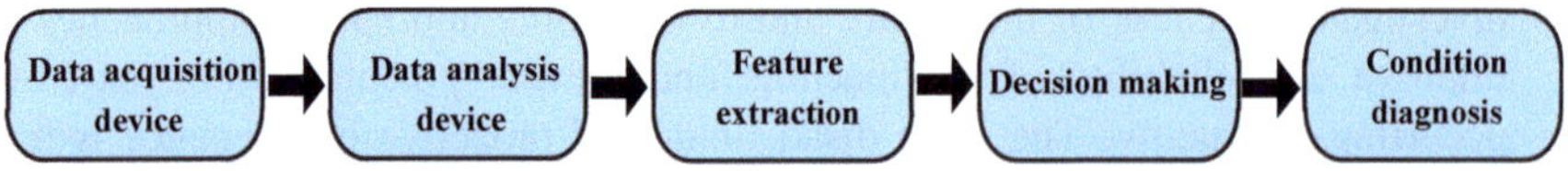

Fig. 1.4 The procedure of the condition monitoring process (Marwala 2012)

criticality and importance. CM is very important for different types of electrical and mechanical machinery and equipment, and can provide many advantages such as:

1. Improved safety and the avoidance of unexpected catastrophic breakdown.
2. Improved reliability.
3. Improved product quality.
4. Increased operational machine life.
5. Increased machine productivity.
6. Reduced maintenance costs.

A general overview of CM is shown in Fig. 1.4 (Marwala 2012). It is essential to employ appropriate sensors with data capturing devices to gather data from the system. Examples of these devices include accelerometers, strain gauges and microphones. Data analysis is carried out using different signal processing techniques, for example, empirical mode decomposition (EMD), or wavelet or Fourier transforms, with which to analyse the data obtained (Marwala 2012).

A specific feature extraction process is used to detect information about the machine. Moreover, the features in the previous step will be interpreted in the next stage (decision making process). In this stage many artificial intelligence techniques, such as neural network or fuzzy logic, can be employed (Marwala 2012). Once a decision is made, fault diagnosis is completed in the final step.

The prediction of the state of health of a machine or structure represents the main aim of condition monitoring. There are five steps that can be used in the estimation procedure as shown in Fig. 1.5 (Marwala 2012). The first step is the detection of whether or not there is a fault in the machine. After that, the nature of the fault is defined using fault classification techniques. The third step is applied to determine the exact location of a fault, and then the degree of seriousness of the fault identified in the fourth step. The final stage in the fault estimation procedure is the prediction of the remaining useful life of the machine, which is the most inaccurate part of the process (Beebe 2004). These are the ideal steps, but the exact procedure used depends on the specific requirements and nature of the system concerned.

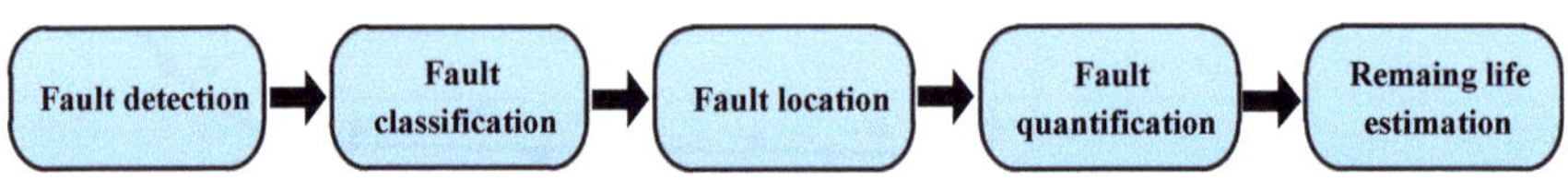

Fig. 1.5 Machine state estimation (Marwala 2012)

1.5 Challenges

Let us imagine that a six degree of freedom robot working in a factory involved in car or aircraft manufacturing, as illustrated in Fig. 1.1. The main job of this robot may be to manipulate a work piece, carrying a tool, or perform spot welding or riveting, performed on the car or aircraft structure at different locations and at fixed time intervals. Normally, such work is performed in a continuous or point-to-point (intermittent) cycle according to a program stored in its control unit. Here the robot arm that holds the tool will move along the desired path, waiting at predefined points for the programmed operation to be achieved by the tool. Having finished the cycle, the robot returns to its initial position, and at the subsequent start signal it repeats the whole series of movements. This repetitive operation has to be accomplished precisely and reliably over time. So, when the robot is first installed it will do its job accurately and meticulously, but after a period of time the robot's performance might degrade. The main causes of degradation in its performance may be because of wear in the robot joint's transmission, or an electrical fault in the motors which are fixed inside the robot joints (as discussed in Sect. 1.2).

Robot arms exhibit very complex dynamic behaviour, and different defects can affect this behaviour. Also, their motion is completely different from that of rotating machines (or other continuously moving machines), for which the majority of present CM systems have been designed. However, in order for the robot to complete the scenario described earlier, each joint in the robot body will move at different angular speeds (and accelerations), needing different torques, and rotating at different angles. When this motion is compared with that of rotating machines, it can be noticed that the latter move at continuous and fixed speed, while the former's movement is discrete and varying over time. Continuously rotating machines emit continuous signals during their operation. Therefore, the CM system will be relatively easy to implement as long as there is a continuous signal during any time period. On the other hand, the signals emitted from the robot's mechanical or electrical parts will be transitory and last for a very short time. Consequently, the challenge here is how to design a reliable and intelligent CM system to be able to deal correctly with the transitory and non-stationary nature of the robot signals for accurate fault detection and diagnosis. Therefore, the signals captured and features extracted have to be analysed and classified in an appropriate way to provide an unambiguous identification of a faulty robot part before a catastrophic failure occurs.

1.6 Aims and Objectives

The majority of the previous research on industrial robots health monitoring is focused on offline monitoring of a limited number of faults, such as backlash in gears (Trendafilova and Van Brussel 2003; Lima et al. 2011; Slamani and Bonev

2013), but does not diagnose the other gear and bearing faults, such as gear tooth and inner (or outer) race bearing faults. Thus, the main aim of this research is to design and develop an online and intelligent condition monitoring system based on wireless embedded technology to detect and diagnose, in addition to backlash faults, the most common faults that could be progressed in the transmission systems (gears and bearings) of industrial robot joints. In the context of this research, intelligence means the ability to automatically acquire signals, process them, extract features, detect the fault and diagnose it and then displaying the result to the operator. To achieve this, the following objectives have been identified:

- To undertake a critical review of recent and past research in the area of condition monitoring and fault diagnosis, with an emphasis on their application to industrial robots. To evaluate the performance of commonly utilized diagnostic techniques in terms of their suitability for robot fault detection, and identify the recent developments in the use of embedded systems in the field of condition monitoring.
- To identify and diagnose the most common faults occurring in industrial robots by using conventional condition monitoring techniques.
- To select appropriate signal processing and feature extraction techniques to deal with the non-stationary nature of the signals from the robot.
- To perform frequency response function (FRF) analyses in order to establish the dynamic characteristics of the robot used in this study.
- To design a data acquisition system, signal analysis, and feature extraction software based on the selected signal processing and feature extraction techniques.
- To develop an experimental procedure for the simulation of different fault scenarios in the robot joints.
- To design an intelligent robot health assessment and fault classification system to be uploaded later onto the embedded system.
- To design an intelligent embedded sensing system for both monitoring and diagnostics. The device has to be wireless with a minimum number of sensors.
- To undertake a series of extensive experiments to validate the performance of the proposed system, after which it can then be implemented on an industrial application.

1.7 Outline of the Thesis

This Chapter provides background information about maintenance strategies and CM. The motivation for conducting this work, its expected challenges, and the aims and objectives of the study are then discussed. A review of previous and recent research on CM techniques applied to different machinery, with an emphasis on industrial robot condition monitoring, is presented in Chap. 2. Chapter 3 discusses the proposed CM algorithm along with the advantages and disadvantages of the different signal processing techniques that have been employed previously for machinery health monitoring. Then, suitable techniques based on the nature of the signal emitted from the robot and the proposed CM algorithm are selected. The dynamic characteristics of the PUMA 560 robot, as used in this study, are found by performing modal analysis on the robot, and this is with a general overview about the PUMA 560 robot are described in Chap. 4.

In Chap. 5 different faults that can occur in the transmission system of the robot with their related frequencies are discussed, along with the design of a data acquisition system based on National Instruments hardware and software, for preliminary evaluation. The focus of Chap. 6 is on analysing the robot system by simulating different types of faults in the robot. Then, features are extracted and saved to be utilized later in the design of the intelligent condition based monitoring (iCBM) system. The work in this chapter is based on the data acquisition system design presented in the previous chapter. In Chap. 7 the iCBM system is designed based on statistical control chart (SCC) and an artificial neural network (ANN). In this chapter, the fundamental principles and application of these two methods are also outlined.

Chapter 8 describes the procedure followed in the design of the wireless embedded system used in this research and the evaluation techniques that were undertaken in the selection process of its component. Chapter 9 then presents a description of flowcharts of the embedded software prepared using the C programming language and discusses the embedded system validation results. Finally, the conclusions of the study along with a summary of the research contributions, achievements and recommendations for future work are provided in Chap. 10.

References

Beebe, R. S. (2004). *Predictive maintenance of pumps using condition monitoring*. New York: Elsevier Advanced Technology.

Girdhar, P. (2004). *Practical machinery vibration analysis and predictive maintenance Oxford Newnes*. Oxford

Halme, J. (2006). Condition monitoring of a material handling industrial robot. In *19th International congress*, 2006 Lulea, Sweden.

Isermann, R. (2005). Model-based fault-detection and diagnosis—status and applications. *Annual Reviews in Control, 29*, 71–85.

ISO 14224:2004. *Petroleum and natural gas industries—Collection and exchange of reliability and maintenance data for equipment*. Geneva: International Organization for Standardization.

Lima, M. F. M., MacHado, J. A. T., & Crisóstomo, M. (2011). Experimental backlash study in mechanical manipulators. *Robotica, 29*, 211–219.

Marwala, T. (2012). *Condition monitoring using computational intelligence methods*. London: Springer.

Slamani, M., & Bonev, I. A. (2013). Characterization and experimental evaluation of gear transmission errors in an industrial robot. *Industrial Robot, 40*, 441–449.

Spong, M. W., Hutchinson, S., & Vidyasagar, M. (2005). *Robot modeling and control*. New York: Wiley.

Trendafilova, I., & Van Brussel, H. (2003). Condition monitoring of robot joints using statistical and nonlinear dynamics tools. *Meccanica, 38*, 283–295.

Wild, P. (1994). *Industrial sensors and applications for condition monitoring*. London: Mechanical Engineering Publications.

Chapter 2
Literature Review

Abstract The thesis title reflects a number of themes, i.e. intelligent system, embedded system, condition monitoring and industrial robot, and previous and recent research in each theme needs to be studied thoroughly; however, there has not been much researches in the field of robot condition monitoring. Thus the purpose of this chapter is to evaluate the condition monitoring methods that have been developed for different machinery with a view to applying the most appropriate one to robots and collectively helps to establish if there is a gap in the area of industrial robot condition monitoring. This chapter will give and describe in its first two sections the necessary background information about the various condition monitoring approaches and techniques. Statistical and Artificial intelligence techniques, such as artificial neural networks (ANN), fuzzy logic system (FLS), genetic algorithm (GA), and support vector machine (SVM), which can be applied to address the issues of fault detection and diagnosis, will also be reviewed. Then, the principle of embedded systems and their application in condition monitoring is reviewed in the third section. The last section of this chapter will discuss the work done in robotics health monitoring and finally the research gap is addressed in the summary section.

2.1 Condition Monitoring Approaches

In machines condition monitoring (CM), fault detection/diagnostic approaches can be classified into two types, dependent on whether the diagnosis evaluation is based on deterministic or stochastic information (e.g., historical, statistical parameters). The first of which has been termed a model-based or "white box" approach, while the second is known as data-driven or "black box" approach (Park and Zak 2003; Kim 2010; Butler 2012). Sections 2.1.1 and 2.1.2 define the principles of these approaches and present a brief overview of previous work based on them.

A.A. Jaber, *Design of an Intelligent Embedded System for Condition Monitoring of an Industrial Robot*, Springer Theses,
DOI 10.1007/978-3-319-44932-6_2

2.1.1 Model-Based Approach

The model-based fault detection approach employs a mathematical model of the system under observation, by assuming that a fault in the system will lead to deterministic changes in the model parameters. The model-based approach relies on comparing the model outputs with the actual system outputs to generate a residual signal, and based upon the properties of the generated residual signal, potential fault conditions are identified and useful information is extracted (Ding 2008). The basic concept of a typical model-based fault detection approach is illustrated in Fig. 2.1.

As indicated in Fig. 2.1, there are two main stages in this approach, the first of which generates the residual which is then passed to the residual evaluation stage. Throughout the fault-free operation, the magnitude of the residual signal should be approximately zero, indicating that the proposed model is accurately describing the current behaviour of the system. If, however, the value of the residual signal diverges from zero, appropriate processing and analysis techniques are applied to it in order to obtain valuable features related to the present fault. The properly processed residual signal is then forwarded to a decision logic routine to map the behaviour of the residual signal onto a specific fault condition.

The most important aspect in the model-based approach for precise fault detection is the requirement of an accurate and robust mathematical model of the system under study. In such models since they are usually derived from first principles, using ordinary differential equations, different elements of the model are related to actual physical properties. Therefore, the main advantage of model-based techniques is the capability of detecting unanticipated faults as well as the replacement of hardware redundancy by analytical redundancy (Vachtsevanos et al. 2006). However, in many real world applications it is almost impractical to apply model-based diagnostic approaches, since many physical processes are too complex to develop accurate model. This will cause mismatch between the process and model outputs which, in turn, lead to large error signals usually giving rise to false alarms (Ding 2008).

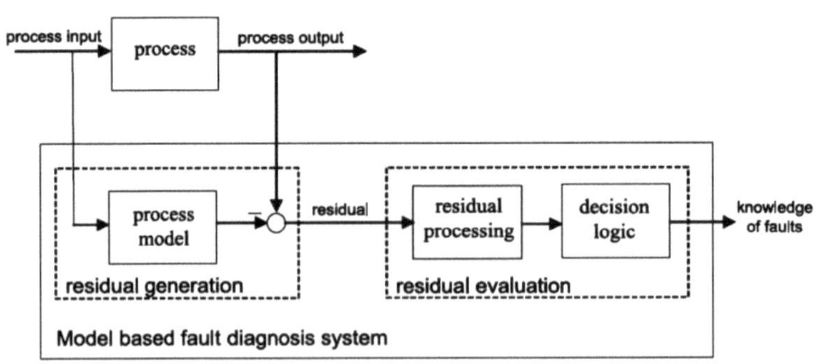

Fig. 2.1 Model-based fault diagnosis flowchart (Ding 2008)

A great amount of work have been undertaken to study the dynamic modelling of, for example gears and bearings, for fault detection. Gears have wide industrial application and unforeseen failures can be enormously damaging, and so research into CM and fault diagnosis in gearboxes is very important. Parey et al. (2006) propose a 6-degree-of-freedom dynamic model for a spur gear pair to simulate and study the effect of lateral–torsional vibration combination on vibration response in the presence of localized tooth defects. The model included two inertial masses representing load and prime mover, two shafts, and bearings, as shown Fig. 2.2. To analyse the simulated signals a multi-resolution signal processing technique was used to break-down the vibration signal to multi-level. Then, statistical parameters such as Crest factor and kurtosis were calculated from each level and utilized to give an early detection of contact surface pitting. Recently, Liang et al. (2015) have developed a dynamic model to simulate the vibration source signals for a planetary gearbox in the healthy and the cracked tooth conditions. The signals were analysed using time- and frequency-domain feature extraction techniques. For model verification experimental work was accomplished and the results were deemed acceptable.

Bearings are of paramount importance in most types of machinery, and bearing failure is one of the foremost causes of breakdown in machines. A generic scheme for health monitoring of rolling element bearings was proposed by Jiang and Zhang (2012). The approach incorporates vibration modelling and online fault dimension estimation. Offline training was applied to obtain the parameters of the vibration model. The trained model is then used online in parallel with a real system (in service) to realize model-based fault detection and diagnosis. To address noise and uncertainties problems associated with fault dimension estimation, an extended Kalman filter was used to increase the accuracy of diagnosis. The proposed system was validated on a case study of rolling element bearing health monitoring and experimental results demonstrated the efficiency of the proposed method. An earlier

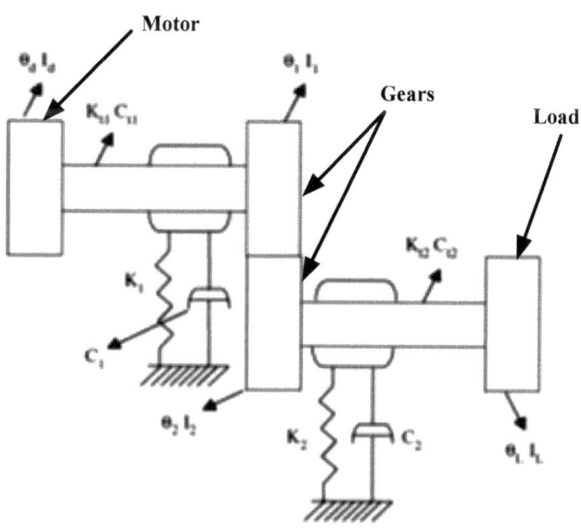

Fig. 2.2 Schematic of modelled spur gear pair (Parey et al. 2006)

paper introduced a dynamic modelling for a gearbox test rig to investigate the effect of a defect in gears and bearings on the vibration signal (Sawalhi and Randall 2008). Two damage types in the gears, spalls and cracks, and inner race, outer race and rolling elements faults in the bearings were adopted in this study. To detect these faults, diagnostic techniques such as spectrum comparison, spectral kurtosis analysis and envelope analysis were applied to the vibration signal for both experimental and simulation work.

2.1.2 *Data-Driven or Model-Free Approach*

Data-driven approaches depend on the features extracted from the measured process data for building a model that represents the process, and these are categorised into two types; statistical based methods and those based on artificial intelligence (AI) techniques (Jardine et al. 2006; Yadav and Kalra 2010). Within these categories, a huge range of techniques have been employed to handle a wide variety of fault detection problems. The suitability for usage a data-driven monitoring strategy is for processes in which an explicit mathematical model is difficult to construct because of shortage of knowledge or information related to the process (Yadav and Kalra 2010). In the following subsections, an overview of these methods and their applications are presented.

2.1.2.1 Statistical Based Approaches

The concept of model-free based fault detection and diagnosis has stimulated the interest of using *novelty detection* for condition monitoring, which focuses on identifying any deviations between the features extracted from the recent measured data and the data measured under normal (healthy) operating conditions. The features obtained from a machine in its undamaged state will have a distribution with an associated mean and variance. However, a variation in the mean and/or variance will appear if the machine is damaged.

Statistical control charts (SCCs) provide a framework for monitoring the distribution of the features and detection if they are inconsistent with the past healthy state, and any change in the distribution characteristics of the features will indicate damage, termed *outlier* analysis. SCCs are one of the earliest statistical fault detection techniques dating back to 1931 (Yadav and Kalra 2010). Starting with the advent of Shewhart control charts for averaging, usually called X-bar ($\overline{X}$) chart, which is normally used in combination with a range chart (*R* chart) or standard deviation chart (*S* chart) (Montgomery and Runger 2014). Further modification to Shewhart charts have resulted in cumulative sum (CUSUM) and the exponentially weighted moving average (EWMA) charts in the early 1950s (Yadav and Kalra 2010). Because these charts are easy to construct, implement, and interpret, they received a large acceptance in the field of machines and processes monitoring.

Techniques based on SCCs can be classified to two approaches: the univariate and multivariate approaches. In the former method each characteristic of interest is monitored independently whereas in the latter the concurrent monitoring of characteristics is accomplished, considering the correlation that may exist among the various characteristics (Yadav and Kalra 2010; Kisić et al. 2013). Despite their acceptance, the main disadvantage of control charts is that their purpose is for the detection of damage, rather than its quantification and location.

Baydar et al. (2001) presented a multivariate statistical methodology for helical gears monitoring. The gathered time-domain vibration signals were employed to form a reference condition model using principle component analysis (PCA). The T-square control chart, type of multivariate SCCs, was adopted as health condition indicator. Researchers concluded that when tooth failures occur, the probability density function (PDF) of the measured signal will change which gives good indication about the health condition. Another paper applied two statistical techniques for wind turbine gearbox CM (Zhang et al. 2012). The first technique was based on data-mining algorithms used to build a statistical model for predicting the jerk indicated by the vibration excitement of the gearbox. This model was utilized in conjunction with experimentally captured vibration signal to produce residual signals. Two control charts, X-bar and EWMA charts, were constructed to evaluate the residual and fault prediction.

Another application, amongst others, of statistical control charts is for roller bearings condition monitoring. Niknam et al. (2013) at University of Tennessee investigate the use of CUSUM chart for detecting bearing failures, such as unbalance, based on acoustic emission signal analysis. Similarly, Zhou et al. (2008) presented an approach for in situ induction motor bearing fault detection by combining noise cancellation and X-bar chart. In this work the motor current signature was analysed to extract features related to bearings deterioration. Two control charts were developed based on Shewhart average chart to identify the initial start point of the bearing defect (Wang and Zhang 2008). These charts are named adaptive moving average chart, and adaptive Shewhart average level charts. Based on these charts the researchers were able to produce warning and action limits, as shown in Fig. 2.3. The findings of this study suggest that the adaptive Shewhart average level chart overcomes the drawback of adaptive moving average charts by working out the limits using all the bearings' data.

More attractive applications of SCC are for induction motors and rotating shafts health monitoring. García-Escudero et al. (2011) proposed a methodology for incipient fault detection in induction motors. They used fast Fourier transform (FFT) and wavelet transform (WT) signal processing techniques to detect significant peaks in the captured current signal. Then, a quality control approach based on multivariate T-square control chart was successfully applied to detect the progressive deterioration of the rotor cage. Another paper proposed the use of CUSMA chart to monitor the misalignment in a rotating shaft (Sun and Chang 2004). The gathered vibration signal was fitted with an autoregressive model and the residuals between the fitted and observed vibration signals used as a monitored parameter. Control chart limits were designed using the healthy baseline residual data. The

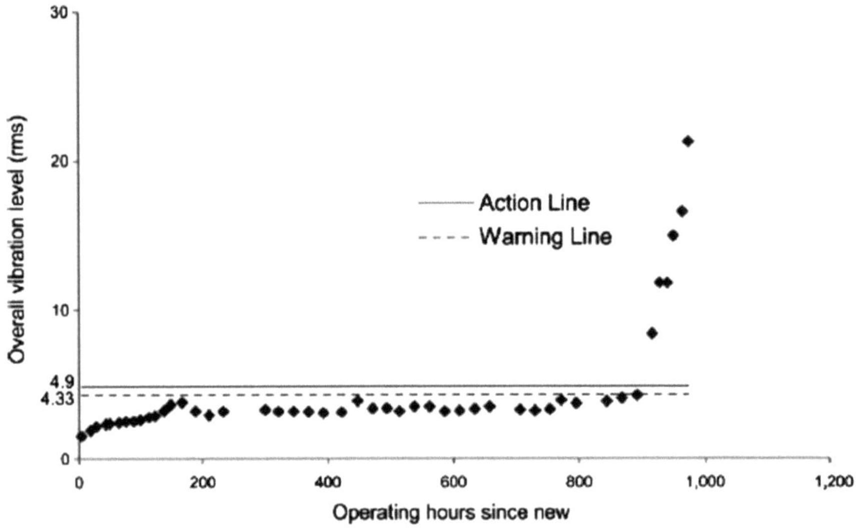

Fig. 2.3 Shewhart rms average chart for bearing fault detection (Wang and Zhang 2008)

results showed that this approach was capable of detecting both mean and variance shifts and also indicate the fault severity.

2.1.2.2 Artificial Intelligence Based Approaches

Artificial intelligence (AI) can be defined as a "computerized approach that employs knowledge, reasoning and/or self-learning to enable machines to perform tasks which humans perform using their intelligence" (Heng 2009). In recent years, AI techniques such as neural networks, fuzzy logic and support vector machines have been widely developed to improve the accuracy and efficiency of fault detection and the diagnosis of machines which can take over some menial and tedious tasks. The intelligent detection and diagnosis begins after data has been collected and important features extracted, as illustrated in Fig. 2.4. The following bulleted sections summarize the fundamental concepts of these methods and their applications to the area of intelligent condition monitoring.

- **Artificial Neural Network (ANN)**: An artificial neural network (ANN) is a computational structure inspired by the data processing and learning ability of biological neurons in the brain. It is composed of simple computation units called neurons, which integrate the functionality of both memory and computation (Samarasinghe 2007; Dreyfus 2005). ANNs can be employed for a variety of tasks, such as function approximation, classification, pattern recognition, clustering, and forecasting (Samarasinghe 2007). For example, an ANN was applied to fault detection and diagnosis in 4-stroke internal combustion engine

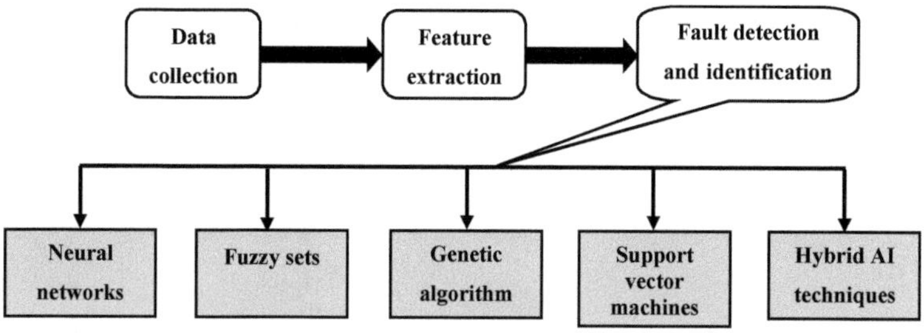

Fig. 2.4 Some of artificial intelligence techniques used for condition monitoring

by depending on a minimum number of sensory data (Chandroth et al. 1999). Cylinder pressure and vibration data were acquired from the engine. By using features of the collected data, two sets of artificial neural nets were trained separately. Experimental work was carried out using a twin cylinder diesel engine, and it was demonstrated that it is possible to detect and diagnose the most common component faults in the engine using either cylinder pressure or vibration amplitude. Such a system would thus require fewer sensors. Neural networks are commonly arranged in layers, and each layer has an array of interconnected elements. Each element receives an input signal, manipulates it, and then output signals are forwarded to the other connected elements in other layers. There are many different forms of neural networks depending on their connection patterns. However, the main classes of neural networks are the feed-forward neural network, whose signal direction is from the input to output layer without any feedback connection. That means the past signals are not used for processing the new signals, as illustrated in Fig. 2.5a. The recurrent (Feedback) Neural Network, which has a feedback connection, and utilizes past signals for identifying the new features, Fig. 2.5b.

There has been significant interest in applying artificial neural networks to identify and diagnose faults in machinery. Kudva et al. (1992) used multilayer perceptron neural networks to deduce the size and location of damage in smart structures using measured strain values at different locations. More recently, and in contrast, Parhi and Dash (2011) applied the same technique for structural monitoring, but vibration signatures were used instead. Both studies achieved acceptable levels of prediction of crack locations.

Lopes Jr et al. (2000) implemented impedance techniques and neural networks to detect, locate, and characterize structural damage. The advantages of smart materials technology and the characteristics of neural network were combined in proposing a self-diagnostic procedure. Experimental investigations were successfully carried out to locate and identify damage in a quarter-scale bridge section. It was concluded that this technique can be applied to complex structures. In another study, a 2-step neural network was used to design a predictive fault detection and diagnosis model for the monitoring of nuclear power plants (Bae et al. 2006). The

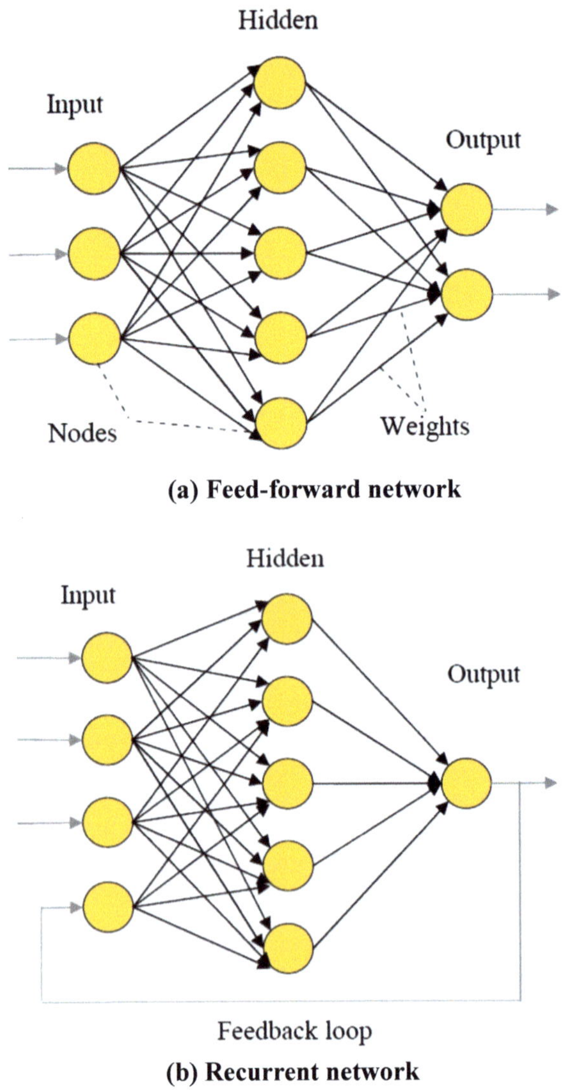

Fig. 2.5 The main categories of artificial neural network (Heng 2009)

main role of the first network was to classify failure type, and then failure severity was determined using the second network. The results showed that this model was suitable for failure detection, but additional work was needed to increase its accuracy. Another study conducted by Zhang et al. (2007) looked at fault diagnosis in a steam turbine generator by applying of an integrated neural network based on combined information. The aim of this method was to overcome the problem of multi-failure mode diagnosis in a single neural sub-network, which normally requires many learning samples. The preliminary diagnosis of faults was implemented using one sub-neural network, after which the sub-neural networks were merged to yield the final decision. It was found that there are many advantages to this approach as system accuracy and reliability are increased, and the uncertainty of system information is reduced.

- **Fuzzy Logic (FL)**: Fuzzy logic or fuzzy set theory is an important technique first introduced by Zadeh in the 1960s to deal with vague, imprecise and uncertain knowledge and data. It is especially suitable for systems with a mathematical model which is difficult to drive. FL is composed of four elements (Marwala 2012): a fuzzy set, which is applied to achieve a flexible representation of the elements in the fuzzy system; a membership function, which shows the level of possibility that an object is an element of a certain class; logical operators, which are used to find new fuzzy sets from the existing fuzzy sets; and fuzzy rules, which show the conditional articulations used to perform the input–output relationships of the system, which can include human descriptive judgments such as:

IF speed is high THEN stopping-distance is long
IF speed is low THEN stopping-distance is short

Decisions in FL can be made with estimated values and incomplete information. A decision might be changed at a later time when extra information is available, or when it may not be correct. For instance, if the input parameter values of a system might be 'fuzzy' or incomplete, the conclusions drawn will be incomplete or incorrect as well (Munakata 2008). The major advantages of fuzzy systems are their robustness and flexibility, since they are not restricted to a true or false approach, and they are ideal where system information is limited and unclear (Lim 2009). The application of fuzzy systems in CM has recently been studied in building reliable monitoring systems. For example, Navarro et al. (2010) successfully designed a FL system for monitoring of electric motor bearings. The researchers used more than one signal to accurately detect bearing failures. These included vibration, stator current, bearing frequency, and acoustic emission. In another paper, a computer system-based fuzzy tool was developed to monitor an induction motor by measuring its vibration signal (Janier and Fazrin Zaim Zaharia 2011). The information received from the vibration sensors was used to determine whether or not faults had occurred and actions would then take place to protect the motor from further damage.

Aliustaoglu et al. (2009) developed a fusion approach based on a two-stage fuzzy system and sensor readings for tool wear monitoring. The machine acoustic emission, thrust force, and vibration signals were used to drive the statistical parameters by applying the first stage of the proposed approach, using a Mamdani fuzzy model, as demonstrated in Fig. 2.6. The output values from the first stage were taken as the input parameters of the second stage, which applies the Takagi-Sugeno fuzzy model. Then, the final decision was made using the threshold function and depending on the output values from the second stage as illustrate in Fig. 2.7. The authors mentioned that the performance of this approach can be improved by using electric motor current as a fourth input parameter to the fuzzy system, in addition to using various classifiers.

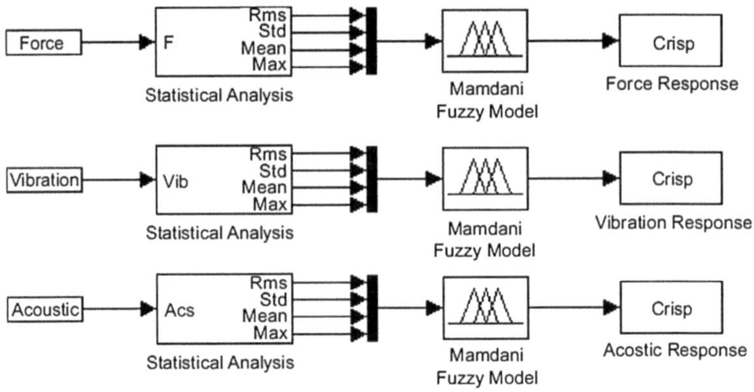

Fig. 2.6 First stage of proposed Fuzzy-sensor model (Aliustaoglu et al. 2009)

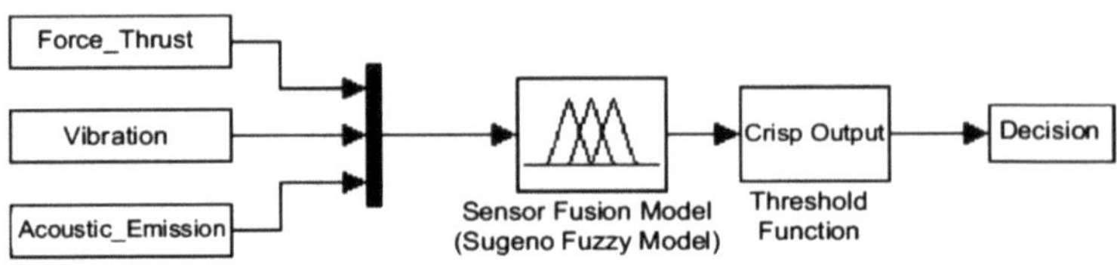

Fig. 2.7 Second stage of proposed Fuzzy-sensor model (Aliustaoglu et al. 2009)

- **Genetic Algorithm (GA)**: The genetic algorithm was first introduced by John Holland in the early 1970s. It can be defined as a computational technique that mimics the genetic processes of biological organisms in order to solve search and optimization problems (Negnevitsky 2005; Munakata 2008; Marwala 2012). To apply a GA to any problem, several key steps have to be followed (Goldberg 1989; Marwala 2012). Firstly, a number of possible individual solutions containing a number of chromosomes are randomly generated. Then, the fitness value of each individual current solution has to be computed, the purpose of which is to evaluate the performance of each chromosome. Once the fitness values have been calculated, a new population will be generated by applying crossover and mutation operations to the individuals. When a convergence criterion is reached, the algorithm stops; and if not, this process is repeated from the second step.

Genetic algorithms have since been adopted in many different disciplines, such as for automatic programming, missing-data estimation, finite-element analysis, and condition monitoring (Marwala 2012). For example, GA was successfully applied in the bearing monitoring process to select the most important features from a large set of vibration signals (Jack and Nandi 2000), where in one set of experiments the GA was capable of selecting a subset of six inputs from a set of 156 features. In a similar study, a GA-based optimization method was applied to select the optimal cutting conditions for each pass of a turning operation, with consideration given to the effect of overall progressive tool wear on machining performance (Pal et al. 2011). The optimization process showed precisely that cutting parameters are not constant when tool wear is taken into account, from which it was concluded that the GA has very good classification accuracy. In other studies, a two-stage process was utilized to detect structural damage (Chiang and Lai 1999; Moslem and Nafaspour 2002), where in the first stage the residual force method was applied to initially locate damage. The GA was then used in the second stage to evaluate the damage in the identified structure.

Differences in the natural frequencies of force vibration are most frequently represented as a potential damage indicator (Ostachowicz et al. 2002). However, changes in the first four frequencies have been used to identify the exact location and magnitude of an added concentrated mass on a simply supported, isotropic plate. A GA model was developed which showed good ability in finding the accurate location and value of added mass. Meruane and Heylen (2011) have presented a technique based on model properties and a GA to detect faults in a tri-dimensional space-frame structure. Two damage scenarios were adapted in this work to verify this technique. The findings showed that this method was capable of detecting and quantifying three simultaneous instances of damages.

- **Support Vector Machines (SVM)**: Support vector machines are a type of artificial intelligence methodology applied mostly for the classification and regression of data. SVMs were first introduced by Vapnik in the late 1990s and are supervised learning methods derived from statistical learning theory, as in most neural network systems. Supervised learning methods refer to machine learning methods which try to generate a clear map between the inputs and outputs in the training data. SVMs are suitable for two-class classification, but there are a number of extensions which enable this technique to be used for multi-class classification problems (Marwala 2012; Lim 2009).

SVMs have gained significant importance recently because of their superior ability to generate an accurate representation of the relationship between the input and output from a small amount of training information (Sharma 2008). For example, if there is a two-class dataset, a SVM will classify them by finding a separating plane that will divide the space containing the data. All points at each side of the hyper-plane will belong to a specific class. The best separating plane can be a linear boundary in the input feature space, however, in some cases a non-linear boundary could be used to separate the target classes where a linear boundary might not be able to separate them adequately, as shown in Fig. 2.8 (Fulcher 2006).

Nowadays, SVMs are applied in many research fields, such as biological sequence analysis, text classification, data mining, facial recognition and mechanical fault diagnosis, and the results are promising (Lim 2009; Zhang et al. 2009). In terms of fault detection, SVMs have been applied to detect the location of damage in rigid structures (Shimada and Mita 2005; Shimada et al. 2006). Changes in the natural frequencies of the structure were used first as training data for the SVMs,

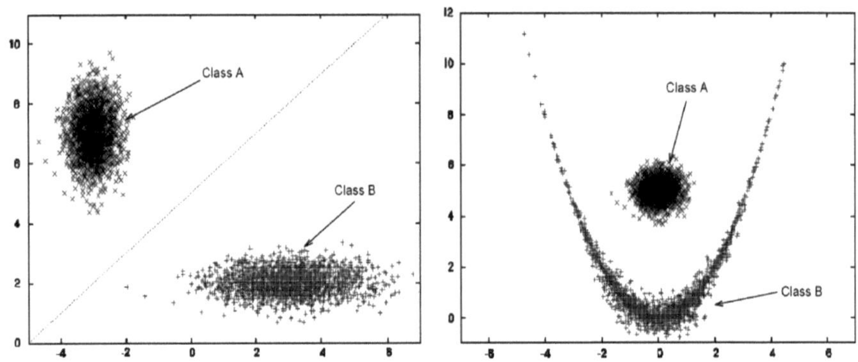

(a) Linear separation (b) Nonlinear separation

Fig. 2.8 SVMs for data classification (Fulcher 2006)

and then to detect damage location. The main goal of this study was to reduce the number of sensors used to collect important data from the structure. The authors pointed out that this technique effectively decreased the possibility of incorrect damage detection. A comparison study of artificial neural networks (ANNs) and support vector machines (SVMs) has been presented which compares their performance in gear fault detection (Samanta 2004). Vibration signals in the time-domain were used in this research for feature extraction. Moreover, the number of nodes in the hidden layer in the case of ANNs and kernel parameters in the case of SVMs were optimized using GAs. The researchers used experimental data for a known machine to train the ANNs and SVMs. The findings showed that the classification accuracy of SVMs is better than that of ANNs without GA, and the performance of both classifiers increased when GA was used. Additionally, Zhong et al. (2010) used SVMs for the intelligent diagnosis of gearbox faults. An experimental test rig was designed to simulate the most common faults occurring in the gearbox, such as imbalance and misalignment. It was concluded that the SVMs are able to precisely recognize different fault types and their severity.

A research study was conducted using SVMs technique to detect and classify of faults in rolling bearings depending on vibration signals in the time-domain (Rojas and Nandi 2006). Four bearing faults were simulated in this study: an inner race fault, outer race fault, rolling element fault and cage fault. It was found that the accuracy of SVMs is minimized if there is a limited amount of training data. Furthermore, this technique has been utilized for the purpose of rotor failure assessment (Yan et al. 2009). It was found that the SVM was reasonable and effective assessment of machinery degradation especially in complicated operating conditions since it does not have limits of input parameters, and the computational required time is also short. On the other hand, Caccavale et al. (2009) reported that "the main drawback of the SVMs is the absence of control over the number of data points selected as SVs by the learning algorithm. This may lead to a heavy computational load in the presence of a large number of training data". From the above it can be concluded that SVMs for condition monitoring is still under development and requires further investigation.

- **Hybrid systems**: A hybrid intelligent system is a combination of at least two of the intelligent approaches mentioned previously to achieve more accurate results and better performance. A hybrid system can combine the advantages of different technologies. The main concept of hybrid systems is to create new approaches where the components complement each other's weakness (Lim 2009; Negnevitsky 2005; Munakata 2008). Recently, there has been an explosive growth in the use of hybrid intelligent systems in condition monitoring, in particular, for instance, in Neural- and Genetic-Fuzzy systems.

Neural networks have the capabilities of learning, memorizing, and recognizing patterns in a way the fuzzy systems do not. In contrast, the strength of fuzzy logic lies in its ability to model the decision-making of humans. So, the synergetic integration of neural networks and fuzzy logic can complement each other (Munakata 2008; Negnevitsky 2005). A growing number of researchers have constructed and examined different forms of neural-fuzzy or fuzzy-neural networks. Yen and Meesad (2001) developed a classification algorithm based on fuzzy-neural networks called the incremental learning fuzzy neural (ILFN) network. This technique has the capability of learning new information without forgetting old information. The authors concluded that this approach in classification is better than many recognized classifiers. Additionally, an evaluation study has been conducted to discuss the usability of three artificial intelligence (AI) methods in lathe turning tool wear estimation (Balazinski et al. 2002). These methods were the feed-forward back-propagation neural network, a fuzzy decision support system, and neural network-based-fuzzy inference system with moving consequents in If-Then rules. All three methods gave similarly acceptable results, but there were differences in the training time required. The neural network and fuzzy logic systems needed a considerable amount of training data. On the other hand, the training time was very short for the neural-fuzzy system, making it easy to optimize and use in industry.

In another study, a neural-fuzzy system was applied for the detection of faults in alternating current (AC) motors (Sainz Palmero et al. 2005). This method was tested using an AC motor, and 15 non-destructive fault types were generated. The results showed good levels of detection and classification. Moreover, the knowledge extracted by a fuzzy rule set had an acceptable degree of interpretability. A multiple adaptive neural-fuzzy inference system (MANFIS) methodology has also been applied to detect cracks in dynamic structures (Parhi and Dash 2011). The input layer of the controller was the fuzzy layer and the other layers were neural layers. The relative deviation of the first three frequencies and mode shapes were used as inputs to the fuzzy layer, and the outputs of this layer were used as inputs to the neural layer. The final findings from the use of this method were relative crack depth and relative crack location, showing good agreement with experimental results collected using an aluminium beam with transverse cracks.

A genetic algorithm has a perfect machine learning capability and satisfactory global search ability, whereas its drawback is chance-dependent outcomes and lengthy computation time. When combined with the benefits of fuzzy logic mentioned earlier, it introduces flexible and robust inference methods under high possibility of imprecision and uncertainty. An improved artificial intelligence technique called a genetic-fuzzy system (GFS) can be developed by the hybridization of a genetic algorithm and fuzzy logic. The genetic-fuzzy system combines the learning ability of the genetic algorithm with the uncertainty representation characteristics of fuzzy logic (Munakata 2008; Pawar and Ganguli 2003).

A genetic-fuzzy system has been utilized for damage detection in a cantilever beam and helicopter blades (Pawar and Ganguli 2003). This method was used to find the existence, location, and extent of damage. In order to calculate changes in beam frequencies because of structural damage, a finite element model of a cantilever beam was applied. These changes in frequencies were used to generate the fuzzy system, and rule-base and membership function optimized by a genetic algorithm. It was concluded that this system allows easy rule generation for different structures. The same technique was used in a similar study by the same research group to detect cracks in a thin-walled hollow circular cantilever beam, which was made of composite material and used as part of a helicopter structure (Pawar and Ganguli 2005). It was found that the effectiveness of this method depends on the number of parameters, such as crack density and noise level. Furthermore, it was observed that the genetic-fuzzy system showed reasonable performance in damage detection and isolation.

Genetic algorithms have also been used for optimizing fuzzy system parameters. This technique has been applied to monitor the performance of a cutting machine (Gallova 2010). A simulation study was conducted along with experimental work for result validation. The findings from the experimental work showed good accuracy with theoretical results, and it was concluded that the proposed technique is suitable for large-scale problems because of the ability of genetic algorithm to extract the most effective features from a considerable number of parameters. Furthermore, this hybrid technique has been used in medical diagnosis applications to achieve correct disease classifications (Di Nuovo and Catania 2007). The authors' main aim of this study was to obtain an efficient diagnostic system and at the same time reliable and easy for practitioners to use. The approach was applied to three real-world benchmarks and compared with relevant work to show its effectiveness.

2.2 Condition Monitoring Techniques

In the past few decades and with the development of sensing technologies, different CM techniques have been utilized for the purpose of gears and bearings (or other machines) health monitoring. These techniques, but not limited to, are vibration; sonic emission; motor current; wear debris and lubricant analysis; and thermal monitoring. Each one of these techniques has its unique advantages and disadvantages and may suit one application but not another. For instance, wear debris monitoring technique requires measuring and analysing the wear particles and contaminants inside the used oil, and needs advanced and expensive laboratory equipment, human expert inspection of the wear debris samples, and is very time consuming (Ebersbach et al. 2006). In this section, a brief description of the first three techniques with a review of previously and recently achieved work based on them will be given, as they are the most applicable techniques to the work reported in this thesis.

2.2.1 Vibration Condition Monitoring

When a defect is developed in a rotating machine while it is in operation, the result is usually an increase in vibration level. Each component of a machine has its own characteristic frequencies determined by its geometry and the rotational speed of the machine. Therefore, by establishing the relationship between the measured frequencies and expected defects, either by theoretical modelling of the machine or by measurement, the defect along with its cause and severity can be determined by performing a detailed vibration signal analysis. Vibration signals analysis has been extensively used for machines fault detection and diagnosis in various industrial applications (Gelman et al. 2011). Gear defect diagnosis based on the analysis of vibration signals using multi-scale statistics was introduced by Loutridis (2008). Experimental investigation to test the capability of this statistical technique was carried out by analysing the vibration data recorded from a single stage gearbox with a pair of spur gears. The fault was simulated on the gear by removing approximately 10 % of the tooth material from the root. Wind turbine gearboxes are one of the most important and fault-critical components in the turbines because of the complicated alternating loads from wind turbulence. Empirical mode decomposition (EMD) signal analysis technique has been used for analyzing the vibration signal of a wind turbine gearbox (Teng et al. 2014), and intrinsic information from the vibration signal of a field gearbox has been extract using EMD.

Huang et al. (2014) published a spur bevel gear fault diagnosis method based on vibration signal analysis by employing wavelet analysis. To test the performance of the proposed method, an experimental study was undertaken of two different faults (tooth breakage and tooth surface wear) in the pinion gear. This paper analysed the variations of gear fault vibration signal using time- and frequency-domain signal analysis methods and then fault attributes were extracted using wavelet analysis. Vibration signal analysis is widely used for planetary gearbox fault detection. Recently, a research proposes two features, namely, accumulative amplitudes of carrier orders and energy ratio based on difference spectra for gear and bearing monitoring of a planetary gearbox (Lei et al. 2015). Vibration data acquired from a gearbox test rig has been used to demonstrate the effectiveness of the proposed features. The vibration data is measured under different motor speeds.

A comparison has been made with those reported in the literature, and has been reported that the proposed features are more successful than others in monitoring and diagnosing gearbox faults.

Randall (2004) used the vibration signature of bearing faults to separate gear from bearing signals in a helicopter gearbox. This technique was based on the different statistical properties of bearings and gears, which were the main factors in the fault diagnosis approach. An illustration was conducted using case history data collected from the US, and Australian Navies. Shao and Nezu (2005) developed an early bearing fault detection technique based on vibration signal de-noising. This technique consists of an adaptive noise-cancelation filter and a wavelet-based estimator. The researchers concluded that, using this method can improve the signal-to-noise ratio when the signal is contaminated by noise and thus faults can be detected efficiently.

A time-frequency analysis technique was adopted for real-time bearing fault diagnosis and prognosis (Aliustaoglu et al. 2008). For frequency analysis, the Fast Fourier transform (FFT) method was used, and experimental work carried out on the bearing-shaft mechanism of an AC electric motor. A schematic diagram of the experimental setup is shown in Fig. 2.9. A current sensor was fixed to the phase line of the motor in order to measure the electric current passing through the driver of the motor, while two accelerometers were placed on the bearing housing to

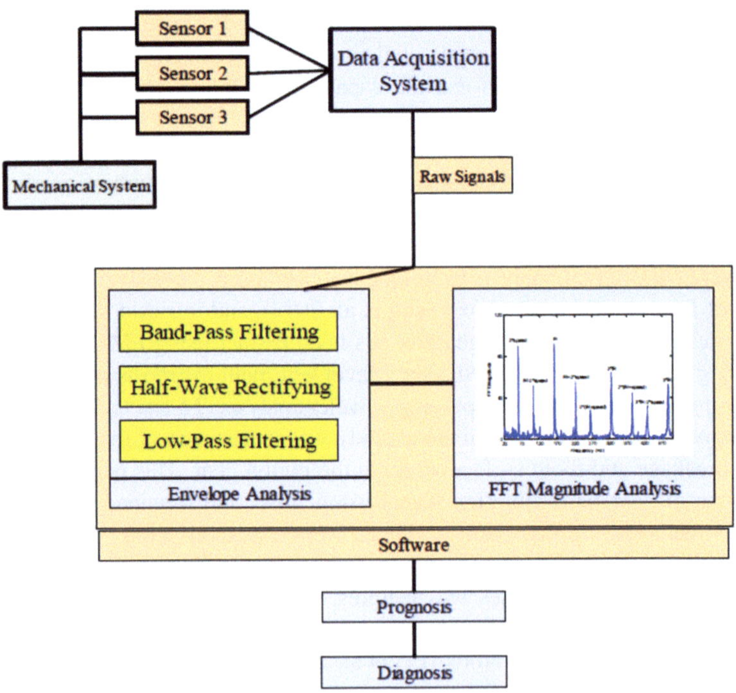

Fig. 2.9 Bearing fault diagnosis and prognosis system (Aliustaoglu et al. 2008)

measure vibration. To analyze bearing status and the progress of any existing faults, vibration and current data were gathered and digitized using a National Instruments data acquisition card. A technique of envelope analysis was applied to separate the modulation signal from the carrier frequency. The authors developed software to perform signal processing task, and six types of defects were defined in this software. The authors claimed that this technique is better than most other advanced techniques, and it could be easily adopted for real-time bearing fault diagnosis.

2.2.2 Noise, Ultrasound and Acoustic Emission (AE) Condition Monitoring Techniques

Acoustics is the study of the generation, propagation, and reception of sound that is heard by a human being (Mohanty 2015). The sounds are classified as desirable or undesirable, which is traditionally known as noise. Human ears are able to hear only sound waves within a specific frequency range, which is known as the audible frequency range (20 Hz to 20 kHz), whereas frequencies above 20 kHz are known as ultrasonic. The acoustic emission (AE) technique deals with signals in the high-frequency range from 100 kHz to 1 MHz (Stamboliska et al. 2015; Mohanty 2015).

Almost all machines under normal operating conditions emit sonic signatures and any variation in these signatures can indicate the start of deterioration of some components. An online gearbox monitoring system based on the LabView program was developed by Wei et al. (2011). This system has the capability to analyze data online and offline, and to query historical data. The authors concluded that the noise detection system can effectively reflect the gearbox's operation status, fault type and location by using spectral analysis. They added that this technique has more advantages than vibration measurement. However, the application of noise measurements in CM is not practical because of the unpreventable noisy background from other neighbouring machines operating in the site which reduces the accuracy of fault detection.

An experimental study was conducted to compare the effectiveness of the ultrasound and vibration measurement technique for the CM of low-speed bearings (Kim et al. 2008). To precisely identify the presence and severity of defects from measured signals, the researchers developed a type of signal processing analysis called the peak ratio (*PR*), suggested by Shiroishi as shown in Eq. (2.1) (Shiroishi et al. 1997):

$$PR = \frac{N \sum_{j=1}^{n} P_j}{\sum_{k=1}^{N} S_k} \qquad (2.1)$$

where, P_j is the amplitude value of the peak located at the defect frequency and harmonics, S_k is the amplitude at any frequency, N is the number of points in the spectrum, and n is the number of harmonics in the spectrum. The modified *PR* (*mPR*) is shown in the following equation, and it depends on disparities between the peak defect frequencies and the average value over the whole spectrum:

$$mPR = 20\log_{10}\frac{\sum_{j=1}^{n}(P_j - A_s)}{A_s} \tag{2.2}$$

$$A_s = \frac{\sum_{k=a}^{b} S_k}{(b-a)} \tag{2.3}$$

where, A_s is the average spectrum amplitude in the frequency band from a to b. It was observed that the ultrasound technique was more effective than common vibration measurements for fault detection. Recently, this opinion has been supported by another research group (Wei et al. 2011).

The AE technique has been increasingly used for condition monitoring of different machinery and structures. For example, Ogbonnah (2007) applied AE and wavelet signal analysis techniques for gear fault diagnosis and prognosis. The result of the gearbox health change over time was presented in statistical properties of amplitude, and corresponding frequency and energy changes. A linear relationship between AE amplitude, gearbox running time, and pit progression was shown in this study. An intelligent health monitoring system for power transmission systems (Onsy 2009), included fault prediction and classification, using AE, vibration and oil debris analysis were combined using fuzzy logic. The aim of this was to monitor two modes of gear fatigue failure; the progression of gear surface and bending fatigue failure in helical gears. The progression of micro-pitting was monitored using the AE average signal level, vibration signal root mean square (RMS), and the mass of ferrous debris, whilst tooth-bending failure was monitored using the AE peaks, the vibration kurtosis and oil debris mass rate.

As mentioned earlier the AE-based technique deals with signals in the high-frequency range, and thus it requires much higher sampling rates than vibration-based techniques. A comparative study for gearbox tooth damage level diagnostics using AE and vibration measurements based on the same sampling rate has been conducted recently (Qu et al. 2014). Partial tooth cut faults are seeded in a gearbox test rig and experimentally tested in a laboratory. It was concluded that the AE signals show more stable performance in fault detection, and reported that the AE-based approach has the capability to differentiate gear tooth damage levels in comparison with the vibration-based approach, since the vibration signals are easily affected by mechanical resonance.

2.2.3 *Motor Current Signature Analysis*

Motor current signal analysis (MCSA) offers a non-intrusive and alternative method to detect mechanical faults through investigating electrical signatures. Provided there is access to the current-carrying conductor to the motor, the drawn current by the stator of the motor can be measured at distant locations from the motor; this represents one of the advantageous of MCSA techniques, since there is no need to mount any transducers or measuring equipment on or near the monitored machine. The MCSA technique was limited to monitoring different faults on induction motors (Kar and Mohanty 2006), including bearings faults.

Schoen et al. (1995) addressed the application of motor current spectral analysis for rolling-element bearing damage detection in induction machines. The study, first, reviewed and found the bearing characteristic frequencies and the modes of failure associated with the construction of the bearings. Then, the relationship between motor current and induced vibration, due to incipient bearing faults, was considered and investigated. This was done by deriving the effects of different bearing faults on the stator current spectrum. The experimental results verified the predicted relationship between the vibration and current frequencies, and confirmed that the stator current signature can be applied to detect the presence of a bearing fault. Another study by Stack et al. (2004) developed a method for detecting progressive motor bearing faults via stator current analysis. The method starts by removing the significant frequency content that are irrelevant to bearing faults by filtering the stator current. The filtered healthy current signal is then used to train an autoregressive model to produce a baseline or reference model. When bearing health is degraded, the deviation in spectral content from its baseline measurement is increased. This increase in spectral deviation was then used as the fault index. A CM technique based on statistical and numerical tools was suggested for detecting the onset of faults in induction motors (García-Escudero et al. 2011). The FFT was used to find the spectrum of the motor current, and a multi-resolution technique using wavelet function was implemented on this spectrum in order to detect the significant peaks. The researchers carried out an experimental study to prove the effectiveness of this approach, concluding that it is very reliable and convenient in detecting failures at their early stages, and it can also take into account the presence of serious anomalous feature measurements.

Nowadays, however, many studies have concentrated on using MCSA for power transmission systems condition monitoring as a replacement for typical monitoring techniques. The MCSA was used as the basis for CM of a multi-stage gearbox by using discrete wavelet transform (DWT) (Kar and Mohanty 2006). By observing the FFT analysis of the captured signals it was concluded that the low frequencies of vibration signatures have sidebands across line frequency of the motor current whereas high frequencies of vibration signature were difficult to detect. A suggestion of applying the discrete wavelet (DWT) to decompose the current signal was made, followed by FFT analysis on some of the DWT results to trace the sidebands of the high frequencies of vibration. In the experimental test rig the faults

(a) One tooth removed (b) Two teeth removed

Fig. 2.10 Simulated faults in the gears (Kar and Mohanty 2006)

was artificially simulated on one of the gears by removing one tooth and then two teeth, Fig. 2.10.

The electromagnetic motor torque estimation can give significant information about the efficiency and health condition of an electromechanical system, and this technique has recently been used for fault diagnosis of gears. The loss of lubrication in a gearbox is considered as a gear failure due to its influences on the vibration and on the electromagnetic estimated torque signatures. By using the electromagnetic torque estimation technique this and other fault types, such as tooth breakage fault in a high-ratio gear in cement kiln drives, have been identified (Kia et al. 2010; Bogiatzidis et al. 2013). Initially, a theoretical validation through a modelling approach to investigate how periodic impulse torque excitation affects the motor current spectrum and how it is expected to be demonstrated at the motor electromagnetic torque has been fulfilled. The effectiveness of this technique has successfully been demonstrated via experimental verification, and validated using vibration signal measurement and analysis simultaneously with the electromagnetic torque analysis.

Apart from aforementioned techniques, several non-destructive and contactless condition monitoring methods have been developed for monitoring machine health and compared with the traditional ones. For instance, a research group applied an infrared thermo-graphic technique to monitor deep-grooved ball bearing with circular weights mounted on them and different lubrication states (Seo et al. 2011). They compared the results from this method with those of the traditional vibration spectrum analysis to evaluate the efficiency of the suggested method. Figure 2.11 shows the test rig, and the infrared camera (Silver 450 M from Cedip Corp). The vibration analyzer shown in Fig. 2.11 was used for spectrum analysis, and the data acquired using this technique was reported to be clearer than that derived using vibration analysis technique.

Another study by Onsy et al. (2012) applied an image registration (IR) technique for online health monitoring of gears system. The main aim of this study was

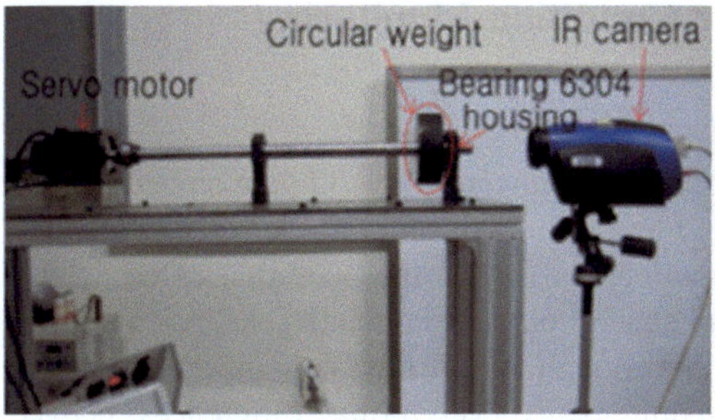

(a) Infrared camera

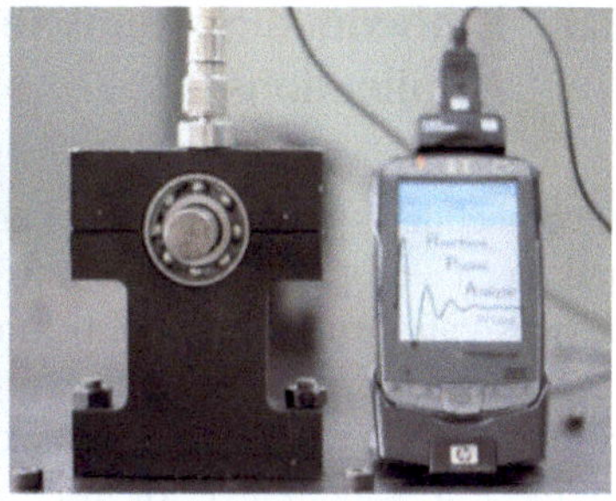

(b) Vibration analyser

Fig. 2.11 Test rig for comparison between the thermo-graphic and vibration analysis techniques for bearing health monitoring (Seo et al. 2011)

Fig. 2.12 Failure index (FI) values of geared system (Onsy et al. 2012)

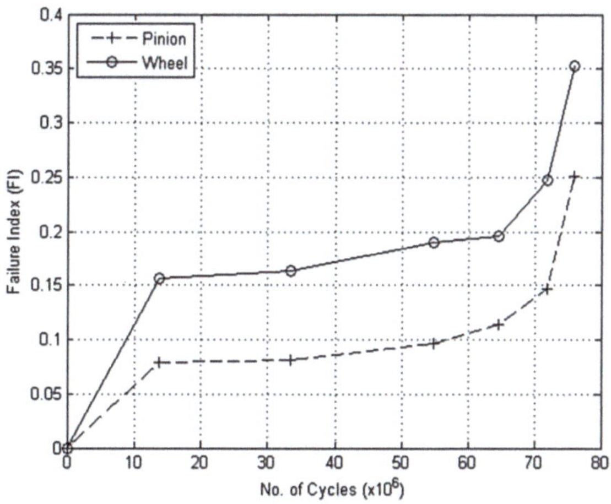

monitoring the progression of micro-pitting and surface scuffing failures. A back-to-back gearbox was designed, and a variable speed electric motor used to drive the system. To evaluate the state of health of this system, the failure index (FI) was found by comparing captured images at different time intervals with reference images taken before running the test. Figure 2.12 shows the values of failure index for pinion and wheel gears versus number of cycles and it can be concluded that the micro-pitting progressed gradually during testing. To check the capability of this technique, the FI results computed using the IR technique were correlated with vibration and oil debris analysis indicators measured for the same test rig and the findings were considered promising.

2.3 Embedded Systems for Condition Monitoring Applications

An embedded system consists of computer hardware with software embedded in it, and has a set of specific functions to be performed, often in real-time. Embedded devices can be used to control, monitor or assist in the operation of equipment, machinery or plant. They differ from general purpose computers such as a personal computer (PC) which are to be flexible enough to perform many different tasks and to meet a wide range of user requirements. The prime differences between embedded systems and PC computers are that the former often do not have displays or keyboards, usually come within larger systems or machines, and have constraints such as small memory, slow CPU or real-time response (Collins 2000).

Embedded systems are found in many applications, including modern cars, airplanes, and mobile robots. Their main merits are low-cost, flexible structure, steady performance, small size, low power consumption, high reliability and integration, and the ability to work in constricted spaces and tough environments (Wang 2009; Sarrafzadeh et al. 2006). Basically, all embedded systems contain a processor and software to execute instructions, and incorporate a memory to store the executable code, as well as input and output devices. Sensors and probe devices can be used to provide inputs, and outputs generally display the changes in the physical world via wire or wireless communications links (M 2002).

Embedded systems are one of the most widely used types of device in many current applications. One research study has described the applications of embedded systems for diagnostic and treatment planning in health care applications for patients with chronic diseases (Srovnal and Penhaker 2007), where the systems have to be portable, non-intrusive, and low in weight and cost in order to be suitable for use. This research also suggests that embedded home care systems could be used as predictive diagnostic systems. Interestingly, proposed applications have expanded to include home safety and environment (Zhai and Cheng 2011). In this study an embedded system was designed to monitor smog percentage and gas parameters, and to collect video information from within a house. Figure 2.13 shows the architecture of the proposed system, which basically contains two

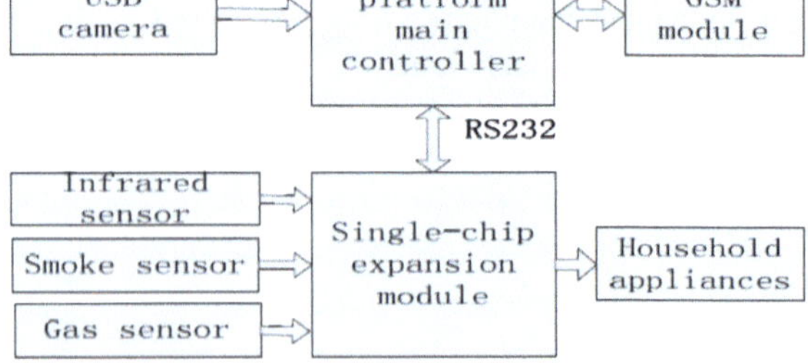

Fig. 2.13 Embedded system for household appliance monitoring (Zhai and Cheng 2011)

controllers (a main controller and an expansion module), and a number of different sensors connected to them. In addition, this system has the ability to communicate remotely with household appliances using a global mobile communications (GSM).

Recent years have witnessed a trend in using the embedded systems for machine fault detection and diagnosis. Generally speaking, there are two categories of embedded systems that can be used for industrial machinery condition monitoring; these are discussed as follows.

2.3.1 Wired Systems

Health monitoring systems are typically fulfilled using wired embedded systems by connecting communication cables directly between the processing and the input/output units. Different varieties of such systems are available nowadays; ranging from simple sensing devices that detect peak-acceleration or peak-strain and inform the user when a certain threshold is exceeded to a more complex system such as a piezoelectric accelerometer with a built-in charge amplifier connected directly to a hand-held, single-channel fast Fourier transform (FFT) analyser. For instance, a microcontroller-based data acquisition system integrated with an accelerometer was used for a milling machine vibration monitoring (Zhang and Chen 2008). The acquired data was sent to and analysed on a PC in real-time manner utilizing software developed in Visual Basic. Time-domain and FFT signal analyses methods were applied for feature extraction and visual interpretation was relied for tool health assessment. Furthermore, a vibration faults simulation system, which involves data acquisition and analysis using LabVIEW-based virtual instrument technology, was proposed to serve as teaching equipment for mechatronics students in the area of CM (Gani and Salami 2004). A test rig was developed to simulate and study most common vibration fault signatures encountered in rotating machines.

An embedded system which implements self-organizing maps using a neural network has been applied to the online detection and classification of faults in electromechanical valves used for flow control (Gonçalves et al. 2009). The aim was to build a proactive maintenance scheme for these valves. A mathematical model of the valve was used to train the map for the fault detection process, and fault classification training was carried out by fault injection based on parameter deviations using the same model. Throughout the online monitoring, the embedded system works to find the best match between the current torque and position, and their values which were calculated using the trained map. The embedded system was prototyped using a Xilinx FPGA (Field programmable gate array) development board. It was found that the embedded system is a hopeful solution for predictive maintenance in these actuators. Cabal-Yepez et al. (2013) has presented a design and implementation of an embedded system that utilizes reconfigurable hardware based on FPGA. This system performs time-frequency signal analysis techniques, such as short time Fourier transform (STFT) and discrete wavelet transform (DWT), on vibration signals captured from an industrial robot for the purpose of early abnormalities diagnosis. To the best of the author knowledge, this paper is considered the only one which focuses in the area of industrial robotics CM based on embedded system. However, there are several shortcomings in the paper, such as: only backlash fault has been considered, lack of intelligence capability since it cannot notify the operator if a fault has developed, and it relies on a wired communication.

A smart sensor network based on Texas Instrument digital signal processor (DSP) type TMS320F2812, AD7656 analogue-to-digital converter (ADC), accelerometer and temperature sensor was used to establish an embedded system for vehicle fault diagnosis (Lijun et al. 2010). The wavelet transform (WT) signal analysis technique for feature extraction was implemented on the system. The communication between the DSP and the other hardware peripherals was enabled using a controller area network (CAN) bus, which is a communication standard designed to permit microcontrollers and devices to communicate with each other in applications without a host computer. Other successful utilizations of DSPs as an embedded system are for medical diagnosis and gearbox vibration signals analysis (Chen et al. 2009b; Lijun et al. 2010). WT was also applied in these two papers, but on this occasion using a TMS320C6713 DSP.

It is clear that the digital signal processing and artificial intelligence algorithms are very powerful and becoming more commonly employed as tools for solving different monitoring problems. Traditionally, these algorithms are implemented using PCs, dedicated DSP chips or FPGAs. These solutions are very efficient in this matter but, on the other hand, they are expensive and large, as in the case of implementing them on PCs. Thus, low cost microcontrollers, such as Arduino and PIC microcontrollers, represent an alternative solution to implement these algorithms. An example is the implementation of ANN on an inexpensive 8-bit PIC microcontroller (Cotton et al. 2008; Tripathy et al. 2014; Rai and Rai 2013). Marandu (2014) has designed an intelligent mechatronic system based on modern version of PIC Microcontrollers, called dsPIC digital signal controller. It has been used for online dental material testing and surface wear monitoring with vibration signal capturing and analysis. LabVIEW software was used to design the graphical user interface (GUI) to send and receive the data from the system. Limitations of these controllers, however, are low memory and central processing unit (CPU) performance.

2.3.2 Wireless Systems

Based on the concept of wireless sensor networks (WSNs), which comprise of a number of battery-powered (or take advantage of nearby power supply if available) sensor nodes, each of which contains different (or the same) sensors types to monitor different variables and transmit the data wirelessly, embedded systems have been extensively used for building different wireless condition monitoring systems.

The typical sensor node should be small size, low power consumption and low cost. WSN solutions are being increasingly employed in CM applications, for example in vehicle fault diagnosis (Shukla et al. 2009). The system comprises a large number of sensors able to communicate with each other through a wireless network, able to get live data from the vehicle, such as oil temperature, wheel balance, and fuel level. The embedded microprocessors gather the data and send them to an external monitoring entity. Another paper has suggested an intelligent diagnosis system combining WSN with a multi-agent system (MAS) (Wu et al. 2011); to satisfy the needs for high sampling rates, high precision, high speed and large amounts of data transmitted from mechanical equipment. The efficiency of the system for a coal preparation plant was investigated, and its practicability was demonstrated.

Other applications of embedded system are found in structural health monitoring. Rad and Shafai (2009) utilized wireless embedded sensors as a successful alternative to fiber optics sensors to assess the state of the infrastructure of bridges in North America. Wireless sensor networks have also shown sufficient potential in data collection when they have been applied to monitor wind turbine blades (Taylor et al. 2011). Here piezotronic accelerometers were used to pick up the signals from blades in both healthy and damaged states, and the sensors were fixed at different locations on the blades and wireless data acquisition utilized. Micro-electro-mechanical-sensors (MEMS) have also been used for condition monitoring. For example, a tiny and very light weight MEMS accelerometer has been mounted on a rotor shaft to monitor its dynamic behavior (Elnady et al. 2011). The accelerometer was connected to a wireless sensor node for the wireless transmission of vibration signals, as shown in Fig. 2.14. Without any added imbalance and at different rotating speeds, vibration measurements such as acceleration values were taken with acceptable performance. It was reported that this technique assisted in reducing the number of sensors needed to monitor the rotating parts.

An embedded system has been applied to helicopter gearbox monitoring (Qin and Hu 2012), with the aim of designing a wireless sensor node fixed to the planetary gears' carrier in order to gather vibration signals to an external receiver through the antenna which extends into the gearbox. The acquired signal was

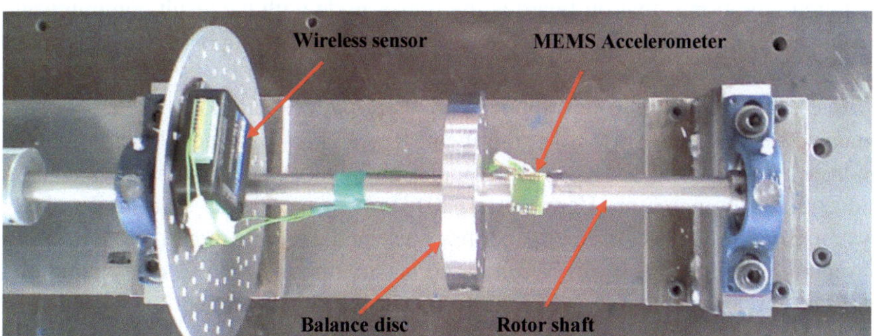

Fig. 2.14 Vibration measurement using MEMS accelerometer (Elnady et al. 2011)

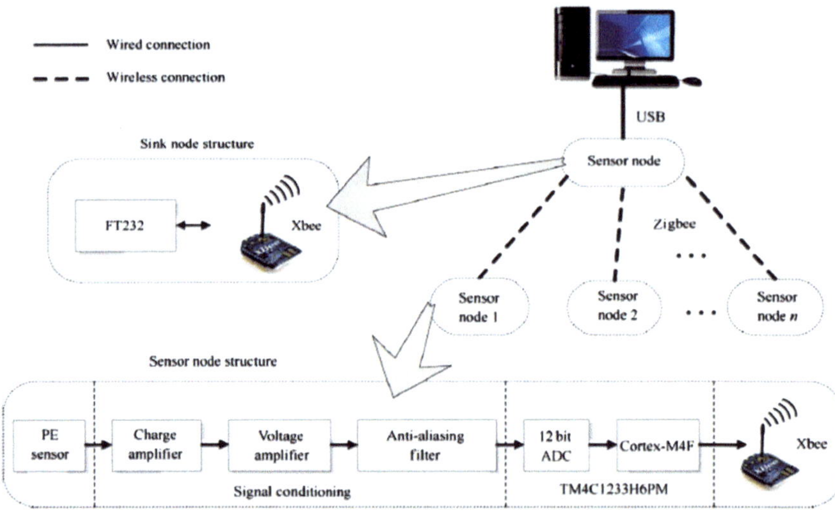

Fig. 2.15 Structure of the wireless bearing CM system (Feng et al. 2015)

analyzed using signal processing methods. An experimental system consisting of a set of planetary gears built using one sun gear and four planetary gears was constructed, and four wireless sensor nodes were installed in the space between each two neighboring gears.

Zigbee is a wireless protocol widely adopted in WSN because of its low cost, low power consumption and applicability to create large scale networks. A research study has implemented the envelope analysis algorithm for wireless bearing CM based on vibration signals measurement (Feng et al. 2015), however, to overcome the limitations of memory size and restricted computational capabilities in the commercially available wireless nodes, the authors have used a 32-bit microcontroller type TM4C1233H6PM from Texas Instruments along with Zigbee wireless module. The power consumption on the wireless node, which represents a considerable problem in the WSN systems, has been reduced by processing the acquired vibration signal on-board at the sensor node with the result communicated to the recipient node. The hardware architecture of the proposed wireless monitoring systems is shown in Fig. 2.15.

Interestingly, the rapid developments in smartphones and portable devices have changed the traditional way of using them. Researchers have developed a scalable android application based on a smartphone to diagnose some types of fault in an industrial air compressor (Verma et al. 2013). They mentioned that the developed system is very reliable. Another paper has presented a remote monitoring system for a rotating machine which can be run based on smartphone or PDA (personal digital assistant) (Wanbin and Tse 2006). In this paper the developers put the capability of informing the concerned user if a fault appears in the remotely monitored machine. Similarly, a paper proposes a real-time method to perform the monitoring of

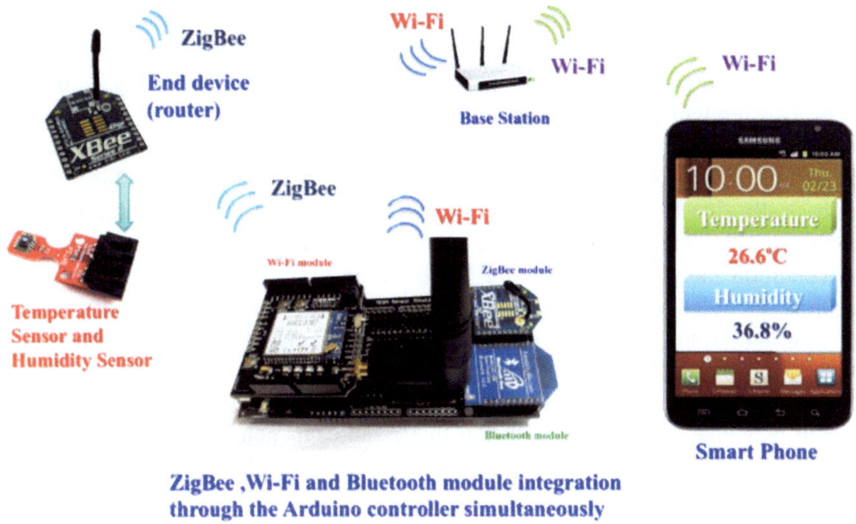

Fig. 2.16 Intelligent embedded system for a factory monitoring

temperature, humidity, air quality and vibrations of operating machinery in a factory zone using smart phones (Lian et al. 2013). The integration of ZigBee and Wi-Fi communication protocol were utilized to build the intelligent monitoring system, Fig. 2.16. By using the ZigBee protocol, the sensors on the factory site transport the real-time sensed data to an integrated embedded controller. The embedded controller was constructed based on an open-source, 32-bit ARM core Arduino Due module. This controller is able to instantly provide numerical results, depending on the received and analysed data, to the smart phones of the factory manager. However, this aspect of CM is lacking of exploration and needs to be further investigated in order to be applied to different machines.

2.4 Applications of Condition Monitoring Techniques in Industrial Robots

After reviewing the previously applied condition monitoring approaches and techniques for a range of machinery, this section explores how these approaches and techniques are applied to industrial robots health monitoring; and to survey the state of the art research with a view to making existing gaps in this area clearer to the readers and researchers. Unfortunately, there is very little (or no) published information regarding the distribution of the robot fault types. Even if available, the most recent one was published in 2000 and it is mainly related to the robot actuator faults (Arvallo and Tesar 2000), and it does not include gearbox faults, which could be a large proportion since robotic gearboxes are frequently overloaded and experience direct effects of shocks. The author has contacted several robot manufacturers regarding this matter, but regrettably no one of these companies has provided information, not surprisingly, since this might be something connected with their reputation in the competitive market.

Industrial robots are extremely complex mechanism and hence the application of condition monitoring for them differs from that of 'simple' rotating machinery. This is basically due to the instantaneous change of geometrical configuration of the robot arm. Previously in this chapter, it has mentioned that there are two approaches to condition monitoring, which are model-based and model-free. Either of these approaches or a combination of both has been adapted in industrial robot condition monitoring. Filaretov et al. (1999) used a nonlinear model to address problems of fault detection and isolation in complex systems, such as in robot manipulators. Algebraic functions were implemented to design the nonlinear diagnostic observer, which was able to dispense with the linearization in nonlinear models to avoid model errors. The robot modeling was conducted using Matlab in discrete time. It was shown that, despite the fact that the use of this model dispenses with linearization, it does not allow some faults to be isolated. In another paper, a model-based fault detection and isolation (FDI) scheme for rigid manipulators was designed which depends on a suboptimal second-order sliding-mode (SOSM) algorithm (Brambilla et al. 2008). In order to make the procedure of FDI possible, an input signal estimator and output observers were adopted and SOSM was used to design the input laws for the observers. Experimental work and theoretical simulations were accomplished with a COMAU SMART3-S2 robot manipulator, and the results showed that the scheme has a good ability to detect and identify faults. On the other hand, the proposed scheme was not able to deal with multiple faults in more than one actuator or sensor, and is also neglected elastic effects in the robot.

Another technique proposed for fault detection and isolation in robot manipulators was based on a new simplified Euler-Lagrange (EL) equation capable of reducing the complexity of the approach (Mohseni and Namvar 2009). The use of this equation allowed the uncertainty in the manipulator's gravity term to be handled. Moreover, the effects of noise and an uncalibrated joint torque sensor could be taken into account. Simulation was conducted using Matlab-Simulink environment to illustrate the performance of this method. A study by Caccavale et al. (2009) presented an approach based on support vector machines (SVM) to detect and isolate fault in a robot's actuators, using an available dynamic model of the manipulator, and trained SVMs offline to compensate for unknown dynamics, uncertainties and disturbances. Furthermore, a radial basis function network was implemented to interpolate unknown actuator faults. Finally, an investigation was performed experimentally using Comau SMART-3S industrial robot to check the effectiveness of the approach.

A model-based fault diagnosis to detect actuator faults in a robot manipulator was introduced by Capisani et al. (2010). Analytical redundancy was achieved using higher order sliding mode unknown input observers (UIO). In addition, the design of the input laws for the observers was based on the super-twisting second order sliding mode control (SOSMC) approach. Simulation and experimental work

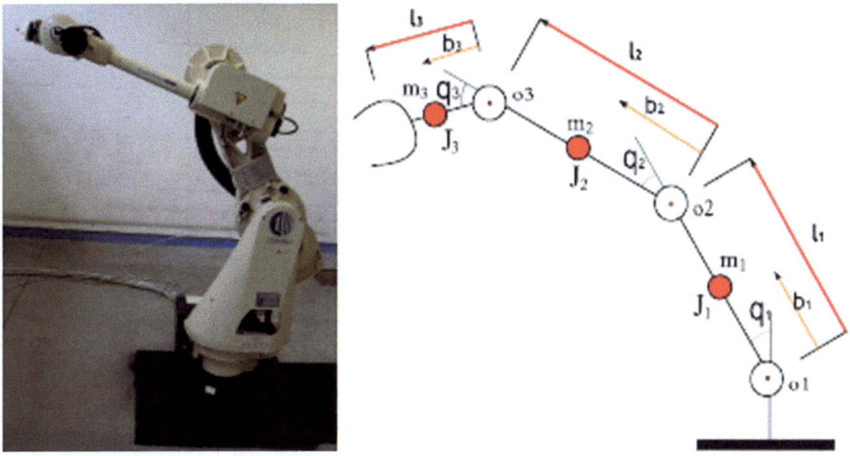

Fig. 2.17 The SMART3-S2 robot (Capisani et al. 2010)

was conducted on a COMAU SMART3-S2 with three links and three joints as illustrated in Fig. 2.17. The results were compared with those of previously proposed approach which depends on sub-optimal second order sliding mode control (SOSMC). It was concluded that the super-twisting approach did not always provide good performance in terms of avoiding false alarms.

Because of the wear process in robot's joints, the friction level will increase, and a study has been conducted to consider the problem of wear estimation in standard industrial robot joints (Bittencourt et al. 2011). A static friction model was used to find the wear level and then this model was extended to take account of the effects of wear. The resulting model illustrates the relationship between friction in the joints and changes in speed, load, temperature and wear. As a result of the experimental and theoretical work, a wear estimator was proposed which was able to distinguish between wear effects under large temperature variations.

Because precise mathematical models for complex systems like a robot are difficult to obtain, model-free methods based on AI or statistical approaches have become prevalent choices for robot health monitoring. The backlash and looseness in the power transmission system of a robot may cause torque variations. However, the electric motor itself generates what is known as a back electromotive force (EMF) when subjected to mechanical load making them acting as a torque transducer (Yuan et al. 2011). The torque variations measurements via current fluctuations on robotic actuators have been applied for robot monitoring (Abdul and Liu 2008; Yuan et al. 2011). The advantage of this technique, as early mentioned, to the robots health monitoring is that the motor current can remotely be measured along the power cables utilizing standard current sensors without supplementary instrumentation on the robot.

Some reported robot fault diagnostic systems are based on acoustic signals analysis. Such systems would have to be able to distinguish the correct information

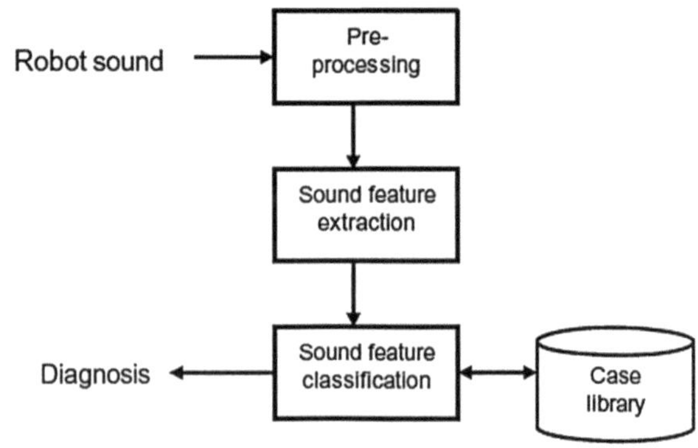

Fig. 2.18 Fault diagnosis steps (Olsson et al. 2004)

from the ambient noise. Case-based reasoning and signal processing were adopted to build an approach to diagnosis the faults in an industrial robot (Olsson et al. 2004). Wavelet analysis was applied to remove noise from the acoustic signals and to extract the most relevant features, which were then sent to the classification component, which uses case-based reasoning to identify the class of faults according to the characteristic of the previous fault cases. Experimental work on an industrial robot was used to assess the performance of this approach, and Fig. 2.18 shows a schematic diagram of the set-up.

A microphone was used to gather sound signals from the robot, and unwanted noise was filtered out in a pre-processing step, after which the important sound features were extracted, and their classification was performed based on previously classified sound descriptions in the case library. The authors reported that "this system is able to successfully diagnose faults in an industrial robot based on a low number of previous examples". The same principle was applied to an industrial robot, but on this occasion the ANN was used for noise analysis and classification (Yildirim and Eski 2010). Noise sensors with data acquisition hardware and feature extraction software were used to prepare the training data for designing the ANN-based noise fault detection of robot manipulator's joints.

Different types of faults in robot transmission systems, such as bearing and gear faults, can cause system degradations and thus lead to development of lost motion (looseness). Vibration analysis algorithms, which rely on measured vibration signal by an accelerometer, are often used and represent the vast majority of utilized techniques for industrial robot health monitoring. Modal analysis, which can give information concerning the dynamic characteristics of machines, involves machine vibration measurement has previously been applied to assess the dynamic characteristics or for fault detection of industrial robots or rotating machinery (Ma et al. 2007; Liguo et al. 2009). Experimental modal analysis was utilized to find the dynamic characteristics of a PUMA 560 robot and use them to detect robots joint faults (Bicker et al. 1989). An experimental programme was accomplished to investigate the changes in the vibration spectra resulting from induced faults in the transmission of the elbow joint on the robot. From the obtained results, the

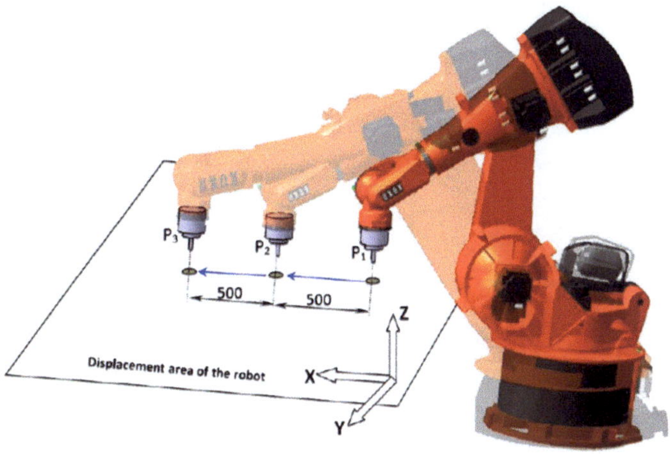

Fig. 2.19 The experimental configuration for impact test (Claudiu et al. 2012)

researchers concluded that backlash can influence the vibration response of the robot elements during normal operations. But, the slight change of peak amplitude at some frequency bands for various backlash conditions was not reliable enough to be used as a criterion for fault diagnosis. Also, at the reversal of motion, the characteristics of backlash can be averaged out over the whole cycle if a conventional spectrum analysis is used to process the vibration signatures. Consequently, the distinct differences between the signatures for different fault conditions cannot be identified accurately.

Recently, modal analysis of KUKA type milling robot has been carried out by Claudiu et al. (2012). The researchers tried to evaluate the robot stiffness at three different and most commonly applied working configurations, as in Fig. 2.19. The first step in this research was to identify the robot self-excited frequencies using impact testing (modal analysis). Then, vibration analysis was conducted first when the robot moving, and after that whilst performing milling process. The result showed that the robot configuration has a significant effect on its stiffness, and therefore on its natural frequencies.

In another study of fault diagnosis in rigid link manipulators, an online learning architecture with a neural network was used for fault detection and isolation by monitoring the behavior of the system (Vemuri et al. 1998). A two-link robotic system was used to show the capability of the neural network in fault diagnosis. Results showed that the learning methodology which was used can provide a model of a fault via analysis of input/output properties as well as detecting its occurrence. A large backlash level in the robot's joints represents a very serious problem. Pan et al. (1998) used vibration signals during normal operation to diagnose joint-backlash on a PUMA 762 industrial robot. Time-domain and frequency-domain analyses were employed to identify features such as probability and density. Artificial neural networks were then used for pattern recognition. The experimental work was performed as shown in Fig. 2.20. One accelerometer was fixed

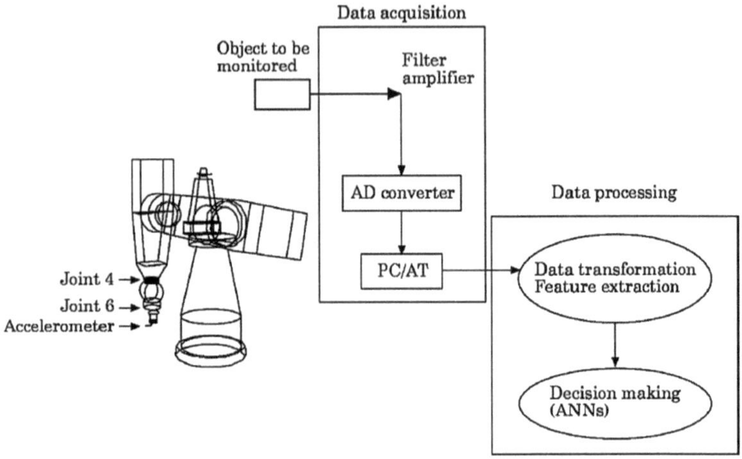

Fig. 2.20 Schematic diagram of the experimental set-up (Pan et al. 1998)

on the robot end effector to measure vibration responses. Additionally, different levels of backlash were artificially contrived in joints 4 and 6 to validate this method. It was pointed out that this technique could be applied in real working environments, and moreover it was inexpensive as only one sensor was used to detect the robot's faults.

Artificial neural networks (ANNs) were used for residual generation and analyzing them in robotic manipulators (Terra and Tinós 2001). For residual analysis, three types of ANN architectures were employed. The first is the radial basis function network (RBFN), which uses position and velocity residuals to identify faults. The second architecture also uses a RBFN, but it utilizes only the velocity residual, and the third is a multilayer perceptron (MLP). A comprehensive simulation study of the PUMA 560 yielded results collected from three joints. It was concluded that the post-failure control of the mechanical manipulator in a hybrid system framework could be included in this work.

Similarly, a technique using only one accelerometer mounted at the robot tip has been applied for the online fault diagnosis in the 4 Degree of Freedom (DOF) SCARA robot (Liu et al. 2009). The tip acceleration was calculated using a dynamic model of the robot, and was used as a reference. By comparing the experimental tip acceleration with the reference, the condition of the robot could be identified. In contrast, another study used more than one sensor for robot joint condition monitoring (Trendafilova and Van Brussel 2003). The objectives were to extract the vital features directly from the measured acceleration signals, and to try to specify defects by finding properties dependent on fault size. Signals were analyzed from the robot joints without error, and subsequently from joints having backlash, using nonlinear dynamics and statistical tools. The proposed system was validated using three robot types: spherical robot arm; SCARA robot arm; revolute robot arm, and on different joints. In order to simulate robot damage conditions, three levels of backlash (small, medium, and large) were generated in the joints by implementing a variety of loads and adjusting the backlash screws. The authors used the pattern recognition principle with nonlinear autoregressive (NAR) analysis for the detection of defect from the data, and acceptable performance was demonstrated. The same technique applied for fault quantification was less effective, however.

Halme (2006) studied the condition monitoring of servomotors and gears in an industrial robot using performance criteria monitoring, which is a model-free method. This study was implemented with a 6 DOF robot (type Fanue R-J2 M-6i) utilized for material handling. This robot has a weighs 290 kg, and is capable of moving 6 kg with a repeatability of ±0.1 mm in space. Acceleration, acoustic emission, and sound sensors were used in order to monitor the accuracy of the robot's path. By comparing different vibration signatures with signals measured over time, deviations in the performance of the robot could be found. However, this method cannot represent an eclectic technique since it always needs to compare signals with references measured at different times and from the same production process.

Another research study used wavelet multi-resolution analysis (WMRA) coupled with a neural network-based approach in order to diagnosis faults in an industrial robot manipulator (Datta et al. 2007). A Matlab-Simulink environment was used to monitor the neural network classifier for a robot used in semi-conductor fabrication. It was concluded that the WMRA is excellent for data reduction and capturing the important properties of signals, although it did not show good performance in distinguishing some of the signals. On the other hand, two neural networks have been used to propose an algorithm for the online monitoring of two-link manipulators (Van et al. 2011). This approach focuses on identifying changes in robot dynamics due to faults. It was noted that this technique was able to provide estimates of fault characteristics.

2.5 Summary

The foregoing literature review has covered a variety of topics, approaches, and techniques applied in the field of CM. The review has shown that two main categorizes of CM approaches are available: model-based and model-free approaches. In model-based CM, a model of the system (or of how it is presumed to behave) is created. Then, the predicted behaviour from this model is compared to the actual behaviour of the machine and any detected significant deviations can be interpreted as indications of faults. Accurate analytical models of industrial robots are often not practicable and difficult to be constructed due to dynamic complexity of the robot and unavoidable uncertainties (Verdonck et al. 2001). Accordingly, model-free approaches based on statistical and artificial intelligence tools are going to be considered for establishing the condition monitoring system of the robot.

It has also been established that many different techniques of machine condition monitoring, such as the analysis of vibration, acoustic emission, wear and thermal measurements, are frequently used. Vibration and acoustic emission analysis represent the most important techniques that have commonly been applied to monitor the status of industrial machines, and such methods have been widely studied by researchers. Because of the high frequency range of the acoustic signals, CM systems based on AE techniques require highly specialized, expensive sensors and extremely high sampling rates, which means a considerable amount of data has to be acquired in order to detect the exceptional events (Ogbonnah 2007; Randall 2011). Also, due to the acoustic signal attenuation during propagation, the AE sensors need to be as close to the source as possible. In contrast, vibration signal respond immediately to manifest itself if any change has appeared in the monitored machine. Consequently, it provides an easy and cost-effective sensing technique to detect faults in machines and for this reason it will be applied for CM of the robot.

Generally speaking, the vast majority of the literature focuses on finding a monitoring system able to take minimum and precise measurements necessary from machines, which can give clear indications of incipient fault modes in a minimum time. Moreover, the issue of feature extraction from data gathered has been a point of debate among many researchers. In short, to design a reliable condition monitoring system, sensors have to be chosen correctly in order to get accurate signals from faulty parts, and appropriate signal analysis techniques have to be employed since these have a significant impact on the sensitivity of the features extracted from the signals captured.

Recently, the field of the condition monitoring of machines has moved from the use of conventional to AI techniques. A wide variety of AI techniques have been applied extensively in monitoring very complex and non-linear systems such as industrial robots, where it is difficult to build accurate mathematical models of the system. However, each of the AI techniques has strengths and weaknesses, and many studies have concluded that combining multiple methods can give better performance in many condition monitoring applications. Nevertheless, despite the large amount of research conducted in the area of AI in condition monitoring, it is still inadequate and requires significant investigation to be performed. Artificial neural networks (ANNs) have broadly been applied in a number of real-world problems of considerable importance and complexity and will be utilized in this research, not only because of their ability to handle the highly non-linear relationships that exist between industrial robot parameters but also because they can deal with large numbers of variables and provide general solutions with significant predictive accuracy (Ogaji 2003).

In the majority of CM systems, with a special focus on industrial robots, it has been noted that for signal acquisition, processing, storage and decision making, data acquisition (DAQ) cards connected to PCs, as their main processing core, were extensively used, which adds considerably to the cost of CM system. Recent developments in electronics and computing have opened new horizons in the area of condition monitoring, and embedded devices other promising solutions, and have shown their practicality in fault detection and diagnosis processes in many of areas. The main aim of using embedded systems is to allow data analysis, which includes feature extraction and diagnostics, to be carried out locally at field level and transmitting the results wirelessly to the base station, which as a result will help to overcome the need for wiring. Furthermore, it seems that there is a serious shortage of studies applying the embedded devices in the industrial monitoring field, and thus they need to be investigated thoroughly, both by researchers, as in this thesis, and by interested companies.

The use of industrial robots has been rapidly increased in a wide range of industrial applications. Therefore, the need for reliable fault detection and diagnosis methods for industrial robots has been increased recently. Most faults in industrial robots occur in their joints, since they have many mechanical and electromechanical parts, such as gears and motors. However, in the above reviewed work in the area of industrial robot CM it is noted that much of the work has been aimed directly at detection of only backlash fault in the gear transmission system. This means that other types of gear and bearing faults have not yet been fully investigated. In this work, an effort to fill part of the gap in the subject of industrial robot CM by not only detection the backlash fault but also other faults, such as gear tooth wear and inner and outer race bearing faults, will be assessed using vibration signal analysis.

In summary, the endeavour of the work in this thesis is to build a wireless and intelligent condition monitoring system for an industrial robot based on an embedded system capable of performing signal capturing, analysis, feature extraction, and then fault detection and diagnosis in real-time operation. The adoption of this route is to try to contribute to the concept of factories-of-future in the 21st century.

References

Abdul, S., & Liu, G. (2008). Decentralised fault tolerance and fault detection of modular and reconfigurable robots with joint torque sensing. In *Proceedings—IEEE International Conference on Robotics and Automation* (pp. 3520–3526).

Aliustaoglu, C., Ertunc, H. M., & Ocak, H. (2008) Applied real time bearing fault diagnosis based on vibration and current analysis. 258–262.

Aliustaoglu, C., Ertunc, H. M., & Ocak, H. (2009). Tool wear condition monitoring using a sensor fusion model based on fuzzy inference system. *Mechanical Systems and Signal Processing, 23*, 539–546.

Arvallo, A. V., & Tesar, D. (2000). *Condition-based maintenance of actuator systems using a model-based approach.* Ph.D Thesis, The University of Texas at Austin.

Bae, H., Chun, S. P., & Kim, S. (2006). Predictive fault detection and diagnosis of nuclear power plant using the two-step neural network models.

Balazinski, M., Czogala, E., Jemielniak, K., & Leski, J. (2002). Tool condition monitoring using artificial intelligence methods. *Engineering Applications of Artificial Intelligence, 15*, 73–80.

Baydar, N., Chen, Q., Ball, A., & Kruger, U. (2001). Detection of incipient tooth defect in helical gears using multivariate statistics. *Mechanical Systems and Signal Processing, 15*, 303–321.

Bicker, R., Daadbin, A., & Rosinski, J. (1989). The monitoring of vibration in industrial robots. In *ASME 12th Biennial Conference on Mechanical Vibration and Noise.*

Bittencourt, A. C., Axelsson, P., Jung, Y., & Brogardh, T. (2011). Modeling and identification of wear in a robot joint under temperature uncertainties. *Automatic Control.*
Bogiatzidis, I. X., Safacas, A. N., & Mitronikas, E. D. (2013). Detection of backlash phenomena appearing in a single cement kiln drive using the current and the electromagnetic torque signature. *IEEE Transactions on Industrial Electronics, 60*, 3441–3453.
Brambilla, D., Capisani, L. M., Ferrara, A., & Pisu, P. (2008). Actuators and sensors fault detection for robot manipulators via second order sliding mode observers. In *2008 International Workshop on Variable Structure Systems* (pp. 61–66). IEEE.
Butler, S. (2012). *Prognostic algorithms for condition monitoring and remaining useful life estimation.* Ph.D Thesis, National University of Ireland, Maynooth.
Cabal-Yepez, E., Garcia-Ramirez, A. G., Romero-Troncoso, R. J., Garcia-Perez, A., & Osornio-Rios, R. A. (2013). Reconfigurable monitoring system for time-frequency analysis on industrial equipment through STFT and DWT. *IEEE Transactions on Industrial Informatics, 9*, 760–771.
Caccavale, F., Cilibrizzi, P., Pierri, F., & Villani, L. (2009). Actuators fault diagnosis for robot manipulators with uncertain model. *Control Engineering Practice, 17*, 146–157.
Capisani, L. M., Ferrara, A., Ferreira, A., & Fridman, L. (2010). Higher order sliding mode observers for actuator faults diagnosis in robot manipulators. In *2010 IEEE International Symposium on Industrial Electronics* (pp. 2103–2108). IEEE.
Chandroth, G. O., Sharkey, A. J. C., & Sharkey, N. E. (1999). Cylinder pressures and vibration in internal combustion engine condition monitoring. *Comadem 99.* Sunderland, UK.
Chen, C. Y., Ke, M. D., & Kuo, C. D. (2009b). Continuous wavelet transformation the wavelet implemented on a DSP chip for EEG monitoring. In *2009 1st International Conference on Information Science and Engineering* (pp. 3633–3636). ICISE 2009.
Chiang, D. Y., & Lai, W. Y. (1999). Structural damage detection using the simulated evolution method. *AIAA Journal, 37*, 1331–1333.
Claudiu, B., Mehdi, C., Alain, G., & Jean-Yves, K. (2012). Dynamic behavior analysis for a six axis industrial machining robot. *Advanced Materials Research.*
Collins, C. M. (2000). *An evaluation of embedded system behavior using full-system software emulation.* Maryland: Master of Science.
Cotton, N. J., Wilamowski, B. M., & Dündar, G. (2008). A neural network implementation on an inexpensive eight bit microcontroller. In *12th International Conference on Intelligent Engineering Systems—Proceedings* (pp. 109–114), INES 2008.
Datta, A., Mavroidis, C., Krishnasamy, J., & Hosek, M. (2007). Neural netowrk based fault diagnostics of industrial robots using wavelt multi-resolution analysis. In *American Control Conference* (1858–1863), USA.
Di Nuovo, A. G., & Catania, V. (2007). Genetic tuning of fuzzy rule deep structures for efficient knowledge extraction from medical data, 5053–5058.
Ding, S. X. (2008). *Model-based fault diagnosis techniques.* Germany: Springer.
Dreyfus, G. (2005). *Neural networks, methodology and applications.* London: Springer.
Ebersbach, S., Peng, Z., & Kessissoglou, N. J. (2006). The investigation of the condition and faults of a spur gearbox using vibration and wear debris analysis techniques. *Wear, 260*, 16–24.
Elnady, M. E., Sinha, J. K., & Oyadiji, S. O. (2011). On-shaft wireless vibration measurement for condition monitoring of rotating machine. In *International Conference on Vibration Problems.* Prague.
Feng, G. J., Gu, J., Zhen, D., Aliwan, M., Gu, F. S., & Ball, A. D. (2015). Implementation of envelope analysis on a wireless condition monitoring system for bearing fault diagnosis. *International Journal of Automation and Computing, 12*, 14–24.
Filaretov, V. F., Vukobratovic, M. K., & Zhirabok, A. N. (1999). Observer-based fault diagnosis in manipulation robots. *Mechatronics, 9*, 929–939.
Fulcher, J. (2006). *Advances in applied artificial intelligence.* Hershey: Idea Group Publishing.
Gallova, S. (2010). *Fuzzy parameters and cutting forces optimization via genetic algorithm approach.* London: Springer.

Gani, A., & Salami, M. J. E. (2004). Vibration faults simulation system (VFSS): A lab equipment to aid teaching of mechatronics courses. *International Journal of Engineering Education, 20*, 61–69.

García-Escudero, L. A., Duque-Perez, O., Morinigo-Sotelo, D., & Perez-Alonso, M. (2011). Robust condition monitoring for early detection of broken rotor bars in induction motors. *Expert Systems with Applications, 38*, 2653–2660.

Gelman, L., Jennions, I., & Petrunin, I. (2011) Detection of chipped tooth in gears by the novel residual technology. In 8th *International Conference on Condition Monitoring and Machinery Failure Prevention Technologies 2011* (pp. 966–977). CM 2011/MFPT 2011.

Goldberg, D. E. (1989). *Genetic algorithms in search, optimization, and machine learning*. Reading: Addison-Wesley.

Gonçalves, L. F., Bosa, J. L., Henriques, R. V. B., & Lubaszewski, M. S. (2009). Design of an embedded system for the proactive maintenance of electrical valves.

Halme, J. (2006). Condition monitoring of a material handling industrial robot. In *19th International Congress*, Lulea, Sweden.

Heng, A. S. Y. (2009). *Intelligent prognostics of machinery health utilising suspended condition monitoring data.* Ph.D, Queensland University of Technology.

Huang, W., Niu, P., & Lu, X. (2014). Spur bevel gearbox fault diagnosis using wavelet packet transform for feature extraction. *Advances in Intelligent Systems and Computing.*

Jack, L. B., & Nandi, A. K. (2000). Genetic algorithms for feature selection in machine condition monitoring with vibration signals. *IEE Proceedings: Vision, Image and Signal Processing, 147*, 205–212.

Janier, J. B., & Fazrin Zaim Zaharia, M. (2011). Condition monitoring system for induction motor using fuzzy logic tool, 3–7.

Jardine, A. K. S., Lin, D., & Banjevic, D. (2006). A review on machinery diagnostics and prognostics implementing condition-based maintenance. *Mechanical Systems and Signal Processing, 20*, 1483–1510.

Jiang, J., & Zhang, B. (2012). Rolling element bearing vibration modeling with applications to health monitoring. *JVC/Journal of Vibration and Control, 18*, 1768–1776.

Kar, C., & Mohanty, A. R. (2006). Monitoring gear vibrations through motor current signature analysis and wavelet transform. *Mechanical Systems and Signal Processing, 20*, 158–187.

Kia, S. H., Henao, H., & Capolino, G. A. (2010). Torsional vibration assessment using induction machine electromagnetic torque estimation. *IEEE Transactions on Industrial Electronics, 57*, 209–219.

Kim, H.-E. (2010). *Machine prognostics based on health state probability estimation.* Ph.D Thesis, Queensland University of Technology.

Kim, E. Y., Tan, A. C. C., Mathew, J., & Yang, B. S. (2008). Condition monitoring of low speed bearings: A comparative study of the ultrasound technique versus vibration measurements. *Australian Journal of Mechanical Engineering, 5*, 177–189.

Kisić, E., Petrović, V., Jakovljević, M., & Đurović, Ž. (2013). Fault detection in electric power systems based on control charts. *Serbian Journal of Electrical Engineering, 10*, 73–90.

Kudva, J. N., Munir, N., & Tan, P. W. (1992). Damage detection in smart structures using neural networks and finite-element analyses. *Smart Materials and Structures, 1*, 108–112.

Lei, Y., Li, N., Lin, J., & He, Z. (2015). Two new features for condition monitoring and fault diagnosis of planetary gearboxes. *JVC/Journal of Vibration and Control, 21*, 755–764.

Lian, K. Y., Hsiao, S. J., & Sung, W. T. (2013). Mobile monitoring and embedded control system for factory environment. *Sensors (Switzerland), 13*, 17379–17413.

Liang, X., Zuo, M. J., & Hoseini, M. R. (2015). Vibration signal modeling of a planetary gear set for tooth crack detection. *Engineering Failure Analysis, 48*, 185–200.

Liguo, Z., Yutian, W., Sheng, Z., & Guangpu, H. (2009). The fault diagnosis of machine based on modal analysis. In *2009 International Conference on Measuring Technology and Mechatronics Automation* (pp. 738–741). ICMTMA 2009.

Lijun, C., Yingtang, Z., Zhining, L., Guoquan, R., & Yiquan, S. (2010). Research on smart sensor network in fault diagnose system. In *2010 International Conference on Computer, Mechatronics, Control and Electronic Engineering* (pp. 23–26). CMCE 2010.

Lim, W. L. (2009). *The application of artificial neural networks for sensor validation in diesel engine condition monitoring and fault diagnosis.* M. Phil., University of Newcastle upon Tyne.

Liu, H., Wei, T., & Wang, X. (2009). Signal decomposition and fault diagnosis of a scara robot based only on tip acceleration measurement. In *2009 International Conference on Mechatronics and Automation* (pp. 4811–4816). IEEE.

Lopes, V., Jr., Park, G., Cudney, H. H., & Inman, D. J. (2000). Impedance-based structural health monitoring with artificial neural networks. *Journal of Intelligent Material Systems and Structures, 11*, 206–214.

Loutridis, S. J. (2008). Gear failure prediction using multiscale local statistics. *Engineering Structures, 30*, 1214–1223.

Ma, H., Li, H., Xie, W., & Chen, F. (2007). Vibration research on winding faults of induction motor based on experiment modal analysis method. In *8th International Power Engineering Conference* (pp. 366–370). IPEC 2007.

Marandu, S. I. (2014). *Design of a mechatronic system for measurement of surface fatigue in dental composites* Ph.D. Thesis, Newcastle University.

Marwala, T. (2012). *Condition monitoring using computational intelligence methods*. London: Springer.

Meruane, V., & Heylen, W. (2011). An hybrid real genetic algorithm to detect structural damage using modal properties. *Mechanical Systems and Signal Processing, 25*, 1559–1573.

Mohanty, A. R. (2015). *Machinery condition monitoring: Principles and practices*. London: Taylor & Francis Group.

Mohseni, S., & Namvar, M. (2009). Fault diagnosis in robot manipulators in presence of modeling uncertainty and sensor noise. In *2009 IEEE Control Applications, (CCA) & Intelligent Control, (ISIC)* (pp. 1750–1755). IEEE.

Montgomery, D. C., & Runger, G. C. (2014). *Applied statistics and probability for engineers*. New York: Wiley.

Moslem, K., & Nafaspour, R. (2002). Structural damage detection by genetic algorithms. *AIAA Journal, 40*, 1395–1401.

Munakata, T. (2008). *Fundamentals of the new artificial intelligence*. London: Springer.

Navarro, L., Delgado, M., Urresty, J., Cusid, J., & Romeral, L. (2010) Condition monitoring system for characterization of electric motor ball bearings with distributed fault using fuzzy inference tools, 1159–1163.

Negnevitsky, M. (2005). *Artificial intelligence: A guide to intelligent systems*. Reading: Addison Wesley.

Niknam, S. A., Thomas, T., Wesley Hines, J., & Sawhney, R. (2013). Analysis of acoustic emission data for bearings subject to unbalance. *International Journal of Prognostics and Health Management, 4*.

Ogaji, S. O. T.-O. (2003). *Advanced gas-path fault diagnostics for stationary gas turbines.* Ph.D. Thesis, Cranfield University.

Ogbonnah, V. (2007). *Condition monitoring of gear failure with acoustic emission.* M.Sc, Blekinge Institute of Technology.

Olsson, E., Funk, P., & Xiong, N. (2004). Fault diagnosis in industry using sensor readings and case-based reasoning. *Journal of Intelligent and Fuzzy Systems, 15*, 41–46.

Onsy, A. (2009). *Intelligent health monitoring of power transmission systems* Ph. D. Thesis, Newcastle upon Tyne.

Onsy, A., Bicker, R., Shaw, B., & Fouad, M. M. (2012). Application of image registration methods in monitoring the progression of surface fatigue failures in geared transmission systems.

Ostachowicz, W., Krawczuk, M., & Cartmell, M. (2002). The location of a concentrated mass on rectangular plates from measurements of natural vibrations. *Computers and Structures, 80*, 1419–1428.

Pal, S., Heyns, P. S., Freyer, B. H., Theron, N. J., & Pal, S. K. (2011). Tool wear monitoring and selection of optimum cutting conditions with progressive tool wear effect and input uncertainties. *Journal of Intelligent Manufacturing, 22*, 491–504.

Pan, M. C., Van Brussel, H., & Sas, P. (1998). Intelligent joint fault diagnosis of industrial robots. *Mechanical Systems and Signal Processing, 12*, 571–588.

Parey, A., El Badaoui, M., Guillet, F., & Tandon, N. (2006). Dynamic modelling of spur gear pair and application of empirical mode decomposition-based statistical analysis for early detection of localized tooth defect. *Journal of Sound and Vibration, 294*, 547–561.

Parhi, D. R., & Dash, A. K. (2011). Application of neural networks and finite elements for condition monitoring of structures. *Proceedings of the Institution of Mechanical Engineers, Part C: Journal of Mechanical Engineering Science, 225*, 1329–1339.

Park, H. G., & Zak, M. (2003). Gray-box approach for fault detection of dynamical systems. *Journal of Dynamic Systems, Measurement and Control, Transactions of the ASME, 125*, 451–454.

Pawar, P. M., & Ganguli, R. (2003). Genetic fuzzy system for damage detection in beams and helicopter rotor blades. *Computer Methods in Applied Mechanics and Engineering, 192*, 2031–2057.

Pawar, P. M., & Ganguli, R. (2005). Matrix crack detection in thin-walled composite beam using genetic fuzzy system. *Journal of Intelligent Material Systems and Structures, 16*, 395–409.

Qin, G., & Hu, N. (2012). Design of embedded wireless sensor and its soft encapsulation for embedded monitoring of helicopter planetary gear set. *Journal of Physics: Conference Series*, 364.

Qu, Y., He, D., Yoon, J., Van Hecke, B., Bechhoefer, E., & Zhu, J. (2014). Gearbox tooth cut fault diagnostics using acoustic emission and vibration sensors—a comparative. *Sensors (Switzerland), 14*, 1372–1393.

Rad, M. F., & Shafai, L. (2009) A wireless embedded sensor for structural health monitoring applications. In *13th International Symposium on Antenna Technology and Applied Electromagnetics and the Canadian Radio Science Meeting, 2009* (pp. 1–4). ANTEM/URSI 2009, 15–18 February 2009.

Rai, N., & Rai, B. (2013). Neural network based closed loop control of dc motor using arduino uno. *International Journal of Engineering Trends and Technology,* 4.

Randall, R. B. (2004). Detection and diagnosis of incipient bearing failure in helicopter gearboxes. *Engineering Failure Analysis, 11*, 177–190.

Randall, R. B. (2011). *Vibration-based condition monitoring*. New York: Wiley.

Rojas, A., & Nandi, A. K. (2006). Practical scheme for fast detection and classification of rolling-element bearing faults using support vector machines. *Mechanical Systems and Signal Processing, 20*, 1523–1536.

Sagar, P. M. (2002). *Embedded operating systems for real-time applications*. Bombay.

Sainz Palmero, G. I., Juez Santamaria, J., Moya De La Torre, E. J., & Perán González, J. R. (2005). Fault detection and fuzzy rule extraction in AC motors by a neuro-fuzzy ART-based system. *Engineering Applications of Artificial Intelligence, 18*, 867–874.

Samanta, B. (2004). Gear fault detection using artificial neural networks and support vector machines with genetic algorithms. *Mechanical Systems and Signal Processing, 18*, 625–644.

Samarasinghe, S. (2007). *Neural networks for applied sciences and engineering*, Taylor and Francis Group, LLC.

Sarrafzadeh, M., Dabiri, F., Jafari, R., Massey, T., & Nahapetan, A. (2006). Low power light-weight embedded systems. 207–212.

Sawalhi, N., & Randall, R. B. (2008). Simulating gear and bearing interactions in the presence of faults. Part I. The combined gear bearing dynamic model and the simulation of localised bearing faults. *Mechanical Systems and Signal Processing, 22*, 1924–1951.

Schoen, R. R., Habetler, T. G., Kamran, F., & Bartheld, R. G. (1995). Motor bearing damage detection using stator current monitoring. *IEEE Transactions on Industry Applications, 31*, 1274–1279.

Seo, J., Yoon, H., Ha, H., Hong, D., & Kim, W. (2011). *Infrared thermographic diagnosis mechnism for fault detection of ball bearing under dynamic loading conditions*. Sanya.
Shao, Y., & Nezu, K. (2005). Design of mixture de-noising for detecting faulty bearing signals. *Journal of Sound and Vibration, 282*, 899–917.
Sharma, S. (2008). *Application of support vector machines for damage detection in structures*. M. Sc, Worcester Polytechnic Institute.
Shimada, M., & Mita, A. (2005). Damage assessment of bending structures using support vector machine. In *Smart Structures and Materials* (pp. 923–930). International Society for Optics and Photonics.
Shimada, M., Mita, A., & Feng, M. Q. (2006). *Damage detection of structures using support vector machines under various boundary conditions*. International Society for Optics and Photonics.
Shiroishi, J., Li, Y., Liang, S., Kurfess, T., & Danyluk, S. (1997). Bearing condition diagnostics via vibration and acoustic emission measurements. *Mechanical Systems and Signal Processing, 11*, 693–705.
Shukla, A. P., Garg, H., Varshneya, G., & Srivastava, A. K. (2009). Real time acquisition of vehicle diagnostic data using wireless sensor network. In *Fifth IEEE Conference on Wireless Communication and Sensor Networks (WCSN), 2009* (pp. 1–5). IEEE.
Srovnal, V., & Penhaker, M. (2007). Health maintenance embedded systems in home care applications.
Stack, J. R., Habetler, T. G., & Harley, R. G. (2004). Bearing fault detection via autoregressive stator current modeling. *IEEE Transactions on Industry Applications, 40*, 740–746.
Stamboliska, Z., Rusiński, E., & Moczko, P. (2015). *Proactive condition monitoring of low-speed machines*. Switzerland: Springer International Publishing Switzerland.
Sun, Z., & Chang, C. C. (2004). Statistical wavelet-based method for structural health monitoring. *Journal of Structural Engineering, 130*, 1055–1062.
Taylor, S. G., Farinholt, K. M., Park, G., Farrar, C. R., & Todd, M. D. (2011). Application of a wireless sensor node to health monitoring of operational wind turbine blades. 45–53.
Teng, W., Wang, F., Zhang, K., Liu, Y., & Ding, X. (2014). Pitting fault detection of a wind turbine gearbox using empirical mode decomposition. *Journal of Mechanical Engineering, 60*, 12–20.
Terra, M. H., & Tinós, R. (2001). Fault detection and isolation in robotic manipulators via neural networks: A comparison among three architectures for residual analysis. *Journal of Robotic Systems, 18*, 357–374.
Trendafilova, I., & Van Brussel, H. (2003). Condition monitoring of robot joints using statistical and nonlinear dynamics tools. *Meccanica, 38*, 283–295.
Tripathy, J. R., Tripathy, H. K., & Nayak, S. S. (2014). Artificial neural network implementation in microchip PIC 18F45J10 8-bit microcontroller. *International Journal of Engineering and Advanced Technology (IJEAT)*, 3.
Vachtsevanos, G., Lewis, F., Roemer, M., Hess, A., & Wu, B. (2006). *Intelligent fault diagnosis and prognostic for engineering systems*. New York: John Wiley.
Van, M., Kang, H. J., & Ro, Y. S. (2011). A robust fault detection and isolation scheme for robot manipulators based on neural networks. In *International Conference on Intelligent Computing* (pp. 25–32). Berlin: Springer.
Vemuri, A. T., Polycarpou, M. M., & Diakourtis, S. A. (1998). Neural network based fault detection in robotic manipulators. *IEEE Transactions on Robotics and Automation, 14*, 342–348.
Verdonck, W., Swevers, J., & Samin, J. C. (2001). Experimental robot identification: Advantages of combining internal and external measurements and of using periodic excitation. *Journal of Dynamic Systems, Measurement and Control, Transactions of the ASME, 123*, 630–636.
Verma, N. K., Sarkar, S., Dixit, S., Sevakula, R. K., & Salour, A. (2013). Android app for intelligent CBM. In *IEEE International Symposium on Industrial Electronics*.

Wanbin, W., & Tse, P. W. (2006). Remote machine monitoring through mobile phone, smartphone or pda. In *Proceedings of the 1st World Congress on Engineering Asset Management* (pp. 309–315). WCEAM 2006.

Wang, Q. (2009). Artificial neural network and hidden space SVM for fault detection in power system. In *International Symposium on Neural Networks* (pp. 391–397). Berlin: Springer.

Wang, W., & Zhang, W. (2008). Early defect identification: Application of statistical process control methods. *Journal of Quality in Maintenance Engineering, 14*, 225–236.

Wei, X. B., Zheng, W., & Lin, R. (2011). Design of LabView-based system of noise measurement on gear box.

Wu, B., Lin, J., & Xiong, X. (2011). Design and implementation of intelligent monitoring and diagnosis system based on WSN and MAS. *CCIS, 238*, 290–297.

Yadav, S. K., & Kalra, P. K. (2010). Condition monitoring of internal combustion engine using EMD and HMM.

Yan, J., Ma, H., Li, W., & Zhu, H. (2009). *Assessment of rotor degradation in steam turbine using support vector machine*. IEEE

Yen, G. G., & Meesad, P. (2001). An effective neuro-fuzzy paradigm for machinery condition health monitoring. *IEEE Transactions on Systems, Man, and Cybernetics. Part B, Cybernetics, 31*, 523–536.

Yildirim, A., & Eski, I. (2010). Noise analysis of robot manipulator using neural networks. *Robotics and Computer-Integrated Manufacturing, 26*, 282–290.

Yuan, J., Liu, G., & Wu, B. (2011). Power efficiency estimation-based health monitoring and fault detection of modular and reconfigurable robot. *IEEE Transactions on Industrial Electronics, 58*, 4880–4887.

Zhai, Y., & Cheng, X. (2011). Design of smart home remote monitoring system based on embedded system. In *IEEE 2nd International Conference on Computing, Control and Industrial Engineering (CCIE), 2011* (Vol. 2, pp. 41–44). IEEE.

Zhang, J. Z., & Chen, J. C. (2008). Tool condition monitoring in an end-milling operation based on the vibration signal collected through a microcontroller-based data acquisition system. *International Journal of Advanced Manufacturing Technology, 39*, 118–128.

Zhang, E., Zhang, H., & Xue, B. (2007) Application of integrated neural network based on information combination for fault diagnosis in steam turbine generator, 1293–1297.

Zhang, Y. F., Ma, B., Zhu, Y., & Zhang, J. L. (2009). Study on condition monitoring of power-shift steering transmission based on support vector machine.

Zhang, Z., Verma, A., & Kusiak, A. (2012). Fault analysis and condition monitoring of the wind turbine gearbox. *IEEE Transactions on Energy Conversion, 27*, 526–535.

Zhong, J., Yang, Z., & Wong, S. F. (2010). Machine condition monitoring and fault diagnosis based on support vector machine, 2228–2233.

Zhou, W., Habetler, T. G., & Harley, R. G. (2008). Bearing fault detection via stator current noise cancellation and statistical control. *IEEE Transactions on Industrial Electronics, 55*, 4260–4269.

Chapter 3
Signal Processing Techniques for Condition Monitoring

Abstract Signal processing plays a significant role in building any condition monitoring system. Many types of signals can be used in the condition monitoring of machines, such as vibration signals as in this research; and processing these signals in an appropriate way is crucial in extracting the most salient features related to different fault types. A number of signal processing techniques can fulfil this purpose, and the nature of the captured signal is a significant factor in the selection of the appropriate technique. This chapter starts with a discussion of the proposed robot condition monitoring algorithm. Then, a consideration of the signal processing techniques which can be applied in condition monitoring is carried out to identify their advantages and disadvantages, from which the time-domain and discrete wavelet transform signal analysis are selected.

3.1 The Proposed Intelligent Condition Monitoring Algorithm

Robots are required to perform a variety of different repetitive tasks and are as designed programmable and configurable machines; and consequently the joints are subjected to continuously varying loads and speeds. Therefore, designing a CM system for a robot being adaptable for different robot tasks is challenging. In this work, to achieve this, it is decided to conduct the robot CM using two stages, as shown in Fig. 3.1. The first stage is only responsible for detection the fault and is performed during the robot movement for accomplishing whatever the task. The vibration signals are captured and features are extracted using time-domain signal analysis technique (as explained later in this chapter). Then, the features are analysed in order to select the most fault-sensitive one (Chap. 6). From the extracted features that are related to the robot healthy state, threshold values are calculated, in order to be used as a baseline reference, using the statistical control chart (SCC) approach (Chap. 7), which is a technique by which a plant (or process) is monitored to investigate whether or not the plant remain in control. The above mentioned steps have to be done offline, before running the monitoring system and

A.A. Jaber, *Design of an Intelligent Embedded System for Condition Monitoring of an Industrial Robot*, Springer Theses,
DOI 10.1007/978-3-319-44932-6_3

the purpose is to compute the threshold values. If the robot is reprogrammed for a different task than the previous, the same steps are needed to be followed, in order to establish different threshold values for the new task. During the online operation the selected (most) fault-sensitive feature will be calculated and compared to the reference thresholds. The result of the fault detection stage will report either the robot is healthy or a fault is developing. If a fault is detected, the robot should be stopped and the second stage of the CM system conducted. In the first stage, the time-domain signal analysis and SCC have been selected because they are relatively computationally easy to implement on the embedded electronic system and the fault category is not known at this stage. From a practical point of view, stopping the robot may affect the whole production line, thus the maintenance team in the factory should find the appropriate time for taking the robot out of service for performing the second stage of the CM system and also maintain it.

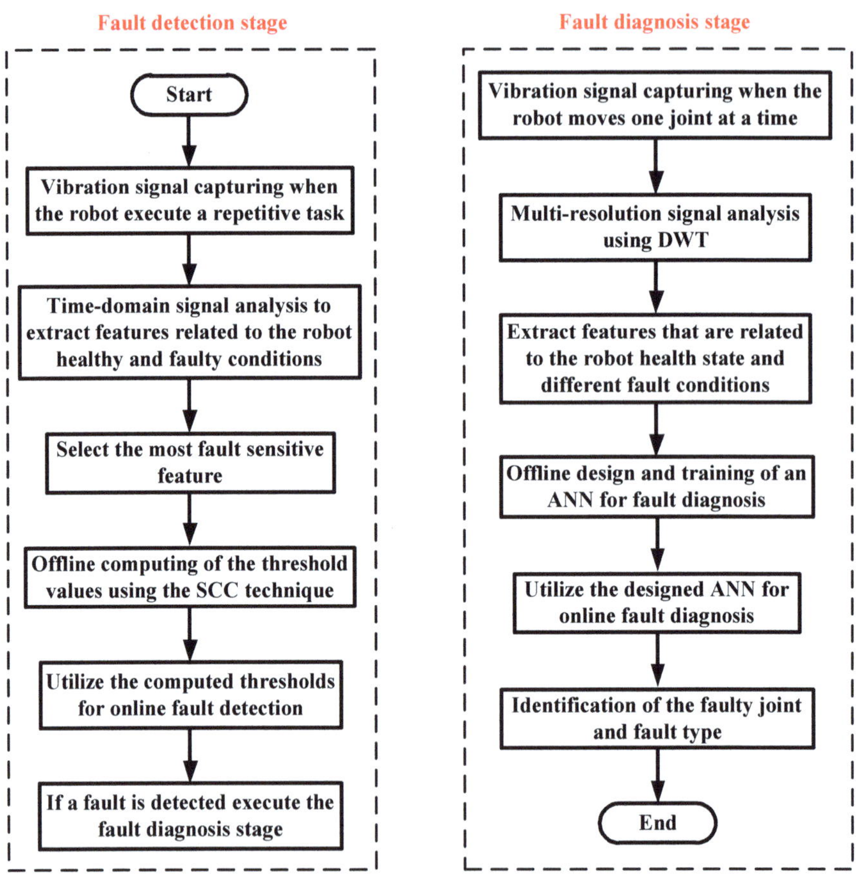

Fig. 3.1 Descriptive flowchart for the proposed intelligent condition monitoring algorithm

The aim of the second stage of the monitoring algorithm is to accurately identify in which joint the fault has occurred and what is its type exactly, for example, backlash, gear tooth wear or bearing fault. To achieve this, the robot will be programmed to move each joint independently in a cyclic movement. The vibration signals are captured and analysed, but this time using multi-resolution signal analysis technique based on the discrete wavelet transform (DWT), since it has been found very appropriate for non-stationary vibration signal analysis, which is the case in industrial robots in which the speed and load on each joint is continuously changing, and can assist in the precise diagnosis of faults, as discussed later. Then the features related to the healthy state and different fault conditions are determined and used for design and training an artificial fault classification system using the artificial neural network (ANN) (Chap. 7). The established ANN is then employed for online fault diagnosis (Chap. 9).

3.2 Basic Concepts of Signals

A signal can be defined as a function that describes a physical variable as it evolves over time. Analogue signals, such as sound, noise, light and heat, represent the majority of signals in nature. Variations in these signals are continuous over time and the processing of analogue signals is called analogue signal processing (ASP). By sampling such continuous signals at repeated time intervals using data acquisition equipment, they can be converted into discrete format, and the processing of the digital (discrete) signal is named digital signal processing. A discrete signal, on the other hand, has values only at specific time periods. The benefits of converting signals from analogue to discrete (digital) form are that it can avoid the degradation and corruption of the signals. Knowing the type of signal to be analyzed has a significant influence on the type of analytic technique chosen. Subsequently, it is necessary to carefully inspect the various types of signal that are encountered in practice. Thus, signals can be classified as shown in Fig. 3.2.

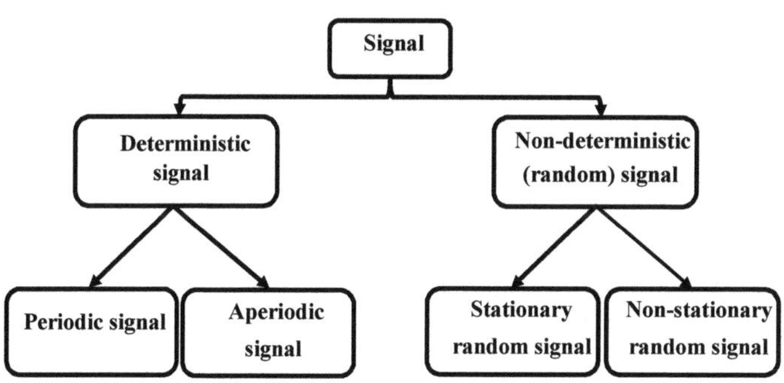

Fig. 3.2 Schematic diagram of signal classification

- **Deterministic signal**: If, after a suitable number of measurements, the signal can be described by an analytical expression and its values can be predicted at any time in the past and future, then it is called a deterministic signal, such as a sinusoid. A deterministic signal may be classified as a periodic signal if the change in the magnitude of the signal repeated at regular time intervals, and if not it is termed an aperiodic signal (Figliola and Beasley 2011).
- **Non-deterministic**: Conversely, non-deterministic or random signals cannot be described by a deterministic mathematical expression and they are more complex than deterministic signals. By determining their statistical properties, random signals can be broken down into stationary and non-stationary parts. Therefore, if the statistical properties of the random signal do not change with time, then it can be called a stationary signal, otherwise, it is named non-stationary (Wilkinson 2008).

However, a majority of the signals emitted from industrial machines are non-deterministic. And when a fault starts to appear in a machine the signals monitored tend to non-stationary in nature. Therefore, a suitable signal processing technique has to be applied to analyse this type of signal, as discussed in the coming sections.

3.3 Signal Analysis Techniques

After a signal is being captured, a large number of signal processing techniques can be utilized to extract the most sensitive and interesting features concerning defects. As a matter of fact, choosing the most suitable method for each specific task represents a major challenge in condition monitoring. Signal processing techniques are classified as using time domain, frequency domain, and time-frequency domain methods. These methods are not totally independent, and in many situations they complement each other. Some of the widely used signals processing techniques are discussed in the following sub-sections to establish their suitability for robot fault detection and diagnosis.

3.3.1 Time-Domain Signal Analysis Technique

The technique used in processing the signal can be classified as a time-domain method if it processes a raw signal directly in the time domain without being transformed into another domain, such as the frequency domain (Li 2006). It is considered one of the cheapest and simplest approaches to implement for fault detection. The purpose of time-domain analysis is to determine the statistical features of the original signal by manipulating the series of discrete numbers. With this technique, however, only the fault can be detected without diagnosing its source. Statistical parameters such as peak value (PK), which represents the maximum amplitude in the signal regardless of sign, can be used to give useful information about the hidden defects represented in the time domain signal. Some of these parameters are illustrated as shown:

Root-mean-square (RMS) is defined as the square root of the average of the sum of the squares of the signal samples, (Eq. 3.1). RMS can be used for measuring the overall level of average power in the vibration signal (Lihui and Gao 2006; Kim et al. 2007).

$$RMS = \sqrt{\frac{1}{N}\left[\sum_{n=1}^{N}(x[n])^2\right]} \tag{3.1}$$

where $x[n]$ is the original sampled signal, N is the total number of samples, and n is the sample index.

Crest factor (CF) is a non-dimensional parameters defined as the ratio of maximum absolute value (or PK value) to the RMS value of the signal, and is given by Lihui and Gao (2006), Kim et al. (2007)

$$CF = \frac{PK}{RMS} \tag{3.2}$$

CF is a normalized measurement of the amplitude of the signal which increases in the presence of a small number of high amplitude peaks, such as in the case for some types of local tooth damage in a gearbox. The sensitivity of CF to the changes in the sharpness of the signal is much higher than the RMS value, and it is much less likely to give false alarms than using the (PK) on its own (Engin et al. 1999).

Skewness (Sk) and Kurtosis (Ku) are also dimensionless parameters and denote the statistical moments of the signal (Shin and Hammond 2008). The distribution shape of the signal can be described using the 3rd moment or skewness, which is a gauge of symmetry of the probability density function (PDF) around its mean. If the distribution is symmetric, its value is zero. The skewness becomes negative if the distribution develops a longer tail left of the mean, and positive if the other way around, indicating that something is going wrong in the monitored system, as shown in Fig. 3.3 (Vachtsevanos et al. 2006). The 4th moment or kurtosis represents a measure of the relative flatness or spikiness of a signal compared to its normal state. Skewness and Kurtosis can be calculated using the following equations:

$$Sk = \frac{\frac{1}{N}\sum_{n=1}^{N}(x[n]-\bar{x})^3}{\sigma^3} \tag{3.3}$$

$$Ku = \frac{\frac{1}{N}\sum_{n=1}^{N}(x[n]-\bar{x})^4}{\sigma^4} \tag{3.4}$$

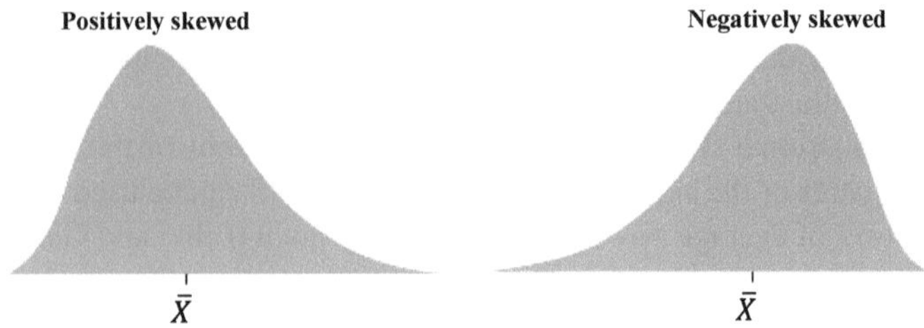

Fig. 3.3 Positive and negative skewness (Vachtsevanos et al. 2006)

where $\bar{x}$ and σ are the signal mean and standard deviation, which represent the first and second moment of the signal respectively, as given by:

$$\bar{x} = \frac{1}{N}\sum_{n=1}^{N} x[n] \tag{3.5}$$

$$\sigma = \sqrt{\frac{1}{N}\sum_{n=1}^{N} (x[n] - \bar{x})^2} \tag{3.6}$$

Every signal distribution has different kurtosis values as shown in the Table 3.1 (Onsy 2009). The monitored signal usually shows a normal pattern with a kurtosis value of approximately 3 if it is healthy. When a fault is developed in the system, the kurtosis value increases indicating that the signal is no longer normally distributed, and therefore, it is useful in identifying the machine nature (Marwala 2012).

A substantial number of papers have employed time-domain technique to identify defects in many applications. Zhen and Zhang (2012) and Li and Frogley (2013) utilized time-domain analysis to detect faults in wind turbine bearings and gears respectively. Piezoelectric accelerometers and capacitive sensors were used to acquire vibration signals from healthy and faulty bearings. Statistical parameters include peak value, average, variance, RMS and kurtosis have been calculated for the bearings and by comparing their results, the status of bearing has been found easily. Although time-domain analysis has many advantages, including straightforward signal processing and simple calculations, it is relatively insensitive tool for early stage fault detection and severely distributed defects if used without being combined with other machine health evaluation techniques (Ghafari 2007). This was verified in two other papers on bearing health monitoring (Williams et al. 2001; Qiu et al. 2003), who concluded that most of the bearing fatigue time is consumed

Table 3.1 Kurtosis values for different signal distribution

Type of signal distribution	**Ku**
Normal-peak	~ 3
Flatter than normal	<3
Sharper	>3

during the development of material accumulative damage, whereas the period of crack propagation and development is comparatively short. The time available for initiating a maintenance action before a catastrophic failure after confirming a defect will be very short if this traditional technique is used.

In contrast, Tseng et al. (2014) and Kamiel et al. (2015) have shown that utilizing the statistical process control (or statistical control chart) techniques combined with the time-domain features have effectively improved the fault detection process, but could not diagnose it. Thus, as it was mentioned previously that the first stage is in charge of only detecting the fault, so the combination of time-domain signal analysis with SCC will be applied at this stage.

3.3.2 Frequency-Domain Signal Analysis Technique

In most applications, signal representation in the time domain is not the most appropriate, since much of the relevant information is hidden in the frequency content of the signal. Frequency or spectral analysis provides additional information about time series data, and can be used to explain the spectra of frequencies which exist in the signal. The parameters of frequency domain analysis are more reliable in damage diagnosis than time domain parameters. However, time-domain signals can be represented by a family of complex exponents with infinite time duration using Fourier transforms (FTs). Additionally, any given time-domain signal can be written as a function of all of the frequencies present within it using Fourier transforms, which allows analysts to concentrate on all or specific frequencies. This is achieved by representing a time-domain signal $x(t)$ by sinusoidal components with infinite time duration (He 2001; Kehtarnavaz 2008), which are given by:

$$X(f) = \int_{-\infty}^{\infty} x(t)\, e^{-j2\pi ft} dt \tag{3.7}$$

where $X(f)$ is the transformed signal, f is frequency, and t is time. To regenerate the time domain signal back from the frequency domain signal, an inverse Fourier transform has to be applied:

$$x(t) = \int_{-\infty}^{\infty} X(f)\, e^{j2\pi ft} df \tag{3.8}$$

However, the continuous-time Fourier transform can only be applied to signals of continuous time and infinite duration. Additionally, in most applications, signals are commonly acquired and sampled at a specific frequency, which is called the sampling frequency (f_s), and converted into a set of digital data points, and therefore it is necessary to use the discrete version of the Fourier transform (DFT) (He 2001; Kehtarnavaz 2008), which is:

$$X[k] = \sum_{n=0}^{N-1} x[n]\, e^{-j(2\pi nk/N)} \tag{3.9}$$

$$x[n] = \frac{1}{N}\sum_{k=0}^{N-1} X[k]\, e^{j(2\pi nk/N)} \tag{3.10}$$

where $X[k]$ and $x[n]$ denote discrete frequency and time signal respectively, k and n represent the frequency and time indices, and N represents the total number of points that are equally spaced.

To perform the DFT, a large number of complex computations are involved. This is a computationally intensive process and not practical when performing real-time signal analysis. Thus, an algorithm that is able to perform rapid calculation of the DFT by greatly reducing the number of computations was developed in the early 1960s (Mohanty 2015). This algorithm is known as the fast Fourier transform (FFT) and commonly used in industry for analysing the data. The FFT algorithm requires the time domain sequence to $x[n]$ have a length of data points equal to the power of 2; which means that 2^m samples are required where m is a positive integer (Mohanty 2015). Generally speaking, the FFT is a useful technique for transforming difficult operations into very simple ones, and for analysing stationary signals, which have spectral content that does not change over time. Also, in many signal processing applications, the Fourier transform represents an adequate analytic method.

However, the Fourier technique can become less effective and inefficient if the analysed signal is non-stationary and transitory, with characteristics that change over time, due to its constant time and frequency resolutions (Tse et al. 2004; Al-Badour et al. 2011). Furthermore, it has a major drawback that when it is used in transforming the signal from the time domain to the frequency domain, all of the information belonging to time will be lost (Al-Badour et al. 2011). Nevertheless, provided that the signals are stationary, the task of distinguishing faulty from normal conditions based on the FFT can be accurately achieved. This is performed by investigating particular estimated frequencies related to some component in the machine, such as gears or bearings. If a fault has developed in these components, the amplitude of these particular frequencies will change or some sideband frequencies will be distributed around them. Therefore, many fault diagnosis studies using this technique have been published, having been successfully applied for the condition monitoring of electrical motors, cutting tools, bearings and gears (Iorgulescu et al. 2009; Ngolah et al. 2011; Hsieh et al. 2012).

Industrial robots, on the other hand, are required to function under a wide range of joint speeds and variable loading within a large working area and varying joint articulation. Also, the typical cycle of robot motion starts with an acceleration from the initial position, then moving at constant speed, and finally deceleration towards the end position, which means movement at a time-varying speed (Bicker et al. 1989; Pan et al. 1998). This motion makes the robot a highly non-linear dynamic system and introduces the non-stationary phenomenon in the captured vibration signal, and this will be more complicated if a fault is progressed in the robot. Using a conventional FFT signal analysis technique to process such signals with transiently nature is not feasible for accurate robot fault diagnosis in second stage. Therefore, several methods of signal processing have been developed to cope with this category of signals, such as joint time-frequency techniques, as discussed in the following section.

3.3.3 *Time-Frequency Signal Analysis Technique*

The signals from faulty parts have a non-stationary nature. However, if the frequency component of the non-stationary signals is calculated using the Fourier transform, the results will represent the frequency composition averaged over the duration of the signal (Sawicki et al. 2009). Consequently, the characteristics of the transient signal cannot be described adequately using the Fourier transform, however, time-frequency analysis has been investigated and applied for the fault diagnosis of machinery because of its capability of signal representation in both the frequency and time domains (Sawicki et al. 2009; Al-Badour et al. 2011). This unique feature of time-frequency analysis techniques means that it is suitable for non-stationary signals. Moreover, time-frequency methods can give interesting information with regard to energy distribution over frequency bands. A number of techniques of time-frequency analysis, such as the short time Fourier transform and wavelet transforms, have been used for fault detection and diagnosis. These techniques will now be discussed to identify the main differences between them and select the best to be used at the diagnosis stage.

3.3.3.1 Short Time Fourier Transform

To overcome the limitations of the Fourier transform technique, Gabor introduced a windowing technique in 1946 known as the short time Fourier transform (STFT). The STFT algorithm is based on the division of the signal into small portions which are assumed to be stationary. Then, a window function is located at the start of the signal and multiplied together. After that, the Fourier transform will be taken for the result of

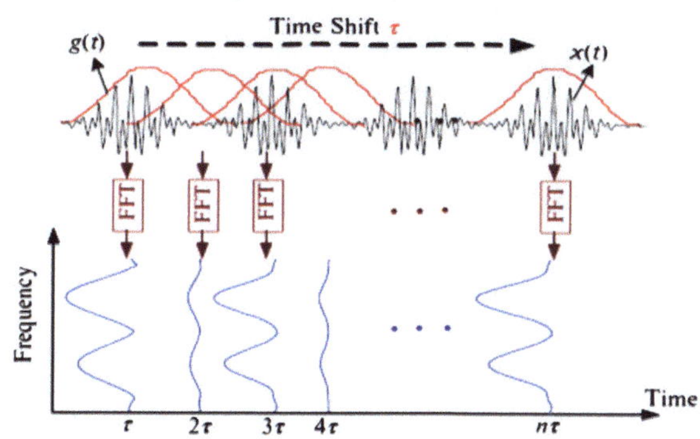

Fig. 3.4 Signal analysed by STFT (Gao and Yan 2011)

this product. Next, this window function is moved to a new segment of the signal and the above-mentioned process is repeated. This sequence is repeated until the end of the signal is reached (Al Kazzaz and Singh 2003). As a result, the STFT outlines the time-domain signal into a two-dimensional time-frequency representation. This can be mathematically expressed and graphically revealed as follow (Fig. 3.4)

$$STFT(t,f) = \int_{-\infty}^{\infty} x(t).g(t-\tau)\, e^{-j2\pi ft} dt \tag{3.11}$$

$STFT(t,f)$ is the Fourier transform of the signal $x(t)$ which has already been windowed by the window function $g(t)$ with respect to the time shift variable τ. Various window types, with each one employed for a particular application, have been developed over the past decades. For instance, the Hann and Hamming windows are utilized for analysing random and narrowband signals (Gao and Yan 2011), whereas a Gaussian window is exploited for analysing transient signals. Selection of the window function has a direct influence on the time and frequency resolutions of the analysed signal. Generally, superior separation of the essential components within a signal can be achieved if high resolution in the time and frequency domains is used. To illustrate the difference between FFT and STFT a LabVIEW programme, designed by Kehtarnavaz (2008) was used, in which three forms of signals were combined to produce a non-stationary signal with 512-input

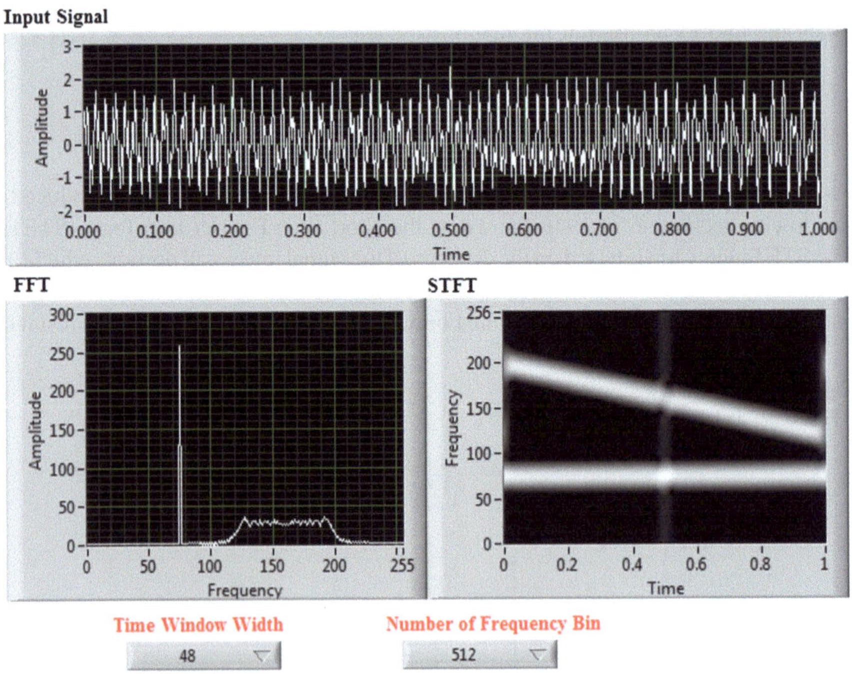

Fig. 3.5 The difference between FFT and STFT for a non-stationary signal

points. The forms of the combined signals are a chirp signal with linearly decreasing frequency from 200 to 120 Hz, a sinusoidal signal of 75 Hz, and an impulse signal located at the 256th sample and having amplitude of 2. The composite signal and its FFT and STFT are shown in the Fig. 3.5.

From Fig. 3.5 it can be observed that in the FFT spectrum graph there is one major peak at 75 Hz, and also there is an indication of presence of a signal from 120 to 200 Hz. However, the impulse signal, which has short time duration, cannot be recognized in the spectrum, although it can clearly be observed in the STFT graph at 0.5 s, which shows the spectrogram for a time increment of 1 s and a rectangular window of width 48 points. Although the STFT provides both the frequency spectrum and the time evolution of the signal, it does have a major drawback: it has a fixed resolution with respect to the time window size at all frequencies, and can be explained as follows.

When the FFT is used it can be noticed that there is no time resolution, but on the other hand the frequency resolution is very high. The reason for this high resolution is related to the fact that the window function used in FFT covers the entire time interval from $\pm\infty$. Conversely, the frequency resolution when the STFT is implemented becomes poorer than the resolution given in the FFT, since the window function has a finite length and therefore only a small segment of the signal will be covered. In order to increase the frequency resolution, the window function has to be wide enough, but that will lead to missing time information as well as violating the stationarity assumption which requires the window to be very small.

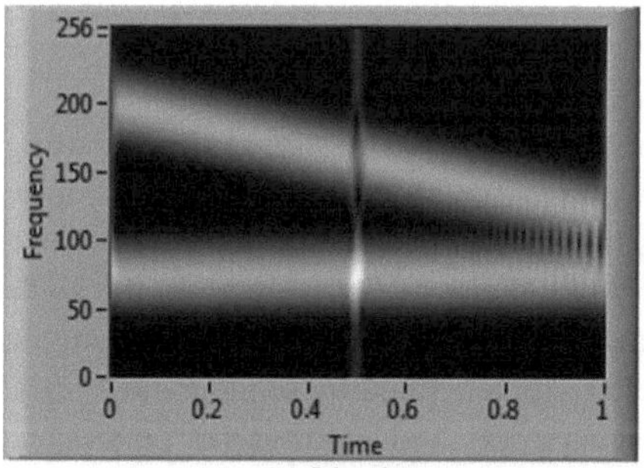

(a) 24 points

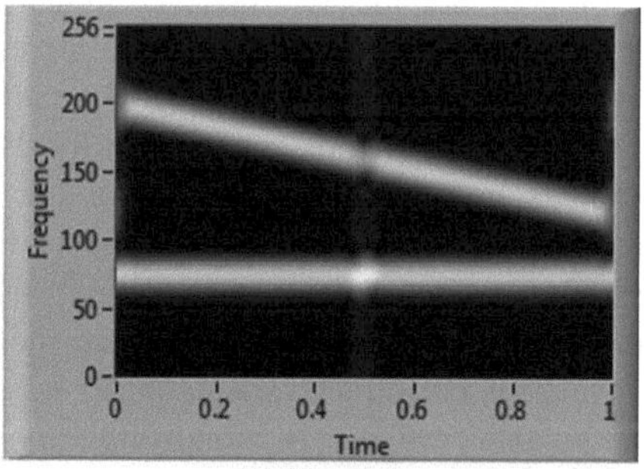

(b) 48 points

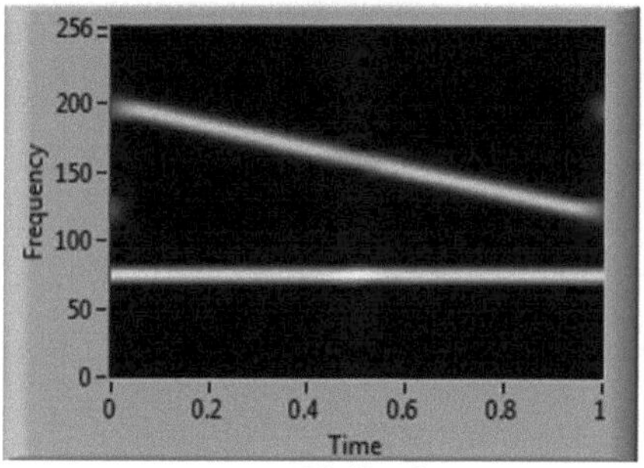

(c) 64 point

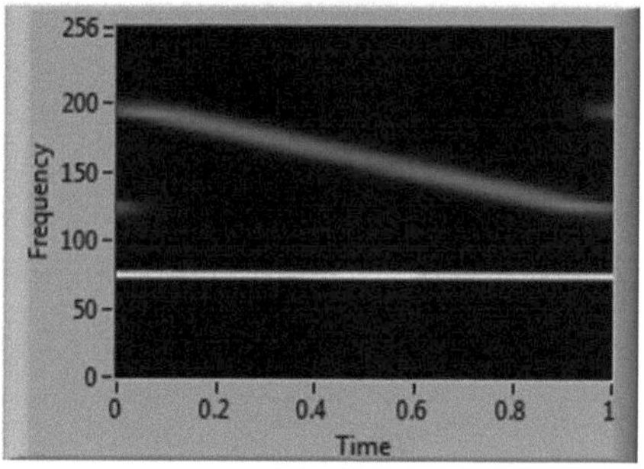

(d) 128 point

Fig. 3.6 STFT with different window widths

Accordingly, there is a trade-off relationship between time and frequency in the STFT. A wide window gives good frequency resolution but poorer time resolution and vice versa (Polikar 1996). This is well illustrated in the Fig. 3.6 below depending on the above analysed signal.

3.3.3.2 Wavelet Transforms

The wavelet transform (WT) was introduced to overcome the resolution limitation of the STFT. The main difference between the WT and the STFT is that the former has varying window lengths, and represents the signal as a sum of wavelets at different scales (Debdas et al. 2011). To clearly understand the differences among the time-frequency resolution of the DFT, STFT and WT, their time-frequency mapping is compared in Fig. 3.7, from which it can be seen that, and as stated earlier, the DFT allows extraction only of the frequency content of a signal and any information concerning time-localization of the frequency components is eliminated. The area of each rectangular box in both STFT and WT has a fixed value (Rajbhandari 2009). However, in the STFT the window has fixed dimensions in both time and frequency axes which offer a constant time-frequency resolution. In WT the window dimensions are not constant, and when the height of the box is greater this corresponds to wide frequency bandwidth, which leads to low frequency resolution, but on the other hand the time resolution is improved. Similarly, if the width is greater, a long time duration is covered providing coarser time resolution in contrast to better frequency resolution. So, the WT behaves rather like a mathematical microscope, as condensing the wavelet corresponds to increasing the magnification of the microscope, which increases more of the signal detail (Rajbhandari 2009).

Complex sinusoids are used in the Fourier transform for signal decomposing, whereas in wavelet analysis a mother wavelet function is utilized. In Fourier analysis sines and cosines are used to fit the signal in order to generate a set of coefficients, however, in wavelet analysis the mother wavelet is fitted on the signal and then the inner product between the analysed signal and a series of daughter wavelets is performed. The daughter wavelets are generated by scaling and shifting the mother wavelet by controlling the scaling (s) and shifting (τ) parameters.

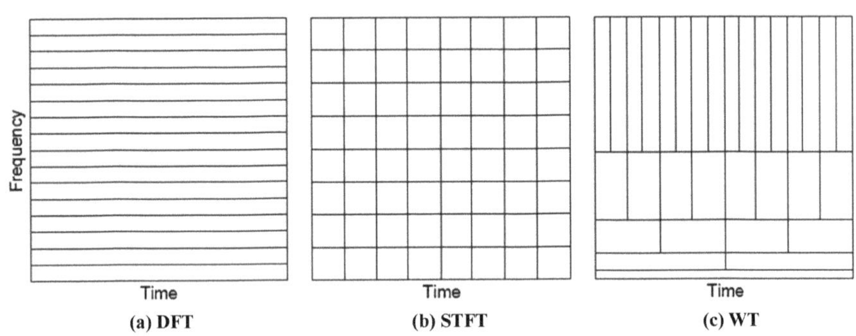

Fig. 3.7 Time-Frequency signal mapping (Kehtarnavaz 2008)

Scaling the mother wavelet is equivalent to stretching or dilating it; although the wavelet is squashed in the vertical axis if it is stretched horizontally, this is to ensure that the energy content in the scaled wavelet is equal to the original mother wavelet (Leavey et al. 2003). In the shifting step, the wavelet is moved along the X-axis until it covers the analysed signal entirely, which can be expressed mathematically as follows (Rajbhandari 2009):

$$WT(\tau, s) = \frac{1}{\sqrt{s}} \int_{-\infty}^{+\infty} x(t)\, \psi\left(\frac{t-\tau}{s}\right) dt \tag{3.12}$$

where $WT(\tau, s)$ is the wavelet transform of the signal $x(t)$ and $\psi(t)$ is the mother wavelet (or the transforming function).

The mother wavelet differs from the infinite sine and cosine functions, as it has a finite start and finish. Mathematically, it can be said that the mother wavelet has "compact support" (Harpen 1998), the importance of which appears in that when the mother wavelet fit to the signal, a localised result will be obtained rather than a global result. A series of coefficients that vary with time will be extracted instead of getting a single coefficient for each sine and cosine as in Fourier analysis, and consequently the wavelet decomposition can accommodate the local and sharp changes in the monitored signal; thus it is suitable for those signals whose spectral content changes over time. Accordingly, it represents the most appropriate method that can be applied for analysing the expected robot vibration signal for precise fault diagnosis and hence it will be adopted in this study.

3.4 Discrete Wavelet Transform (DWT)

The above Eq. (3.12) is called the continuous wavelet transform (CWT) and offers greater accuracy in signal analysis; however, theoretically it is infinitely redundant, which means a significant amount of unnecessary information is produced when it is implemented, and such it is impractical (Giaouris et al. 2006). The redundancy problem is a result of the continuous scaling and shifting of the mother wavelet. This increases the required computational time, power and memory, making the CWT impractical in many situations, particularly when applying real-time wavelet analysis on an embedded system which is the case here. To reduce required power and time it is appropriate to remove any unnecessary information and reduce the number of wavelets without loss of the essential information. The discrete wavelet transform (DWT) was developed to achieve this, in which the mother wavelet is not continuously scaled and shifted, but is instead only at discrete steps along the signals. By using DWT, the original signal is often decomposed into several signals each with a specific frequency band each of which can be handled as an independent signal on which separate analysis can be implemented. The strength of the

DWT is that filters with different cut-off frequencies are utilized to analyse the signal at different scales. First, the signal is passed through a high-pass (HP) filter to analyse high frequencies, and then it is passed through a low-pass (LP) filter to analyses low frequencies. Using digital techniques such as this, a time-scale representation of a digital signal can be obtained.

Another type of wavelet analysis called complex wavelet transform and is represented by the dual-tree complex wavelet transform. It is an alternate, complex-valued extension and enhancement to the standard DWT, and has important properties that provides multiresolution, sparse representation and the capability to reduce the aliasing effects, which is caused by the overlap of opposing-frequency pass-bands of the wavelet filters (Loutas and Kostopoulos 2012; Qu et al. 2016). Two parallel DWTs with different low-pass and high-pass filters in each scale are used for decomposition and reconstruction in the dual-tree implementation. The two DWTs use two different sets of filters, with each satisfying the perfect reconstruction condition. However, the drawback of this transform is that it is exhibits redundancy compared to the standard DWT at the expense of extra computational power; therefore, it was not considered in this study due to the expected computational limitation of the embedded system.

3.4.1 Multi-resolution Analysis Using DWT

Generally speaking, by using the DWT, a multi-resolution analysis can be performed at different frequency bands with different resolutions by decomposing the time domain signal (Debdas et al. 2011; Sawicki et al. 2009). Two sets of functions are employed in the DWT, called the wavelet function and the scaling function, which are associated with the HP and LP filters respectively. At the first level, the original signal $x[n]$ is decomposed by passing it through both of these filters and emerges as two signals, each one having the same number of samples as the original signal, and are termed as coefficients. In order to keep the total number of coefficients in the produced filtered signals equal to the original signal samples they are then down-sampled by a factor of 2, by keeping only one sample out of two successive samples. Thus, the extracted signal coefficients from the HP filter and after down sampling are called the detail coefficients of the first level (cD_1). These coefficients contain the high frequency information of the original signal, whilst, the coefficients that are extracted from the LP filter and after the down sampling process are called the approximation coefficients of the first level (cA_1). The low frequency information of the signal is hidden in these coefficients. This can be expressed mathematically as (Vivas et al. 2013):

$$cD_1[k] = \sum\nolimits_n x[n] * h[2k - n] \tag{3.13}$$

$$cA_1[k] = \sum_n x[n] * g[2k - n] \tag{3.14}$$

where $h[n]$ and $g[n]$ are the high- and low-pass filters respectively. After obtaining the first level of decomposition, the above procedure can be repeated again to decompose cA_1 into another approximation and detail coefficients, as articulated in Eqs. 3.15 and 3.16 (Vivas et al. 2013). This procedure can be continued successively until a pre-defined certain level up to which the decomposition is required to be found.

$$cD_1[k] = \sum_n cA_{l-1}[n] * h[2k - n] \tag{3.15}$$

$$cA_1[k] = \sum_n cA_{l-1}[n] * g[2k - n] \tag{3.16}$$

where $cD_1[k]$ and $cA_1[k]$ are the DWT coefficients at level l, and $cA_{l-1}[n]$ is the approximate coefficient at level $l - 1$.

At each decomposition level, the corresponding detail and approximation coefficients have specific frequency bandwidths given by $\left[0 - F_s/2^{l+1}\right]$ for the approximation coefficients cA_l and $\left[F_s/2^{l+1} - F_s/2^l\right]$ for the detailed one cD_l where F_s is the sampling frequency (Sawicki et al. 2009; Vivas et al. 2013). However, at every level, the filtering and down-sampling will result in half the number of samples (half the time resolution) and half the frequency band (double the frequency resolution). Also, due to the consecutive down sampling by 2, the total number of samples in the analysed signal must be a power of 2 (Ghods and Lee 2014). By concatenating all coefficients starting from the last level of decomposition, the DWT of the original signal is then produced, and it will have the same number of samples as the original signal. A schematic diagram illustrates how the multi-level decomposition is performed shown in Fig. 3.8. The number of decomposition levels is identified by the lowest frequency band needed to be traced,

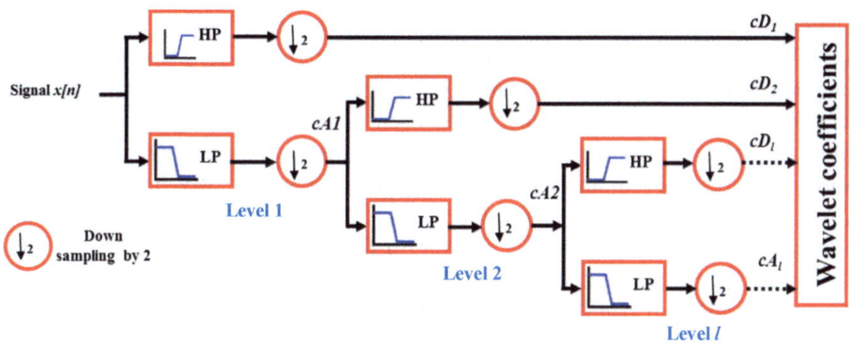

Fig. 3.8 Multi-level signal decomposition using DWT

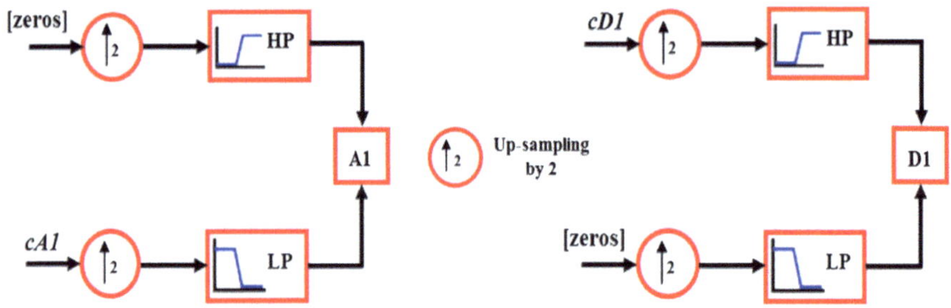

Fig. 3.9 Reconstruction the approximation and detail signals with zero padding

and a higher number of decomposition levels are required if very low frequency band is investigated. However, the highest decomposition level that can be achieved is up to that the individual details consist of a single sample (Misiti et al. 1997).

Once the approximation and detail coefficients are computed to different levels of decomposition, it becomes possible to reconstruct the approximation and detail signals at each level, in order to extract features, such as standard deviation and mean, related to the frequency bands in each level. Each signal, however, will have the same number of samples as the original signal but with a definite frequency band. This can be achieved by up-sampling the approximation (or details) coefficients by two, since they were produced previously by down sampling by 2, and then passing them through high- and low-pass synthesis filters. For instance, to reconstruct the approximation signal of the first level (A_1), just the approximation coefficients at this level are required and a vector of zeros is feed in place of the detail coefficients. Similarly, the first-level detail signal ($D1$) can be constructed using the analogous process. The concept of signal synthesising is illustrated in Fig. 3.9.

3.4.2 Selection the Optimum Mother Wavelet

There are available a number of commonly used wavelet families for performing the DWT. Any discussion of wavelets starts with *Haar* wavelet, which is the first, simplest, and resembles a step function. However, to find the optimum wavelet function for this research, a survey has been conducted to uncover the different types of mother wavelets researchers have used for the purpose of fault diagnosis. Some examples of common wavelets families previously used are *Daubechies (dbN), Coiflet (coifN)* and *Symlets (symN)* (Fig. 3.10), where *N* is the order number in the wavelet family (Sawicki et al. 2009; Elbarghathi et al. 2012). The *N* value also identifies the number of filter coefficients in each wavelet order; for instance, the wavelet *dbN and symN* have *2 N* coefficients in each order. Generally, the use of different wavelets to analyze the same signal would lead to different results, and to date no generic theoretical procedure has been published describing on how to

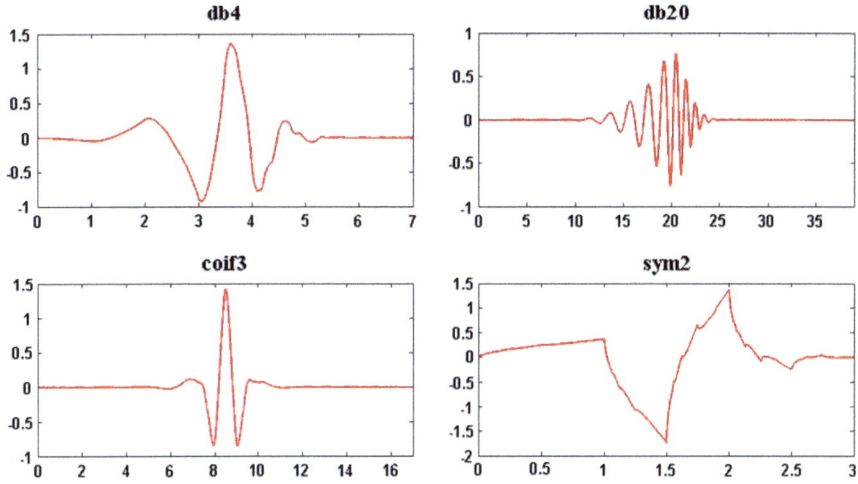

Fig. 3.10 Examples of mother wavelets

select the optimum wavelet family (Kankar et al. 2011; Karthikeyan et al. 2012; Loutas and Kostopoulos 2012).

The selection in many cases is achieved by trial and error. Indeed, the wavelet function is considered appropriate for analyzing the signal under study if there is a significant similarity between the signal and mother wavelet (Ngui et al. 2013), and based on this several quantitative methods have been proposed to measure the similarity between the signal and the mother wavelet. For example, Bouzid (2013) has proposed calculating the cross correlation coefficient between the signal and the mother wavelet. The wavelet that maximises this coefficient considered as the optimum mother wavelet. So, an important question now raised is which wavelet family should be utilized for analyzing the robot vibration signals in the second stage.

From the reviewed work it has been observed that the majority of researchers are performing either off- or on-line condition monitoring using PC platforms. In this case there is no need for concern about computers' memory or processing power, since they are designed for conducting daunting tasks such as this. In this research, however, it is intended to achieve on chip wavelet analysis in conjunction with intelligent fault classification system, and thus the above mentioned factors need to be carefully considered. Therefore, the number of wavelet's filters coefficients in each order of specific wavelet family has to be counted. Some wavelet functions, such as *db10* or *sym7*, have many coefficients in their filters, which will raise the execution time required for real-time wavelet analysis because of the increased computational burden on the embedded system. Also, higher order wavelet function will generate higher number of coefficients from the analysed signals that may case exceed the available system memory (Chen et al. 2009; Loutas and Kostopoulos 2012).

The mother wavelet selection has been limited to the lower order families and hence there is no need to apply further quantitative methods as the remaining

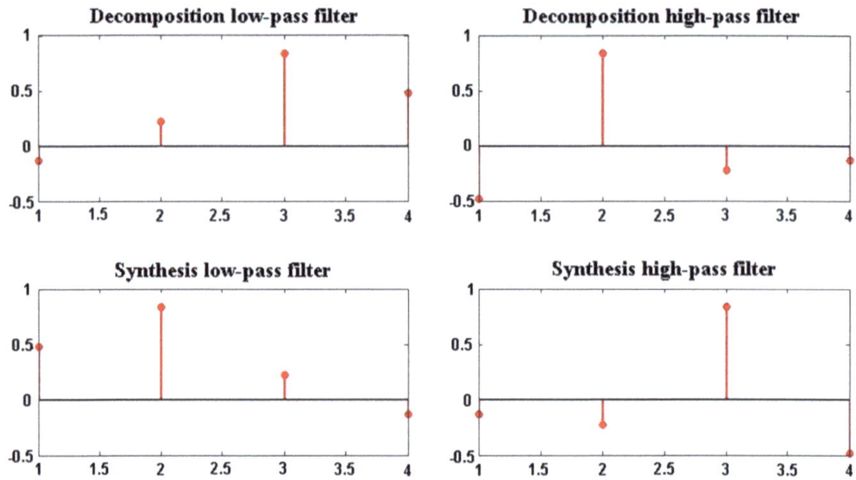

Fig. 3.11 Filter coefficients of Daubechies order 2 (db2) mother wavelet

options are very few. *Daubechies* and *Symelet* families are recognized as very effective in vibration signal analysis and have various wavelet orders, thus, in this project *Daubechies's* second order (*db2*), which is the same as *Symelet's* second order (*sym2*), has been selected. It has four filter coefficients and Fig. 3.11 shows the low and high pass decomposition and synthesis filters extracted from the Matlab software. Extracted features using this wavelet showed high sensitivity to different robot faults, as will be explained later in Chap. 6.

3.5 Summary

In this chapter, an intelligent condition monitoring algorithm composed of two stages that can be used for robot fault detection (first stage) and diagnosis (second stage) has been proposed. An outline of three conventional signal analysis techniques that are commonly utilized in developing condition monitoring systems has been provided, in order to choose the appropriate techniques for the robot fault detection and diagnosis; these techniques are time-domain, frequency-domain and joined time-frequency domain. The advantages and disadvantages along with a brief theoretical background for each method were discussed. Time-domain analysis represents the simplest signal processing technique; it can provide an efficient fault detection performance if it is used with other fault evaluation methods. Subsequently this will be used in combination with statistical control chart (SCC) technique in the first stage for robot fault detection.

Frequency-domain signal analysis based on fast Fourier transform (FFT) is a valuable and widely used technique for analysing signals that have spectral content that do not change over time (stationary signals), but its effectiveness is reduced if applied for analysing signals that have characteristics which change over time (non-stationary signals). The limitations in Fourier transform have been overcome by using time-frequency signal analysis techniques such as short time Fourier transform (STFT) and wavelet transform (WT). The main advantage of these techniques over the Fourier transform is their ability in revealing the non-stationary and random components within the signals of interest. However, STFT is based on fixed window size which means it has fixed resolution for all the frequencies in the signal, which is not appropriate when a non-stationary signal is investigated.

Wavelet transform represents an efficient method of time-frequency analysis and was introduced to surmount the drawback of the STFT, since it uses variable window size to get high frequency resolution at low frequencies and high time resolution at high frequencies. By applying wavelet analysis, the signal can be analyzed down to its sub-band frequencies and it is increasingly being utilized for fault diagnosis. Hence, it will be adapted in this thesis for analyzing the robot transitory vibration signals in order to diagnose the fault in the second stage. Thus, more focus has been placed on examining the practical use of the WT as an efficient signal processing technique for health monitoring. The differences between the continuous wavelet transform (CWT) and the discrete wavelet transform (DWT) are discussed and it was concluded that DWT is more appropriate to be implemented, along the with intelligent classification system, using an embedded system. Before it is applied using the embedded system a preliminary robot vibration analysis will be undertaken in Chap. 6, in order to extract the salient signal features and use them for designing the intelligent embedded system.

References

Al Kazzaz, S. A. S., & Singh, G. K. (2003). Experimental investigations on induction machine condition monitoring and fault diagnosis using digital signal processing techniques. *Electric Power Systems Research, 65*, 197–221.

Al-Badour, F., Sunar, M., & Cheded, L. (2011). Vibration analysis of rotating machinery using time-frequency analysis and wavelet techniques. *Mechanical Systems and Signal Processing, 25*, 2083–2101.

Bicker, R., Daadbin, A., & Rosinski, J. (1989). The monitoring of vibration in industrial robots. In *ASME 12th Biennial Conference on Mechanical Vibration and Noise*, 1989.

Bouzid, O. M. (2013). *In-situ health monitoring for wind turbine blade using acoustic wireless sensor networks at low sampling rates*. PhD Thesis, Newcastle University.

Chen, C. Y., Ke, M. D., & Kuo, C. D. (2009). Continuous wavelet transformation the wavelet implemented on a DSP chip for EEG monitoring. 3633–3636.

Debdas, S., Quereshi, M. F., Reddy, A., Chandrakar, D., & Pansari, D. (2011). A Wavelet based multiresolution analysis for real time condition monitoring of AC machine using vibration analysis. *International Journal of Scientific and Engineering Research, 2*.

Elbarghathi, F., Wang, T., Zhen, D., Gu, F., & Ball, A. (2012). Two stage helical gearbox fault detection and diagnosis based on continuous wavelet transformation of time synchronous averaged vibration signals. *Journal of Physics: Conference Series, 364*.

Engin, S. N., Gulez, K., & Badi, M. N. M. (1999). Advanced signal processing techniques for fault diagnostics—a review. *Mathematical and Computational Applications, 4*, 121–136.
Figliola, R. S., & Beasley, D. E. (2011). Theory and *design for mechanical measurements*. New York: Wiley.
Gao, R. X., & Yan, R. (2011). *From fourier transform to wavelet transform: A historical perspective*. Wavelets: Springer.
Ghafari, S. H. (2007). *A fault diagnosis system for rotary machinery supported by rolling element bearings*. Waterloo: Doctor of Philosophy.
Ghods, A., & Lee, H. H. (2014). A frequency-based approach to detect bearing faults in induction motors using discrete wavelet transform. In *Proceedings of the IEEE International Conference on Industrial Technology* (pp. 121–125).
Giaouris, D., Zahawi, B., El-Murr, G., & Pickert, V. (2006) Application of wavelet transformation for the identification of high frequency spurious signals in step down DC–DC converter circuits experiencing intermittent chaotic patterns. In *The 3rd IET International Conference on Power Electronics, Machines and Drives* (pp. 394–397), 4–6 April 2006.
Harpen, M. D. (1998). An introduction to wavelet theory and application for the radiological physicist. *Medical Physics, 25*, 1985–1993.
He, J. (2001). Modal analysis, Oxford Boston. OxfordBoston: Butterworth-Heinemann.
Hsieh, W. H., Lu, M. C., & Chiou, S. J. (2012). Application of backpropagation neural network for spindle vibration-based tool wear monitoring in micro-milling. *International Journal of Advanced Manufacturing Technology, 61*, 53–61.
Iorgulescu, M., Beloiu, R., & Cazacu, D. (2009). Vibration monitoring for electrical equipment faults detection using fast fourier transform. 2009. 34–38.
Kamiel, B., McKee, K., Entwistle, R., Mazhar, I., & Howard, I. (2015). Multi fault diagnosis of the centrifugal pump using the wavelet transform and principal component analysis. *Mechanisms and Machine Science*, 555–566.
Kankar, P. K., Sharma, S. C., & Harsha, S. P. (2011). Fault diagnosis of ball bearings using continuous wavelet transform. *Applied Soft Computing Journal, 11*, 2300–2312.
Karthikeyan, P., Murugappan, M., & Yaacob, S. (2012). ECG signal denoising using wavelet thresholding techniques in human stress assessment. *International Journal on Electrical Engineering and Informatics, 4*, 306–319.
Kehtarnavaz, N. (2008). *Digital signal processing system design: LabVIEW-based hybrid programming*. Amsterdam: Elsevier.
Kim, E. Y., Tan, A. C. C., Yang, B.-S., & Kossel, V. (2007). Experimental study on condition monitoring of low speed bearings: Time domain analysis. In *5th Australasian Congress on Applied Mechanics*.
Leavey, C. M., James, M. N., Summerscales, J., & Sutton, R. (2003). An introduction to wavelet transforms: A tutorial approach. *Insight: Non-Destructive Testing and Condition Monitoring, 45*, 344–353.
Li, C. J. (2006). *Signal processing in manufacturing monitoring. Condition monitoring and control for intelligent manufacturing*. London: Springer.
Li, R., & Frogley, M. (2013). On-line fault detection in wind turbine transmission system using adaptive filter and robust statistical features. *International Journal of Prognostics and Health Management, 4*.
Lihui, W., & Gao, R. X. (2006). *Condition monitoring and control for intelligent manufacturing*. London: Springer.
Loutas, T., & Kostopoulos, V. (2012). Utilising the wavelet transform in condition-based maintenance: A review with applications. In BALEANU, D. (ed.).
Marwala, T. (2012). *Condition monitoring using computational intelligence methods*. London: Springer.
Misiti, M., Misiti, Y., Oppenheim, G., & Poggi, J.-M. (1997). Wavelet toolbox for use with MATLAB, MathWorks.
Mohanty, A. R. (2015). *Machinery condition monitoring: Principles and practices*. London: Taylor & Francis Group.

Ngolah, C. F., Morden, E., & Wang, Y. (2011). An intelligent fault recognizer for rotating machinery via remote characteristic vibration signal detection. 2011. 135–143.

Ngui, W. K., Leong, M. S., Hee, L. M., & Abdelrhman, A. M. (2013). Wavelet analysis: Mother wavelet selection methods. *Applied Mechanics and Materials.*

Onsy, A. (2009). *Intelligent health monitoring of power transmission systems*. Ph.D Thesis, Newcastle upon Tyne.

Pan, M. C., Van Brussel, H., & Sas, P. (1998). Intelligent joint fault diagnosis of industrial robots. *Mechanical Systems and Signal Processing, 12*, 571–588.

Polikar, R. (1996). *The wavelet tutorial*. United States: Rowan University.

Qiu, H., Lee, J., Lin, J., & Yu, G. (2003). Robust performance degradation assessment methods for enhanced rolling element bearing prognostics. *Advanced Engineering Informatics, 17*, 127–140.

Qu, J., Zhang, Z., & Gong, T. (2016). A novel intelligent method for mechanical fault diagnosis based on dual-tree complex wavelet packet transform and multiple classifier fusion. *Neurocomputing, 171*, 837–853.

Rajbhandari, S. (2009). Application of wavelets and artificial neural network for indoor optical wireless communication systems. PhD PhD Thesis, University of Northumbria at Newcastle.

Sawicki, J. T., Sen, A. K., & Litak, G. (2009). Multiresolution wavelet analysis of the dynamics of a cracked rotor. *International Journal of Rotating Machinery*, 2009.

Shin, K., & Hammond, J. K. (2008). *Fundamentals of signal processing for sound and vibration engineers*. New York: Wiley.

Tse, P. W., Yang, W. X., & Tam, H. Y. (2004). Machine fault diagnosis through an effective exact wavelet analysis. *Journal of Sound and Vibration, 277*, 1005–1024.

Tseng, C. L., Wang, S. Y., Lin, S. C., Chou, J. H., & Chen, K. F. (2014). A diagnostic system for speed-varying motor rotary faults. *Mathematical Problems in Engineering*, 2014.

Vachtsevanos, G., Lewis, F., Roemer, M., Hess, A., & Wu, B. (2006). *Intelligent fault diagnosis and prognostic for engineering systems*. New York: Wiley.

Vivas, E. L. A., Garcia-Gonzalez, A., Figueroa, I., & Fuentes, R. Q. (2013). Discrete wavelet transform and ANFIS classifier for brain-machine interface based on EEG. In *2013 6th International Conference on Human System Interactions, HSI 2013* (pp. 137–144), 2013.

Wilkinson, M. R. (2008). *Condition monitoring for offshore wind turbines Eng. D.*, University of Newcastle upon Tyne.

Williams, T., Ribadeneira, X., Billington, S., & Kurfess, T. (2001). Rolling element bearing diagnostics in run-to-failure lifetime testing. *Mechanical Systems and Signal Processing, 15*, 979–993.

Zhen, C., & Zhang, Y. (2012). Fault diagnosis for wind turbines based on vibration signal analysis.

Chapter 4
PUMA 560 Robot and Its Dynamic Characteristics

Abstract This chapter describes the PUMA 560 industrial robotic system that will be used as the demonstrator throughout this project; highlighting its main components and how they interact with each other. VAL II is a programming language utilized with the used robotic system, hence, included in this chapter is a brief synopsis of some of the important VAL II commands that are needed to programme the robot, referent to the project. In order to design a reliable monitoring system for any machine, some of its characteristics have to be known, such as machine natural frequencies. Thus, the robot natural frequencies are obtained and utilized for preliminary health evaluation and fault detection. To obtain these frequencies an experimental modal analysis (EMA) is performed, which consists of: exciting the robot and then measuring what is called frequency response function (FRF) between the excitation and response. The FRF measurement is normally achieved using specific software to find the natural frequencies from it. Explanation of the EMA theory and how FRFs are calculated, used measurement equipment and software setup for the experiments are also discussed in this chapter.

4.1 PUMA 560 Industrial Robot—General Overview

The PUMA 560 robotic manipulator, produced by Unimate, is an old (>25 years), but functional, multi-joint robot and was used in this project, due to the availability of the spare parts that are required for simulation of the different faults (as discussed in Chap. 5). It is a PC controlled, serial manipulator designed for use in industrial applications, and has six revolute joints/degree of freedom (DOF) with three major axis of motion (X, Y, and Z) and resembles the human arm in function. Each of the robot's joints is actuated by a DC brushed permanent magnet servo motor. Positioning of the end effector, which is normally a device fixed at the end of the robotic arm, is achieved by the coordination of the first three joints, which are named: waist (joint 1), shoulder (joint 2) and elbow (joint 3), allowing the robot to move the end effector into any position with maximum reach of 1 m. Orientation of the end effector is important when it approaches its final position and this is

A.A. Jaber, *Design of an Intelligent Embedded System for Condition Monitoring of an Industrial Robot*, Springer Theses,
DOI 10.1007/978-3-319-44932-6_4

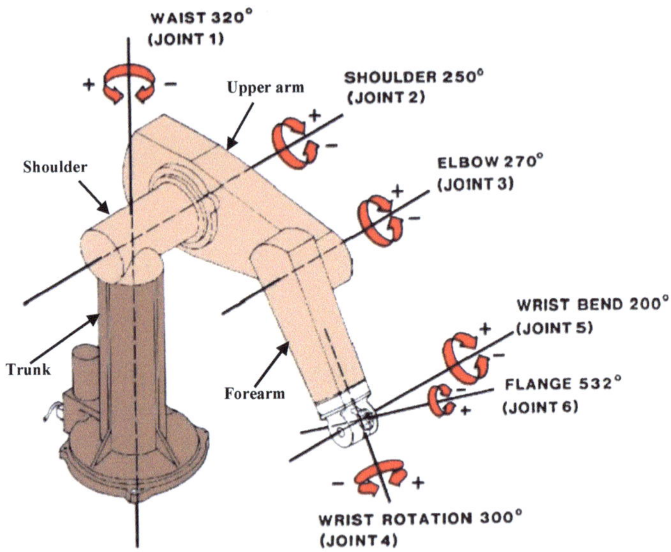

Fig. 4.1 PUMA 560 robot member representations (Rutherford 2012)

achieved using three wrist joints (4, 5 and 6) where the tool can be independently manoeuvred. This gives the robot six degrees of freedom, as shown in Fig. 4.1 where the name of each joint with its maximum range of rotating is indicated. Electromagnetic brakes are equipped with the first three joints (waist, shoulder, and elbow), which lock the motors to prevent collapsing when the power is removed from the robot (Potgieter et al. 2005).

The control unit of the PUMA 560 is the most important part of the robotic system, it controls all operations of the manipulator and any connected device. There is two-way communication to each of the joints motors where sensory information on position and speed is fed back to the control unit; outputs are sent to the motors relative to the task's requirements. External digital input/output ports are available allowing other sensors and actuators to be controlled from this unit. There are a number of ways to position and orientate the manipulator end effector; but generally, one method is used at a time. Connected to the robot control unit is a **teach pendant** used to manoeuvre the robot around the workspace via different modes. In **joint** mode, each individual motor can be driven one at a time, produces rotational motion of the limbs in clockwise or anticlockwise direction. In **world** mode the end effector is moved in a straight line along X, Y and Z axes, relative to the world coordinates, which are fixed on the base of the robot, or moving tool coordinates. Once the operator is satisfied with the end effector position and orientation, then this can be stored within the control unit memory by giving it a code name.

All the operations running through the control unit are programmed using VAL II programming language, which was developed for PUMA robots. Complex programs can be developed by combining some commands, such as 'wait' and 'move', with conditional statements such as 'if-then-else'. A simple VAL II programme below demonstrates this; the function of each line in the programme is shown in the {} brackets, but would not be included when writing a VAL programme.

```
1 SPEED 20 ALWAYS              {Set the arm speed to 20% monitor speed}
2 OPENI                        {Open gripper jaws}
3 IF SIG (1001) == 1 THEN      {If input signal 1 equal to 1 then...}
4 MOVE PICK                    {Move to position Pick}
5 CLOSIE                       {Close gripper jaws}
6 MOVE PLACE                   {Move to position Place}
7 ELSE                         {If signal 1 not equal to 1...}
8 MOVE EMPTY                   {Move to position Empty}
9 END                          {End of the program}
```

To enable real-time control, VAL II has the ability to interface with external devices. Through the available I/O module, the control unit may issue commands to the devices or accept data from them. The module is organised as a set of 32 output lines (OX) and 32 input lines (WX), with each line having its own address. To monitor an input line the address must be preceded by 10 in the programme code (see line 3 in the above code). High and low voltages on the lines are considered by the system as logic 1 and 0 respectively. This facility has been used in this thesis.

4.2 Modal Analysis

Modal analysis is a technique to estimate the natural frequencies of dynamic structures and their associated mode shapes. It can be performed using two approaches: computationally and experimentally. In the first method, which is also known as the analytical (or theoretical) modal analysis, the dynamic behaviour (natural frequencies and mode shapes) of the structure is obtained from its mathematical model through the numerical analysis. Finite element modelling (FEM) is commonly utilized to investigate the dynamic behaviour of structures. For instant, a research was undertaken to theoretically study the dynamic characteristic of KUKA robot (Pupăză et al. 2014). The three dimensional robot modelling was done using computer aided design (CAD) software and the modal analysis was performed using ANSYS FEM software. Other research was also applied modal analysis approach based on ANSYS software for gearbox fault diagnosis (Liguo et al. 2009).

In the experimental model analysis (EMA), also recognized as modal testing, an appropriate external excitation force is applied to the structure, and simultaneously the structure vibration response is measured and analysed to extract the modal parameters. The relationship between the input and output (the excitation and response) is described using what is known as the frequency response function (FRF), as discussed in the following section. There are two commonly applied excitation methods in modal analysis; the impact type by using an instrumented hammer with a force transducer at its tip, and the random (or sinusoidal) excitation type by using an electromagnetic shaker driven by a signal generator. In term of its uses, Elosegui (1994) conducted an experimental modal analysis of a PUMA 560 robot with the main aim is to find its natural frequencies. In the area of structural health monitoring (SHM), modal analysis has been effectively used to study the effects of a crack in a structure on its natural frequencies (El-Kafrawy 2011). The researcher applied experimental and theoretical modal analysis to validate the results.

In this chapter EMA will be applied to determine the natural frequencies of the PUMA robot. These frequencies can be utilized for condition monitoring of the robot, as their values are subject to change if a fault is present in the system; this is discussed in Chap. 6. However, to obtain all the robot's natural frequencies, this test was accomplished with the robot in different configurations and in three axes (X, Y and Z), as explained later.

4.2.1 Frequency Response Function (FRF)

Frequency response function (FRF) is defined as the ratio between the system response to its excitation force, and represents the outcome of the modal analysis. It can be calculated by dividing the Fourier transform of the response $x(t)$ on the Fourier transform of the excitation $y(t)$, as shown (Mohanty 2015).

$$H(f) = \frac{X(f)}{Y(f)} \tag{4.1}$$

where $H(f)$ is the FRF and $X(f)$ and $Y(f)$ are the FFT of the $x(t)$ and $y(t)$, respectively. However, to obtain a noise-free result the FRF can be estimated by dividing the cross-power spectrum of excitation and response $(S_{xy}(f))$ on the auto-power spectrum of the excitation $(S_{yy}(f))$, as follow.

$$H(f) = \frac{(S_{xy}(f))}{(S_{yy}(f))} \tag{4.2}$$

where $S_{xy}(f)$ and $S_{yy}(f)$ can be formulated in Eqs. 4.3 and 4.4 below, t is the measured time record, and X^* is the complex conjugate of $X(f)$.

$$S_{xy}(f) = \frac{1}{t} X^*(f) Y(f) \tag{4.3}$$

$$S_{xx}(f) = \frac{1}{t} X^*(f) X(f) \tag{4.4}$$

The peaks in the FRF correspond to the natural frequencies of the system. Thus, to find the natural frequencies of the PUMA 560 robot, an experimental modal analysis will be performed to find the FRF. In this test the robot will be excited in X, Y and Z directions, using an impact hammer, which is easier to excite the robot's natural frequencies and by using it the robot does not need to be mounted on a shaker (Brandt 2011); the responses from the robot in the three axes are then captured using an accelerometer. This analysis can be accomplished using different dedicated, commercially available data acquisition systems for vibration and modal analysis, along with their driver software. From the peaks of the established FRF the robot's natural frequencies can be extracted. The amplitudes associated with the natural frequencies represent the energy required to excite these modes.

4.2.2 Coherence

In modal analysis it is important to check if the captured response is from the impact force or if it is noise. To this end the correlation between the response and the excitation signals can be checked. This will also help in identifying where to locate the accelerometer on the robot to get an accurate modal result. Accordingly, for data quality assessment, the coherence function in the frequency domain ($\gamma_{xy}^2(f)$) can be used, which shows how much the output is connected to the input. Its value ranges from zero to one, where a value of one corresponds to a perfect correlation between the impact and the output signal and there is no influence of noise, whereas zero means no coherency. The coherence between the input and response signals can be calculated with the following formula (Mohanty 2015):

$$\gamma_{xy}^2(f) = \frac{S_{xy}^2(f)}{S_{xx}(f) * S_{yy}(f)} \tag{4.5}$$

4.2.3 Description of the Measuring Equipment

In the EMA measuring equipment provides the required input and output data. To give a better understanding of the experiments, the measuring equipment, which consists of: an impact hammer, an accelerometer and a dynamic signal analyser, are described in this section.

Impact hammer: this device produce an impulsive excitation force that causes the structure to vibrate followed by free decay of vibration. The hammer contains a force sensor mounted on its striking face to measure how much force is applied to the structure. The force sensor contains a piezoelectric crystal that generates charge as a result of the deformation in its shape, due to the impact applied. Usually the impact hammer has a hole at its striking head for using a variety of impact tips (soft, medium, and hard) and also a removable extra mass secured on the other side to increase the input force magnitude. The main functions of the tip are to transfer the force of the impact to the sensor and also to protect sensor face from damage. Selecting a suitable tip type is very important for accurate modal analysis; however, the hardness of the impact tip and the added extra mass to the hammer influence the excitation frequency bandwidth. A hammer with a hard tip excites a wide frequency spectrum bandwidth, whereas a soft-tipped hammer generates concentrated excitation energy in a low frequency band. Here a PCB integrated circuit piezoelectric (ICP) impact hammer model 086C03 that has its own built in charge amplifier has been used. Its sensitivity is 2.25 mV/N and its mass and length are 160 g and 216 mm, respectively. The hammer can be connected to the data acquisition analyser through a standard BNC jack connector.

Accelerometer: the dynamic response from the structure, due to the applied impact, is measured using a piezoelectric accelerometer, accelerometers are the most popular and most commonly used in modal analysis applications, which their working principle is also based on the deformation in the shape of the piezoelectric crystal. A single axis ICP accelerometer type PCB 352C68 has been utilized to measure the acceleration from the robot in different axes. The integrated amplifiers in the accelerometer and impact hammer are directly fed from the signal analyser. This accelerometer has a standard sensitivity of 100 mV/g and measurement range is 0.5 Hz–10 kHz. The accelerometer comes with a standard stud mount, which would require drilling and semi-permanent attachment to the robot arm. In order to avoid drilling the arm, the accelerometers were mounted using an additional bracket, which is attached to the mounting stud on the accelerometer. This bracket allows for wax or any other adhesives to be applied to the bottom of the accelerometer in order to affix it without modifying the tested structure.

Data Physics Quattro analyser: for data acquisition, the Data Physics Quattro, which is a USB powered vibration and sound analyser, has been used. This analyser offers 4 channels analogue input up to 54 kHz, and can be connected to a laptop using USB 2.0. The analyser samples the voltage signals coming from the accelerometer and the force transducer and converts them into equivalent acceleration and force depending on the sensitivity information of the sensors; also, it has the capability to power ICP sensors. For vibration signal acquisition and analysis the SignalCalc 240 from Data Physics Corp, which is especially designed to work with the Data Physics Quattro analyser, has been used. The experimental setup for this analysis is shown in Fig. 4.2. The datasheets for all aforementioned devices are provided in Appendix A.

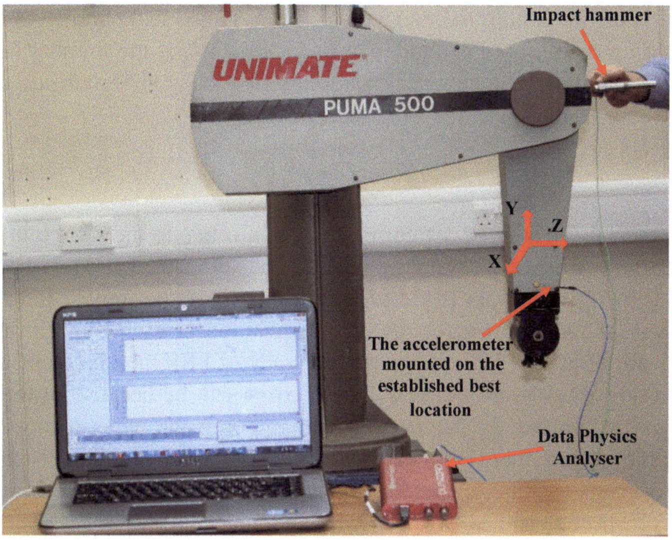

Fig. 4.2 The experimental set up

4.3 Experimental Considerations

A robust result from the EMA depends on number of factors, including, but not limited to, the used testing equipment, setting of the utilized dynamic signal analyser software and the skill of the person doing the experiments. The following subsections discuss some of parameters that need to be considered prior to conducting the modal testing.

4.3.1 Settings of SignalCalc ACE Software

The frequency resolution, which is coupled with the number of measurement lines, represents the first parameter to be set in the software. Higher frequency resolution implies longer time, more data, and large data size; thus a compromise needs to be made. For this analysis, the number of spectral lines has been set to 6400 *lines*. Hence, by using the following equations the frequency resolution (ΔF) can be determined.

$$\Delta F = 2.56 * \frac{f_{max}}{N} \quad (4.6)$$

$$\Delta F = \frac{f_s}{N} \quad (4.7)$$

where f_{max} is the maximum frequency, which is assumed to be known and its value is set based on the number of modes to be captured, N is the number of samples, and f_s is the sampling frequency. The sampling frequency in SignalCalc software is always given by:

$$f_s = 2.56 * f_{max} \tag{4.8}$$

The trigger also has to be set correctly, which refers to the source that initiates the data collection process. Many options can be set in the software, but the most appropriate for this analysis is the ***input*** option where the impact force from the hammer acts as input to trigger the data acquisition at each impact, after which N samples are acquired. To increase the accuracy of FRF measurement, the effects of random errors has to be reduced. These errors can be because of the contaminated noise or induced by the person who is doing the experiments. However, averaging the measurements represents a very useful method to get reliable results. So, after the signals are captured the software then computes the auto- and cross-power spectra, and adds them to a cumulated average. Different types of averaging, such as peak hold, exponential, and stable, can be set up in the software. In this analysis a ***stable*** averaging method is used; by which a pre-set number of constituents are summed, each one is weighted with equal importance and then the average is automatically normalized to the number of constituents recently acquired (Data Physics Corporation 2006). The software also has the capability of halting after the time-domain signal has been collected, but before being included in the average, for a manual investigation process where the user can decided whether to include or reject the current impact in the averaging.

The response signal from the test is a free decay that has to approach zero within the observation time interval; however, a leakage in the signals may occur if the measured signals do not decay to zero. Therefore, to minimize the effects of leakage, the measured signals are multiplied by a window function before they are transformed to the frequency domain. The software has different window types, but for transit signals the appropriate one is the ***rectangular*** window (Data Physics Corporation 2006). Since the impulse signal excites the structure for a very short time period, it is important to make sure that the entire impulse and response signals are captured. In SignalCalc software a pre-trigger delay can be specified, which allows the analyser to begin data sampling before the trigger point occurs. A plastic hammer tip is chosen, as it was found that this delivers adequate energy to excite frequencies within the bounds of interest. Finally, the person doing the experiment has to ensure that the impact axis is parallel to the accelerometer axis, and also there is no double hit, which may occur due to the structure flexibility that bounce back at the hammer (Brandt 2011).

4.3.2 Measurement System Calibration

Calibration can be conducted in one of several ways. Comparison with a reference standard device, that is considered to be more precise than the instrument being calibrated, represents the most common calibration approach (Mohanty 2015). However, the accelerometer used here was purchased specifically for this research and was already calibrated, whereas the impact hammer and Data Physics analyser were several years old. Thus, to test the measurement system accuracy prior to utilize it on the robot, modal analysis was performed to find the natural frequencies of simple cantilever beam test rig. The idea of this procedure is to mathematically calculate the first four natural frequencies of the beam, based on the beam theory (Wahab 2008), and then compare them with the experimentally established ones from the test rig. As observed in Table 4.1, the experimental and theoretical results are identical, presenting a maximum percentage discrepancy of 1.26 % at the 4th mode. The small differences between the result can be attributed to that the theoretically calculated frequencies are based on an assumption that one end of the cantilever beam is properly fixed, but in practice this may not be always the case due to the flexibility and damping in the support.

4.3.3 Coherency Checking

In order to have a reliable result from the hardware and software the selection of response measuring position should be made properly. After conducting the fundamental experimental settings and validation, it remained to check if the signal-to-noise ratio (SNR) is sufficient; established by examining the coherence function that is measured from different accelerometer locations. A pre-modal analysis testing was carried out to investigate the best accelerometer location. The coherence assessment has been done for three different accelerometer's locations which are near to the robot wrist, elbow, and shoulder, as these locations contain the robot's gears so the response will properly propagated through them giving good signal-to-noise ratio. The accelerometer was also attached to the robot's links, but it was found that the noise percentage in the captured signal was very high, and the coherence was very low. Some of results of coherence function are shown in Fig. 4.3.

Table 4.1 Comparison between the experimental and theoretical natural frequencies of a cantilever beam

Mode	Theoretically	Experimentally	Error between theoretical and experimental result %
1	64	64.06	0.09
2	400.9	400	0.22
3	1122.4	1120.31	0.18
4	2199.66	2171.8	1.26

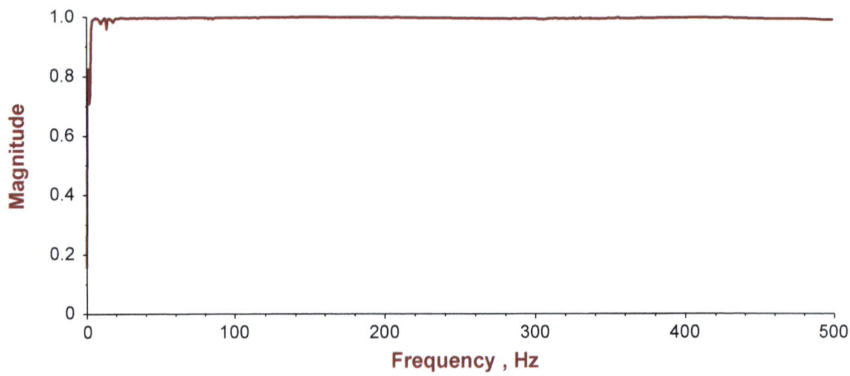

(a) The coherence when the accelerometer on the wrist

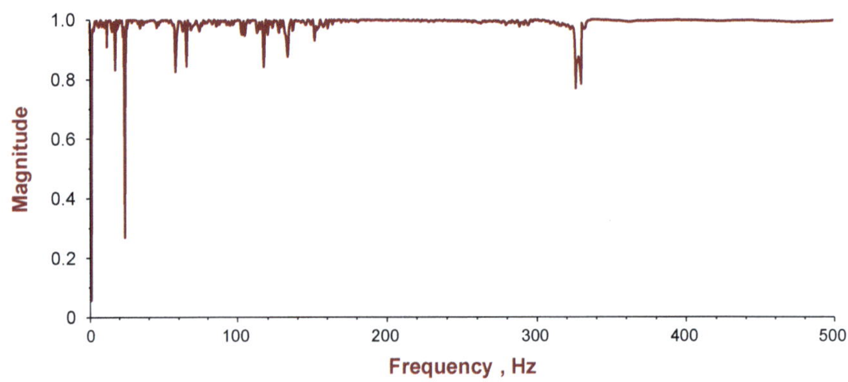

(b) The coherence when the accelerometer on the elbow

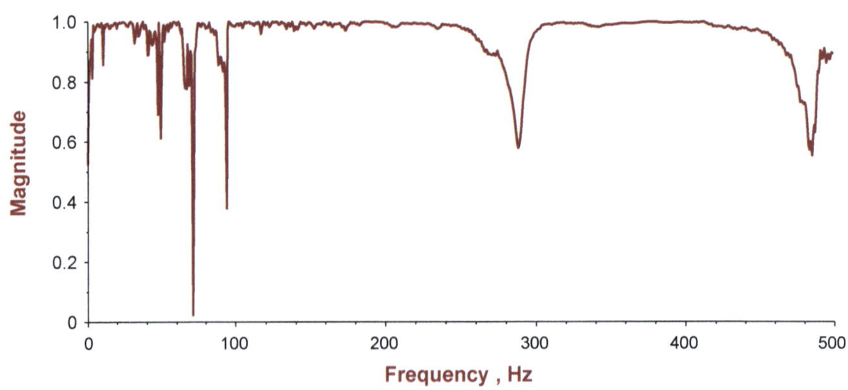

(c) The coherence when the accelerometer on the shoulder

Fig. 4.3 Coherence function when the accelerometer on different locations

From the results it was observed that, in comparison to the elbow and shoulder locations, when the sensor is located at the first location (near to the wrist as shown in the experimental set up in Fig. 4.2) the coherence gets very high. Also, in selecting the best position to put the accelerometer not only has the coherence to be checked, but also the number of excited modes from that position; the higher the number of modes the better the accelerometer location is. Thus, from the preliminary result it was established that a higher number of modes were captured when the accelerometer near to the arm wrist, and therefore will be used. Generally, if a structure is excited in a point located farther from the sensor, the coherence will be very poor; therefore, the point of impact has also been selected carefully.

4.4 Results of FRF Analysis

Since the arm robots are configurable machines, they have different structural stiffnesses for different configurations; and consequently some frequencies appear in one configuration and not in another. So, the FRF analysis was carried out when the robot in four different configurations and when the accelerometer is near to the wrist, which is the best established response capturing location. These configurations represent the standard configurations of this robot, as shown in the Fig. 4.4. In the first one the upper arm is horizontal and the forearm is hanging down; the second is called *stretch* where the entire arm is straight and horizontal; third is the robot *zero* position, which is same as the first configuration just the forearm is up, and the last one is known as the *ready* position where the arm is straight and vertical (Corke 2011).

Before conducting the modal testing it was observed that there is a dithering problem in the robot structure when the arm power is on and does not move. This can be related to the joint motors which produce the required torques to hold the arm static. An FFT analysis was accomplished to investigate if the robot control system excites some natural frequencies or not; it was established that there are no significant frequencies that need to be considered in the spectrum. Furthermore, before achieving modal analysis on the robot structure, the FRF was established in the base of the robot, to find its natural frequencies and monitor them if they will be excited when the robot arm is impacted. However, the base natural frequencies were discovered to be very high in comparison to the robot arm frequencies, and therefore they were not detected when the robot excited.

To measure the FRF for the robot in X, Y and Z axes, first the accelerometer was put in-line with X-axis and the test was conducted, after which it was fixed parallel to Y- and Z-axes and the FRF procedure was performed, respectively. Every time the test was performed, the impact direction was keen to be in-line with the accelerometer, in order to excite just the frequencies in the direction of interest as the displacement is being maximized in that direction. Also, prior to doing the final tests several trials were performed in order to locate the best excitation point that gives the best coherence between the input and response; and was found that it is

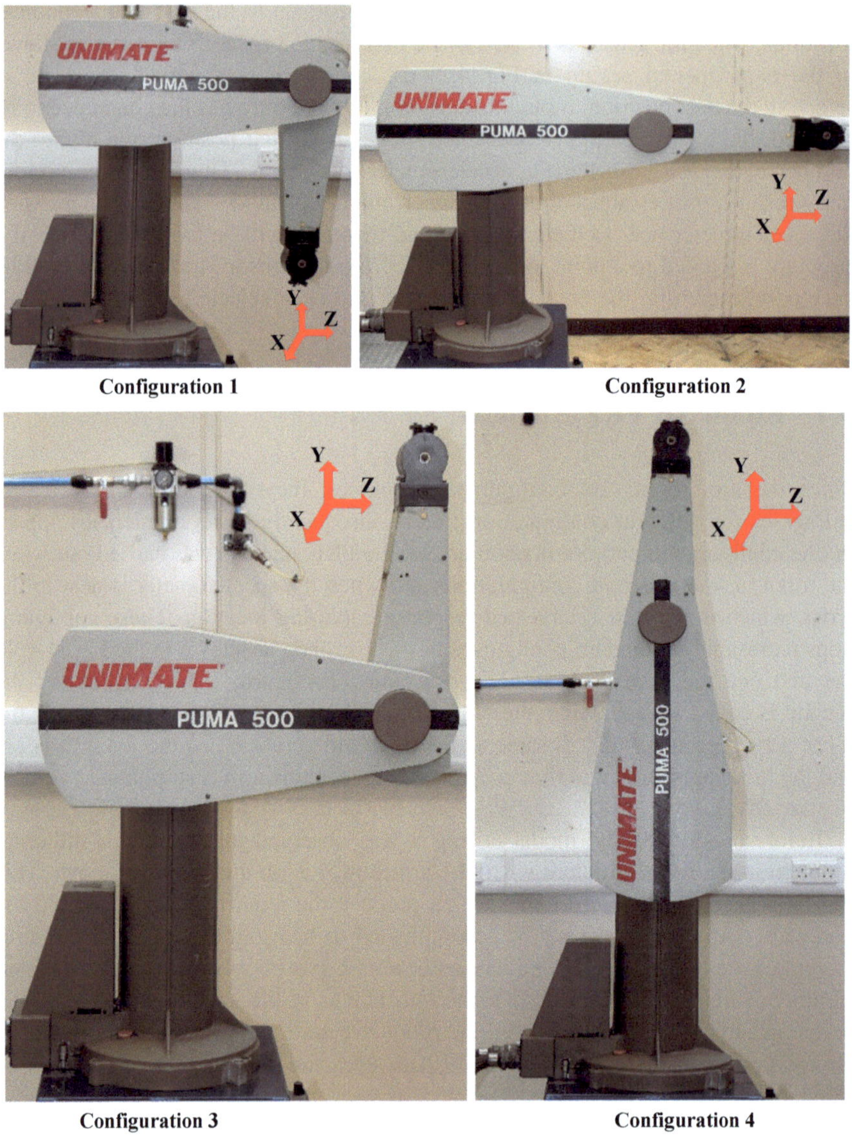

Fig. 4.4 The tested robot configurations

best to hit the structure at the elbow (joint 3) area. The Fig. 4.5 shows the FRF results when the robot in configuration 1. The extracted natural frequencies from the FRF analysis when the robot in different configurations are shown in the Table 4.2.

From the above table it can be seen that some frequencies appear in one configuration and not in another; for instance, the frequency 356.25 Hz can be

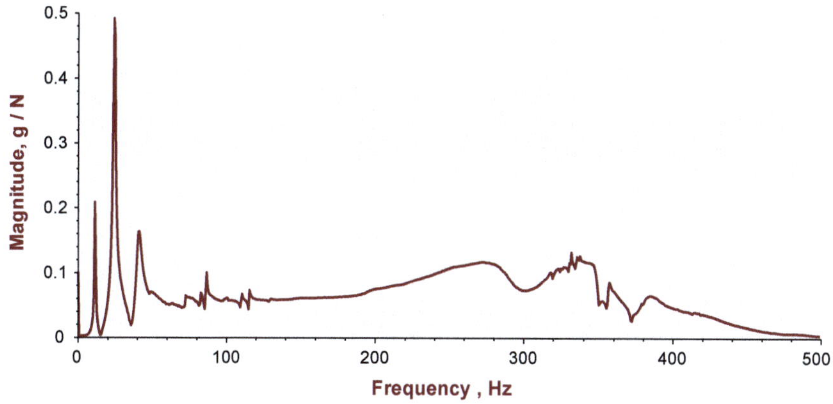

(a) X-axis

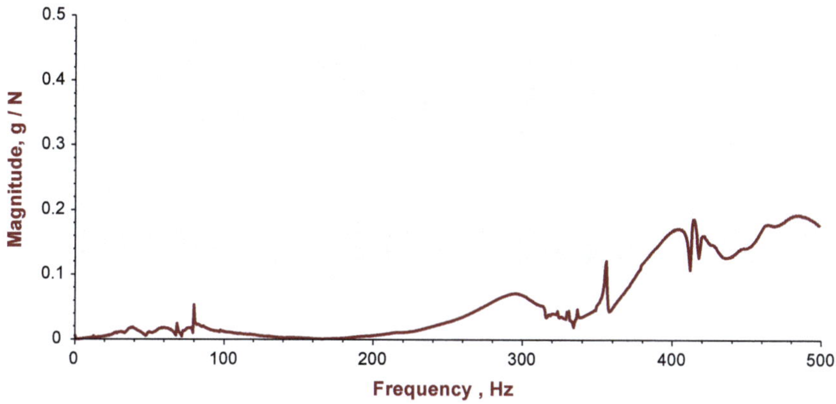

(b) Y-axis

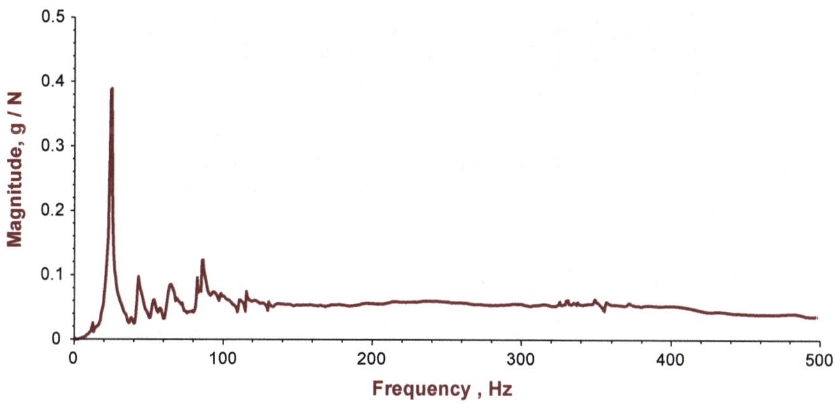

(c) Z-axis

Fig. 4.5 FRF of the robot in configuration 1

Table 4.2 Estimated robot natural frequencies when it is at different configurations

Configuration	Axis	Natural frequency								
		1st	2nd	3rd	4th	5th	6th	7th	8th	9th
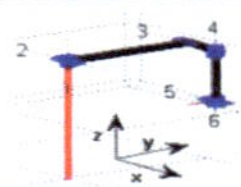	X	12.5	27.37	42.2	86	110.6	115.6	356.25	–	–
	Y	66.87	80.6	356.25	414.4	–	–	–	–	–
	Z	12.5	27.37	42.2	53.75	66.87	86	115.6	–	–
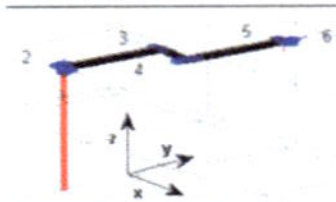	X	9	12.5	17.19	42.2	86	110.6	115.6	–	–
	Y	12.5	17.19	27.37	42.2	–	–	–	–	–
	Z	12.5	20	66.87	86	110.6	115.6	130.6	–	–
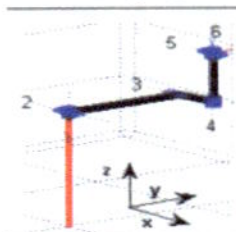	X	9	12.5	35	80.6	86	110.6	115.6	–	–
	Y	12.5	86	110.6	115.6	130.6	197.5	–	–	–
	Z	12.5	35	66.87	80	–	–	–	–	–
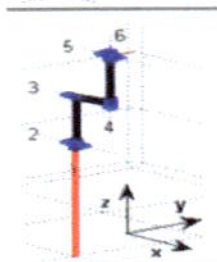	X	12.5	27.37	42.2	86	110.6	115.6	–	–	–
	Y	12.5	20	27.37	35	86	97.5	110.6	115.6	130.6
	Z	42.2	66.87	197.5	356.25	–	–	–	–	–

Table 4.3 Estimated robot natural frequencies compared to another paper result

Mode	Estimated natural frequencies (Hz)	PUMA 560 natural frequencies from (Elosegui 1994)	Mode	Estimated natural frequencies (Hz)	PUMA 560 natural frequencies from (Elosegui 1994)
1	9	5.46	10	86	80.32
2	12.5	12.67	11	97.5	–
3	17.19	–	12	110.6	–
4	20	20.12	13	115.6	–
5	27.37	25.23	14	130.6	–
6	35	–	15	197.5	158.69
7	42.2	40.14	16	356.25	–
8	53.75	51.08	17	414.25	–
9	66.87	67.21	–	–	–

observed only when the robot in the first and forth configuration, while the 414.4 Hz frequency was detected only in the first configuration. In contrast, other frequencies, such as 12.5, 66.87 and 86 Hz, emerged in all of the tested configurations. Due to the torsional stiffness of the transmission systems industrial robots have flexible joints, and the effect of this flexibility accompanied with the gravity field is clearly present in X and Z axes, where higher number of lower natural frequencies are excited especially in the first three configurations; in these configurations fewer frequencies were excited in Y-axis, as it is outside the effect of joints flexibility and gravity field. In the fourth configuration fewer frequencies were stimulated in Z-axis, since the applied torque due to the robot mass will be neglected as the centres of the arm and mass are coincident. It can be concluded that the technique of using different robot configurations has helped in extracting most of the robot natural frequencies. The order (sequence) of the frequencies will be selected depending on their emersion in the table and from lower to higher frequency, regardless of the configuration since during performing a task the robot configuration is changing instantaneously, exciting different frequencies. Also, to validate the results, the natural frequencies from Elosegui (1994) research, which used the same robot but different excitation method and configuration, have been utilized for comparison, as shown in Table 4.3.

It can be seen from Table 4.3 that a higher number of frequencies are excited in the current study with acceptable correlation between the results. The difference between the results may be because of the excitation method, the method of mounting the robot on the ground, the robot configuration in the previous study or the age of the robot. The robot used here is very old in comparison to the same robot model in 1994. It has been observed from previous research conducted in the area of robot modal analysis, the robots have been configured in arbitrary positions, and their natural frequencies extracted. The disadvantages of that are: firstly not all the robot's natural frequencies will be extracted, these natural frequencies might be not accurate, and finally it will be difficult for other researchers to obtain the same results from the same analysis on the same robot. Consequently, after doing this test, it can be said that the frequency response analysis can be undertaken easily by using the above procedure to find the natural frequencies of any robot.

4.5 Summary

A general overview of the used PUMA 560 robot along with its VAL programming language was introduced in the first section of this chapter, followed by the background to experimental modal analysis (EMA) and how it can be employed to extract the system natural frequencies, including the use of frequency response analysis for fault detection. The test equipment, setup of the data acquisition software and measurement system calibration was explained. Considerations, such as finding the best locations to excite the robot and measure the vibration response on the robot structure, were discussed. To correctly identify the robot natural frequencies modal analysis was carried out with the robot in four different configurations. The established frequencies were compared with the results of another research study, which utilized the same robot model but different experimental setup, and an acceptable correlation between both results was observed, giving some confidence in the current findings. These frequencies may be applied for robot fault detection as it will be investigated in the coming chapters. The procedure developed in this chapter could be adopted to establish the dynamic characteristics of other industrial robots.

References

Brandt, A. (2011). *Noise and vibration analysis: Signal analysis and experimental procedures*, New York: Wiley.

Corke, P. (2011). *Robotics, vision and control*, Berlin/Heidelberg: Springer/GmbH & Co. KG.

Data Physics Corporation (2006). *SignalCalc dynamic signal analyzer*. User Manual Data Physics Corporation.

El-Kafrawy, A. (2011). Crack detection by modal analysis in 3D beams based on FEM. *International Journal of Mechanics and Materials in Design, 7*, 265–282.

Elosegui, P. (1994). Measurement of the dynamic modael of a PUMA 560 robot using experimental modal analysis. *Journal of Mechanical Design, 116*.

Liguo, Z., Yutian, W., Sheng, Z., & Guangpu, H. (2009). The fault diagnosis of machine based on modal analysis. *2009 International Conference on Measuring Technology and Mechatronics Automation, ICMTMA 2009*, pp. 738–741.

Mohanty, A. R. (2015). *Machinery condition monitoring: Principles and practices*, Taylor & Francis Group.

Potgieter, J., Zyzalo, J., & Diegel, O. (2005). Reconfigurable mechatronic robotic plug-and-play controller. In V. Kordic, A. Lazinica, & M. Merdan (Eds.), *Cutting edge robotics*. Germany: Pro Literatur Verlag.

Pupăză, C., Constantin, G., & Negrilă, Ş. (2014). Computer aided engineering of industrial robots. *Proceedings in Manufacturing Systems, 9*, 87–92.

Rutherford, J. (2012). *Using the PUMA 560 robot.*

Wahab, M. A. (2008). *Dynamics and vibration: An introduction*, New York: Wiley.

Chapter 5
Robot Hardware, Transmission Faults and Data Acquisition

Abstract In this chapter joint three of the PUMA 560 robot was selected to simulate different transmission faults. The mechanical construction of this joint was comprehensively assessed to establish its assembly/disassembly. A discussion about different types of faults that may appear in the joint's gears and bearings is also included, and the physical simulation of several faults, with different severities, in the gearbox of the selected joint are presented. For vibration analysis a suitable accelerometer needs to be selected, and where to locate and attach it on the robot are important, therefore an optimum accelerometer location will be identified and fixed. Data acquisition software based on LabVIEW graphical programming and Matlab was designed using National Instrument hardware.

In this chapter joint three of the PUMA 560 robot was selected to simulate different transmission faults. The mechanical construction of this joint was comprehensively assessed to establish its assembly/disassembly. A discussion about different types of faults that may appear in the joint's gears and bearings is also included, and the physical simulation of several faults, with different severities, in the gearbox of the selected joint are presented. For vibration analysis a suitable accelerometer needs to be selected, and where to locate and attach it on the robot are important, therefore an optimum accelerometer location will be identified and fixed. Data acquisition software based on LabVIEW graphical programming and Matlab was designed using National Instrument hardware.

5.1 Mechanical Construction of the PUMA 560 Elbow Joint (3)

As stated in Chap. 4, each joint in the robot is actuated by a DC brushed motor and a transmission gearbox; and consequently faults can progress in different components of these joints. However, due to the limited time of the PhD study only one of

A.A. Jaber, *Design of an Intelligent Embedded System for Condition Monitoring of an Industrial Robot*, Springer Theses,
DOI 10.1007/978-3-319-44932-6_5

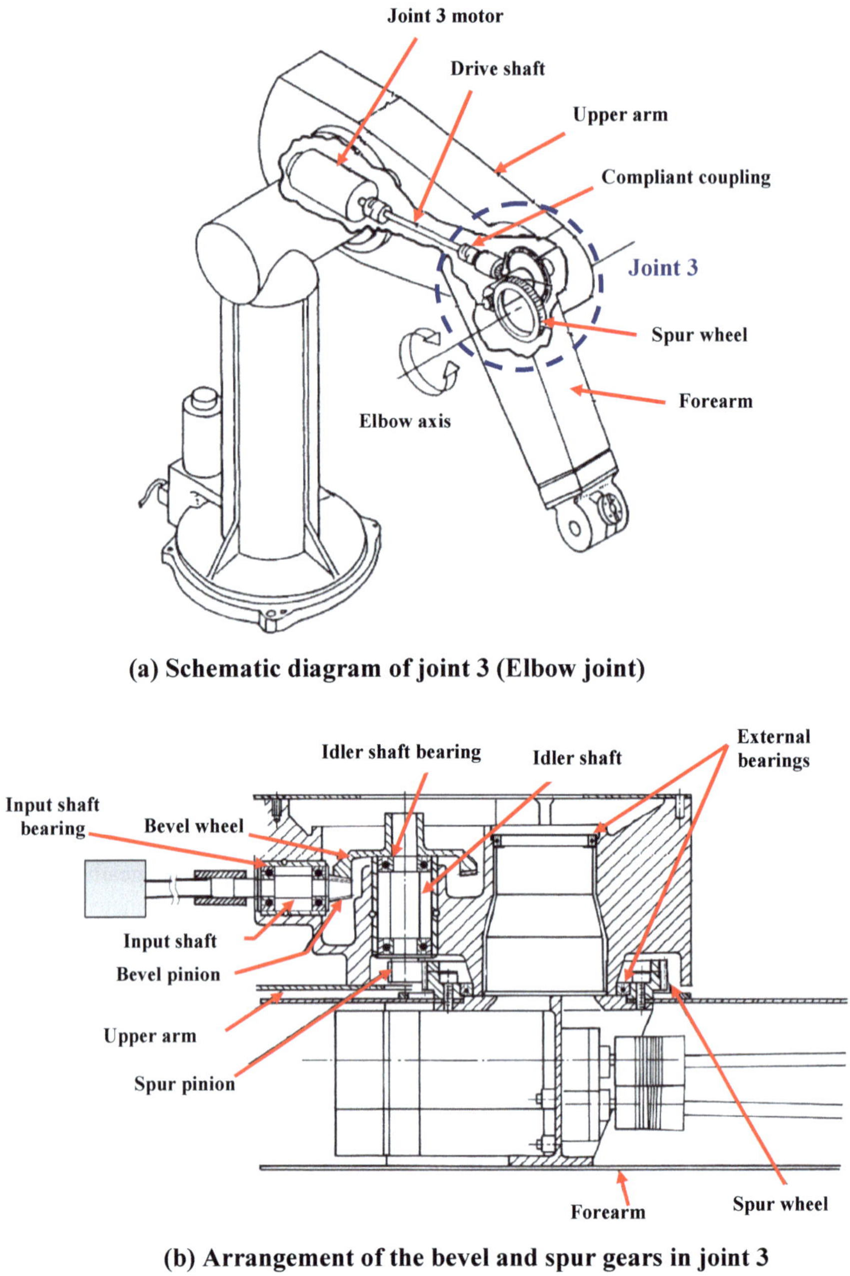

Fig. 5.1 **a** Schematic diagram of joint 3 (*Elbow joint*) **b** arrangement of the bevel and spur gears in joint 3

the robot joints (joint 3—elbow) is considered for simulating different fault types on the PUMA 560 robot.

From the literature it has been established that the elbow joint (joint 3) is the most critical joint in the PUMA robot, and consists of two-stage gear train system. The mechanical construction of this joint is illustrated in Fig. 5.1. The gear train is housed in the end of the upper arm and connected via a drive shaft and compliant couplings to a DC motor housed at the top of the upper arm (Fig. 5.1a). The flexible couplings are there to ensure a smooth transmission of motion while permitting some mechanical misalignment and relative movement between the motor and the gear train. A bevel pinion on the input shaft meshes with a bevel wheel fixed on one end of an idler shaft, Fig. 5.1b. A spur pinion at the other end of the idler shaft engages with a spur wheel fixed to the forearm, which rotates the forearm around the elbow axis. A number of bearings are used to carry the input and idler shafts. Also, there are two external bearings used to support the forearm in the upper arm.

5.2 Gears General Overview

A wide range of types and sizes of gears are used for different machinery. Gears usually operate in pairs; allow power to be transmitted from one shaft to the other without slippage and the direction and speed of rotation of the driven shaft to be changed. The pair gears may have the same or different teeth number; the gear with higher teeth number is called the wheel or gear, while the one with fewer teeth is the pinion. Speed of rotation is increased when the gear drives the pinion and decreased when the pinion drives the gear. Thus, the speed ratio is given by:

$$speed\ ratio = \frac{number\ of\ gear\ teeth}{number\ of\ pinion\ teeth} \tag{5.1}$$

The most common frequencies associated with gears are known as gear mesh frequencies. These frequencies are calculated depending on the number of teeth and the rotational speed as in the equation:

$$Gear\,mesh\,frequency = gear\,speed * no.\,of\,teeth \tag{5.2}$$

The main gear types are demonstrated in Fig. 5.2. To transmit power between two parallel shafts spur gears, which have teeth cut parallel to the shaft, are used. These gears produce only radial forces on their bearings and no axial forces (Radzevich 2012). Helical gears are similar to spur gears but their teeth are cut at a helix angle, which ensures the load is transferred progressively along the length of the tooth and distributed over a larger tooth area than in spur gears. Both axial and radial forces are imposed on the bearings of helical gears, due to their teeth not being parallel to the carrying shaft. Bevel gears are conical in shape and commonly used for transmitting power between intersecting shafts; most often the angle

(a) Spur gears **(b) Helical gears**

(c) Bevel gears **(d) Worm gears**

Fig. 5.2 Main gear types (Childs 2014)

between shafts is 90°. Bevel gears also impose both axial and radial forces on their bearing. Teeth in bevel gears can be straight or cut in spiroid to produce spiral bevel gears, which are smoother and quieter in operation than the straight bevels. Worm gears allow power transmission between two perpendicular shafts. They are screw-like gears and usually meshed with spur or helical gears.

Generally, typical quality requirements for the majority of industrial gears are: high efficiency, low backlash, large reduction in few steps, low friction, high torsional stiffness and low weight (Pettersson 2008). All the gears used in the PUMA 560 robot are of type spur and bevel gears. The number of teeth and speed ratios in joint 3 are presented in Table 5.1.

Table 5.1 Number of teeth in the gears of joint 3

Parameter	First stage (bevel gears)		Second stage (spur gear)	
	Pinion	Wheel	Pinion	Wheel
Number of teeth	13	96	11	80
Speed ratio	7.38		7.27	

5.3 Transmission Error in Gears

Transmission error (TE) is defined as "the difference between the actual position of the output gear and the position it would occupy if the gear drive were perfect" (Welbourn 1979). TE is considered to have a very strong correlation to resulting gear vibration and noise (Hiroaki and Nader 2012; Henriksson 2009). Several different techniques and instruments have been previously used to measure gears transmission errors, such as strain gauge on the drive shaft, torsional vibration transducers, magnetic signal methods, tachometers, rotary encoders systems and tangential accelerometers. The TE concept is illustrated in Fig. 5.3, and its value can be calculated using Eq. 5.3 (Hiroaki and Nader 2012).

$$TE = \left(\theta_{gear} - \frac{R_{pinion}}{R_{gear}}\theta_{pinion}\right) \tag{5.3}$$

where θ_{gear}, θ_{pinion} and R_{pinion}, R_{gear} are angular displacements and pitch circle radiuses of pinion and gear, respectively. Theoretically, when two gears mesh, the rotation of the output gear would be a function of the gear ratio and input rotation, if these two gears have perfect involutes and an infinite stiffness. However, gears will

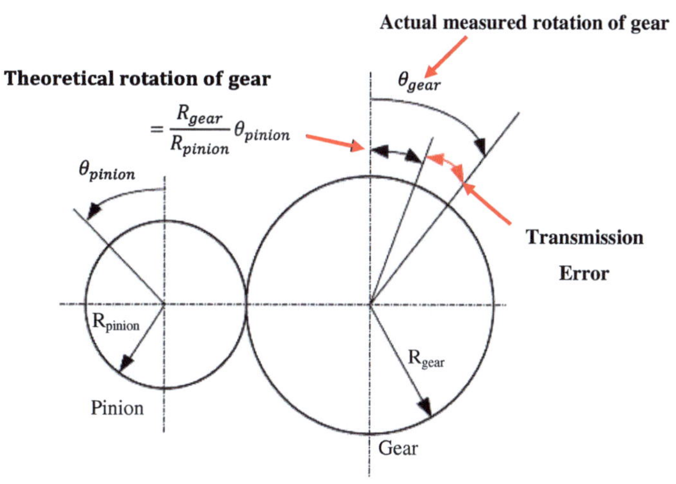

Fig. 5.3 Explanation of transmission error

normally not have a perfect involute shape because of unpremeditated modifications, such as manufacturing errors, and intended shape modifications. Additionally, there will be a motion error of the output gear relative to the input gear due to finite gear mesh stiffness. In short, there are three main sources for TE: the elastic deformation of the gears and associated components, gear geometrical errors and mounting errors (Hiroaki and Nader 2012). Also, varying gear mesh stiffness, unequal load distribution at gear mesh and changeable loads at the meshing position caused by output loads are another sources for TE (Ibrahim 2011).

5.3.1 Mounting Errors

In this sub-section the mounting errors that may occur in the transmission system of joint 3 will be discussed, however, these errors can be generalized for other gearboxes. In order to introduce some faults in the robot gearbox, as will be explained later in this chapter, the transmission in the elbow joint was dismantled. In addition to unavoidable transmission errors, mounting errors are possible if the gears are not correctly installed on reassembly, and thus care is required.

Gears mesh problems, such as poor teeth contact, high noise and vibration, cannot be avoided if at least one of the mating gears is not correctly mounted, and, factors, including misalignment of shafts, clearances in bearings and eccentricities further influence the contact pattern of the gear teeth. For example, in Fig. 5.4 which is a schematic diagram for joint 3 gearbox, if the centre distance between the spur gear pair decreases, the backlash value also decreases, and may result in interference which will prevent the gears from rotating properly. Furthermore, the accuracy of the two parallel axes will be composed of a parallelism error and a shaft offset error, which will influence the tooth contact in the tooth rotating direction.

For spur (and helical) gears, as explained previously, the best mesh for a gear set normally requires only correct adjustment of the centre distance and shafts parallelism and no further adjustment is needed. However, because the bevel gears are conical they can be assembled in an almost infinite number of positions and in spite of these positions that may satisfy the desired backlash value, they may still perform poorly. This makes the challenge of bevel gears adjustment quite complicated. In the practice all types of bevel gears have an optimum position defined by the conic intersection, which provides the optimum transmission performance. The manufacturer determines this position by running a number of tests for each individual gear pair, using marking blue, and permanently indicates it on the gear surface, to ensure smooth running and optimum load distribution between mating bevel gears; and also parameters, such as axial mounting position, shaft angle, and shaft offset, have to be taken in the consideration.

Bevel gear mounting distance (as shown in Fig. 5.4) is "the distance from a locating surface on the back of one gear to the centreline of a mating gear" (Wasilewski 1994). This represents the most important parameter for proper operation, and normally its optimum value can be established experimentally by

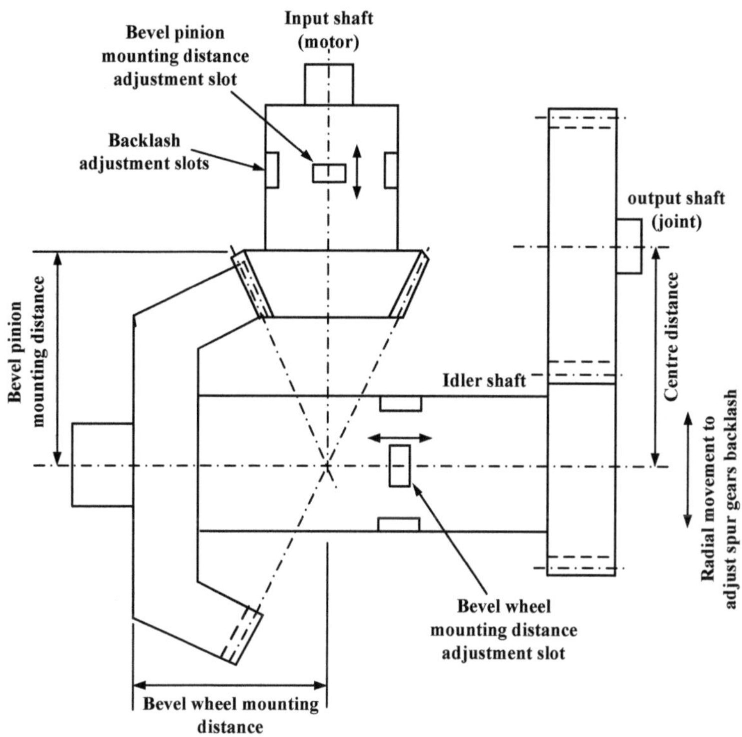

Fig. 5.4 Schematic diagram for the bevel and spur gear sets in joint 3

undertaking a number of gear rotations to find the best contact pattern. In most bevel gear pairs, the pinion mounting position has a larger effect on the contact pattern than the wheel position (Wasilewski 1994), and two types of error (positive and negative) are associated with mounting distances as illustrated in Fig. 5.5 (Kohara-Gear-Industry 2015). When the error is positive, the pinion contact pattern will move towards the tooth root, while the contact of the mating gear will move toward the top of the tooth. Conversely, the contact of the pinion will move toward the top and that of the gear will move toward the root if the mounting distance of the pinion has a negative error. These errors can be corrected by axial adjustment of the pinion and wheel gears. However, mounting distance error will cause a change in backlash; positive error increases backlash whilst negative backlash decreases it.

Figure 5.6 shows another category of mounting errors called shaft angle errors. If the shaft angle error is positive, the tooth contact will move toward the toe end, while the tooth contact will move toward the heel end if the shaft angle error is negative (Fig. 5.6a). Shaft alignment or offset error occurs if the pinion axis does not intersect with the gear axis; however, a crossed contact will be produced if the gears have an offset error, as shown in Fig. 5.6b.

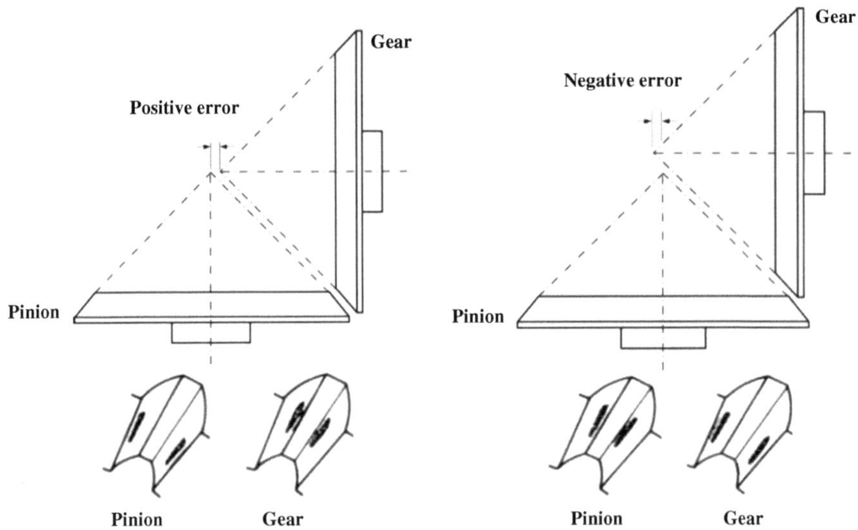

Fig. 5.5 Positive and negative mounting distance error

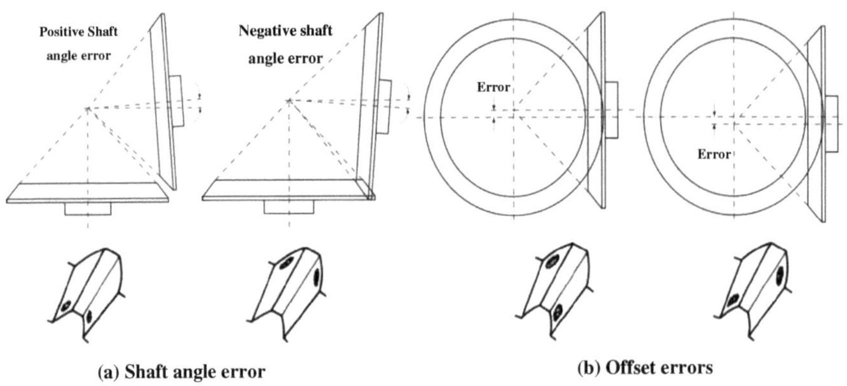

Fig. 5.6 Shaft angle and offset errors in bevel gears

5.4 Common Gear Faults

The main fault types in a robot gear are similar to those occurring in gearboxes (Halme 2006). When a gear is working under high load conditions, failures tend to occur, localised faults in gears tend to develop rapidly once initiated; and the earlier identification of a gear failure gives a clear indication about potential problems in the system prior to failure. A summary of the main gear failures is provided for completeness.

5.4.1 Backlash in Gears Pair

Backlash is defined as "the clearance between the non-working flanks of the teeth of a gear pair when the working flanks are in contact" (British-Standard 2007). High backlash between gears teeth can lead to reduction in system stability, high impact force and generate undesired vibration and noise, which together will decrease the accuracy and repeatability of the robot. However, it is not recommended to totally eliminate backlash between gear teeth, as it is essential to avoid interference between the mating gears and to provide better lubrication of the tooth surfaces. Two methods can be followed to achieve the desired clearance, the first of which is to reduce the tooth thickness and the second method, which is the most common, is to adjust the centre distance between the shaft axes of the meshed gears in spur and helical gears, or by adjusting the mounting distance in bevel gears, as explained earlier in Sect. 5.3.1 (Sommer 2011; Wasilewski 1994). This can be achieved by using anti-backlash gears, precise mechanical design, which is the case in the robot under study.

Later in this chapter, in addition to the backlash fault, joint 3 requires disassembling in order to simulate other kinds of faults in the gears and bearings. Information related to centre distance in the spur gears and mounting distance of bevel gears are not provided by the manufacturer, and the degree of backlash that gives the best gear mesh in spur and bevel gear sets has not been provided. The robot manual does not include procedure of re-installing these gears makes the problem of resetting these gears back very complex, since the mentioned parameters are vital for gear mesh adjustment. To overcome this problem, a considerable amount of time was spent practicing assembling joint 3 using a spare faulty robot.

5.4.2 Scuffing

If the oil-film between the gear teeth breaks down, scuffing failure, which is a mode of tooth surface failure, will occur, and allowing metal to metal contact leading to a localized welding. Overloading, misalignment and high operating temperature are considered the main causes of disruption of the oil-film. When the scuffing occurs, the tooth surfaces will appear slightly rough, dull, and torn in the direction of sliding (Yeýilyurt 1997), as shown in Fig. 5.7.

5.4.3 Abrasive Wear

The presence of hard material, such as fine metal particles or sand, in the lubricant leads to abrasive wear. During the normal operation of the meshed gears, the teeth

Fig. 5.7 Scuffing in gear teeth (www.novexa.com)

are separated from each other by a thin oil-film, however, if a particle with dimension larger than the thickness of the oil-film and hardness near or above the hardness of the gear tooth surface passes through the mesh, the tooth surfaces will be affected. The resulting damage is seen as grooves on the tooth surfaces in the direction of sliding (Smith 1983; Yeýilyurt 1997).

5.4.4 *Pitting or Surface Fatigue*

Pitting is a surface fatigue failure happens when the contact stresses exceed the fatigue tolerance limit and is indicated by the development of cavities in the contact zone of the gear teeth, as shown in Fig. 5.8. After a certain period of operation with repeated variation of load, surface fatigue cracks (or pitting cracks) will be initiated. These cracks propagate until they intersect other surface cracks, and finally a small area of metal on the tooth surface will spall off. After that, pitting will spread rapidly on the tooth surface as the remaining healthy areas must carry the extra load which is previously supported by the damaged areas.

5.4.5 *Tooth Breakage*

All gears failures which have been discussed up to now represent progressive failure types. That means during the progression of the failure the gear set will continue transmitting the power. However, tooth fracture is the most insidious mode of gear failure and leads to either impairment of the drive system or instant

Fig. 5.8 Gear tooth pitting (www.novexa.com)

loss of serviceability. Bending fatigue represents the most common cause of gear tooth fracture. That is because of the cyclic bending stress, introduced during power transmission process due to repetitive loading. However, when this stress exceeds the yield strength of the gear material at the critical section, near the tooth root, a small crack will initiate and then propagates rapidly until part (or the entire tooth) breaks off. As a result, the remaining part of the tooth will have to withstand a greater cyclic loading, and is susceptible to further breakage (Radzevich 2012).

5.5 Rolling Bearings General Overview

Rolling element bearings represent the most commonly used component for minimizing the friction between stationary and moving parts of any rotating machine. Depending on the shape of the rolling elements, bearings can be classified into either ball or roller bearings (Juvinall and Marshek 2012). The main components in all types of rolling element bearings are the rolling elements, an inner ring/race, an outer ring/race, and a cage. The inner race is usually mounted on the rotating shaft of the machine and so it will be in rotation with the shaft, while the outer race is usually mounted on the stationary housing of the machine and does not rotate; and this is the case in joint 3 of the robot. The raceways in the inner and outer race in which the rolling elements are moving will have different forms depending on the shape of the rolling elements (ball or roller). The cage is used to keep the rolling elements separated at equal distances and thus prevents undesired contact and from rubbing against each other. The load in bearings is transferred via the rolling

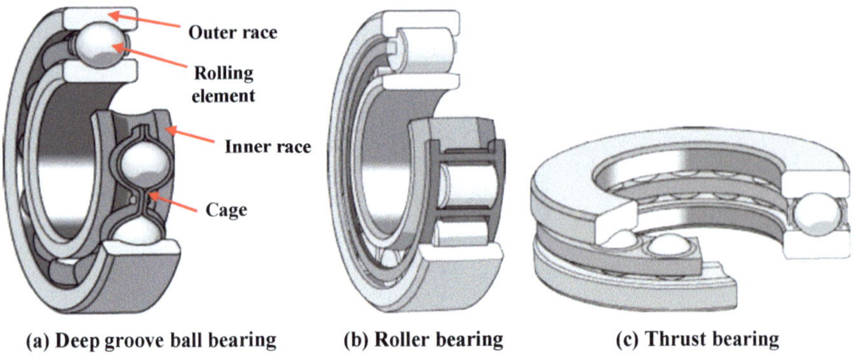

Fig. 5.9 Different types of rolling bearings (Harris and Kotzalas 2007b)

elements, and in ball bearings is transferred via the balls over a very limited point contact area between the inner and outer races. Consequently they are carrying lower radial load capacity than that of the roller bearings, where the rollers transfer the load via a line contact with the raceways. Rolling bearing are also further classified based on their load carrying capability into: radial bearings whereby the load is primarily carried radially; thrust or axial-contact bearings whereby the loads is only carried axially; and angular-contact bearings which carry combined axial and radial loads (Juvinall and Marshek 2012). However, all the bearings in joint 3 of the robot are of type deep-groove ball bearings; this type of bearings can support radial load as well as axial load (Harnoy 2002). The bearings types and their components are illustrated in Fig. 5.9.

5.6 Common Rolling Bearing Faults

Faults in bearings can occur at two different stages and because of different reasons. Factors such as inappropriate design, misalignment, overload, or contact corrosion represent the most common causes that may develop distributed faults, such as roughness and waviness, at the early stage of their service life. In the second stage due to wear, fatigue or intrusion of contamination material localized faults, such as pits and spalls, will be progressed. Table 5.2 provides a description and images of some common bearing faults, however, a comprehensive evaluation of different rolling bearing faults with their cases can be found in a book authored by Harris and Kotzalas (2007a) and on the Emerson Company website (http://www.emersonbearing.com/technical-toolbox).

Table 5.2 Fault types in rolling bearings (Harris and Kotzalas 2007a)

Fault type and description	Fault photo
Corrosion pitting (rusting): this fault happens if there is a moisture or corrosive material in or enters the lubricant making it no longer provides adequate protection.	
Pitting due to electric current passing through the bearing: this fault happens in applications that use electric motor, as in the robots. Electrical current may pass through the bearing if it is not electrically insulated. This will lead to generation of tiny pits; progression of them, due to continuous operation, causing corrugation of the surfaces	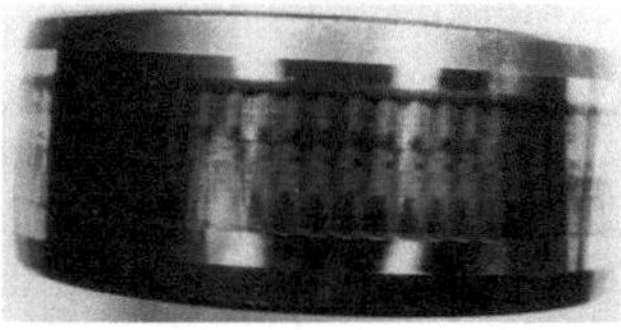
Indentations: this can be developed because of mechanical shocks, hard particle contaminates that may enter to the free space in the baring and then get trapped between the rolling elements and raceways, or due to assemble misalignment.	
Micro-pitting: if the lubricant film between the contacting surfaces in rolling bearings is insufficiently thick, the surface asperities come into contact with each other leading to high frictional shear stress on the surface. This will cause plastic flow of surface material and so pits on the surface.	
Surface-initiated fatigue: cracks on the surface of contact will occur if the cyclic stress exceeds the endurance strength limit of the material. The cracks will propagate until they meet leading to remove tiny spall from the surface.	

5.7 Rolling Bearing Characteristic Frequencies

If a fault is progressed in any of the four main bearing components due to any of the failure mechanisms, a high-level short duration impulsive force is experienced each time one of the balls rolls over the flaw, and causes the natural frequencies of the bearing components to become excited each time (Shrivastava and Wadhwani 2012). Because of this phenomenon, a unique frequency characteristic is generated for each of the bearing components depending on whether the fault exists on the outer raceway, inner raceway or on the rolling element. These frequencies can be employed for detection of bearing faults and they are calculated based on the bearing's geometry and rotational speed (kinematic considerations). These frequencies can indicate what component is failing in the bearing; therefore, the calculation is very useful. The mathematical analysis for calculating these frequencies is presented in this section, but the specific values will be calculated in the next chapter. These frequencies are commonly used for the health evaluation of rotating machines; therefore, the purpose for reproducing them here is initially to explore if it is possible to detect them on the robot using time-frequency signal analysis and secondly for preliminary robot health evolution. Figure 5.10 below shows the different positions of localized defects in rolling bearings.

Consider a rolling bearing mounted on a rotating shaft with a general configuration and both outer and inner races rotating (Fig. 5.11) at constant speeds ω_o and ω_i, respectively. The different fault frequencies that may occur in any rolling element bearing system are found using Eqs. 5.4–5.7. These equations assume pure rolling contact, rollers are equal in diameter and no slippage between the bearings and load producing shaft (Randall 2011).

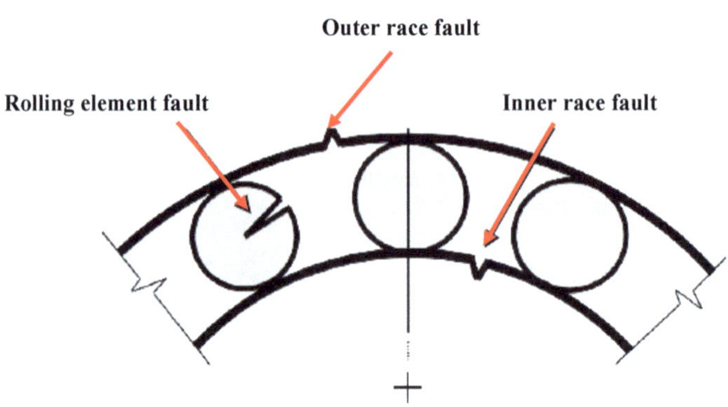

Fig. 5.10 The main types of faults in bearings

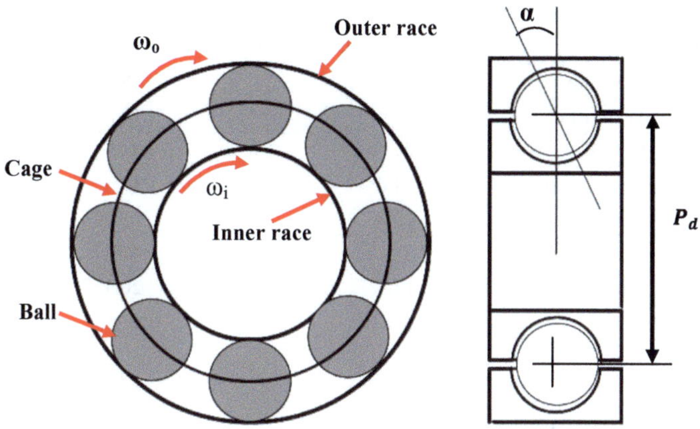

Fig. 5.11 General bearings configuration

Though, the fundamental train (or cage) frequency (FTF) and ball spin frequency (BSF), which are related to the faults in the cage and rolling elements, are:

$$FTF = \frac{1}{2} \cdot \left[\omega_i \left(1 - \frac{B_d \cdot \cos \alpha}{P_d} \right) + \omega_o \left(1 + \frac{B_d \cdot \cos \alpha}{P_d} \right) \right] \tag{5.4}$$

$$BSF = \frac{P_d}{2 \cdot B_d} (\omega_i - \omega_o) \left(1 - \frac{B_d^2 \cdot \cos^2 \alpha}{P_d^2} \right) \tag{5.5}$$

where P_d is the pitch diameter, B_d is the ball diameter, and α is the contact angle. The frequency at which any rolling element passes a specific point on one of the races is called the ball pass frequency. So, the ball pass frequency of outer race (BPFO) and the ball pass frequency of inner race (BPFI) are:

$$BPFO = \frac{N_b}{2} \cdot (\omega_i - \omega_o) \cdot \left(1 - \frac{B_d \cdot \cos \alpha}{P_d} \right) \tag{5.6}$$

$$BPFI = \frac{N_b}{2} \cdot (\omega_i - \omega_o) \cdot \left(1 + \frac{B_d \cdot \cos \alpha}{P_d} \right) \tag{5.7}$$

where N_b is the number of rolling elements. However, for most of applications including this robot, the outer race is fixed in the housing (i.e. $\omega_o = 0$) and the inner race rotates at the same speed as the shaft. To compute these frequencies, the specifications of the robot's bearings are needed; thus they are presented in Table 5.3.

Table 5.3 Robot's bearings specification

Parameter	Input shaft bearings	Idler shaft bearings
Outer diameter (D) (mm)	24	28
Bore diameter (d) (mm)	9	12
Width (B) (mm)	7	8
Inner race diameter (d_1) (mm)	14.4	17
Outer race diameter (D_1) (mm)	19.8	23.2
Ball diameter (mm)	3.4	4.75
Pitch diameter (mm)	17.1	20.1
Number of balls	7	8

5.8 Fault Simulation in the Robot

It is accepted that the best way to introduce a fault in the robot is by operating the robot to execute a specific task for a considerable amount of time until a fault is developed. This method, however, has not been followed in this work primarily because it may resulting other joints of the robot failing. Secondly, it is time consuming in a time restricted study. The faults that will be introduced here are not truly representational of real faults in the robot, since in a practical situation the majority of faults in gears and bearings are progressive. However, this methodology has important advantages such as different types of fault can be produced in different components of the joint 3 gearbox and the degree of fault severity can be controlled. Three types of common gearbox faults are introduced in joint 3 of the robot; how these faults are simulated and their location in the gearbox are discussed in the following subsections.

5.8.1 *Backlash Fault Simulation*

The input and idler shafts in joint 3 are carried inside the robot structure using hollow and eccentric shafts. There are three slots on the surfaces of these hollow shafts, as shown in the Fig. 5.12, utilized to adjust the backlash in the bevel and spur gears pair. Two of these slots, diametrically positioned, are used for changing the centre distance in the spur gear pair and for changing the mounting distance of bevel wheel gear, and are used to varying the backlash in each gear set. These changes can be achived by using adapted grub-screws which fit into the slots. The circumfrentional slot in the two shafts is used to adjust these gears axially to control the contact pattern between meshed gears, and is acomplished using an eccentric grub-screw, shwon in Fig. 5.12.

Three levels of backlash were introduced in the spur gears, starting from very high level (large clearance between the gears = 0.92 mm), then intermediate level (small clearance = 0.8 mm) and interference (0.56 mm), where in the healthy

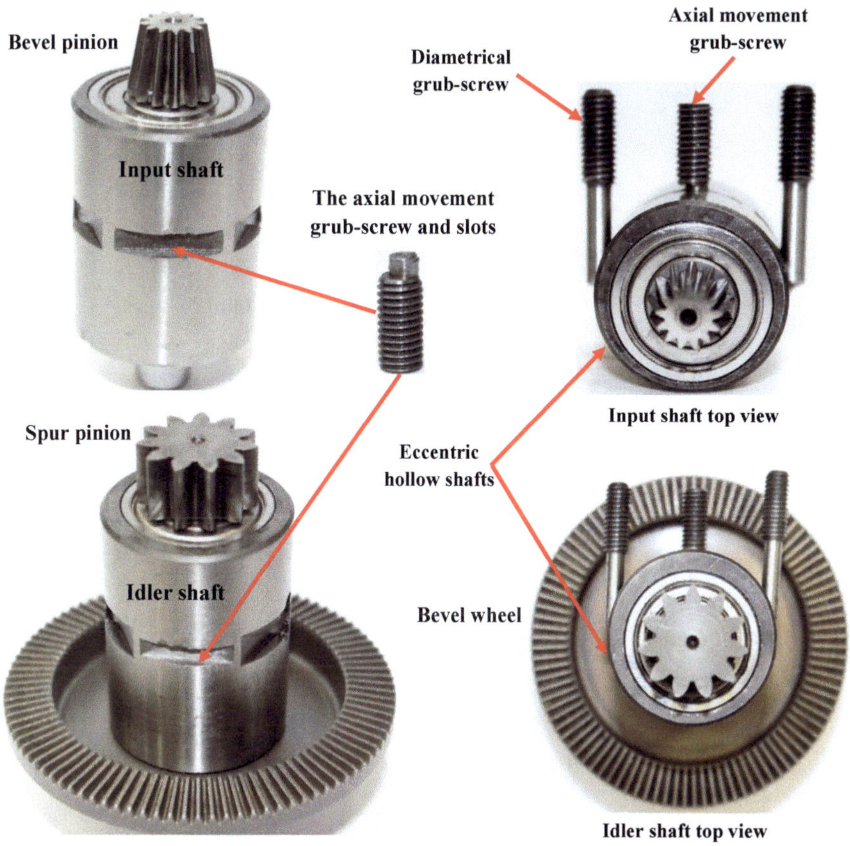

Fig. 5.12 Input and idler shafts and their backlash adjustment grub-screws

condition the backlash is 0.68 mm. It was found that a total of approximately five grub-screw turns are required to move from high backlash level to interference level. Also, depending on the shaft and grub-screw dimensions, simple calculations was done and established that every one screw turn is roughly equivalent to increasing or decreasing the backlash by 0.072 mm. This is an important finding, since it helps to get a preliminarily indication about the number of turns required to simulate different backlash values in this thesis as will be shown in the coming chapters. The interference case should not develop during the normal operation of the robots, but it could occur if the gear mesh are assembled/adjusted incorrectly after maintenance. The optimum backlash setting was first recorded before introducing any fault type on the robot, using a dial-gauge indicator, as shown in Fig. 5.13, and each time the backlash value is calibrated with respect to this optimum value using the same measurement technique (Bicker et al. 1989).

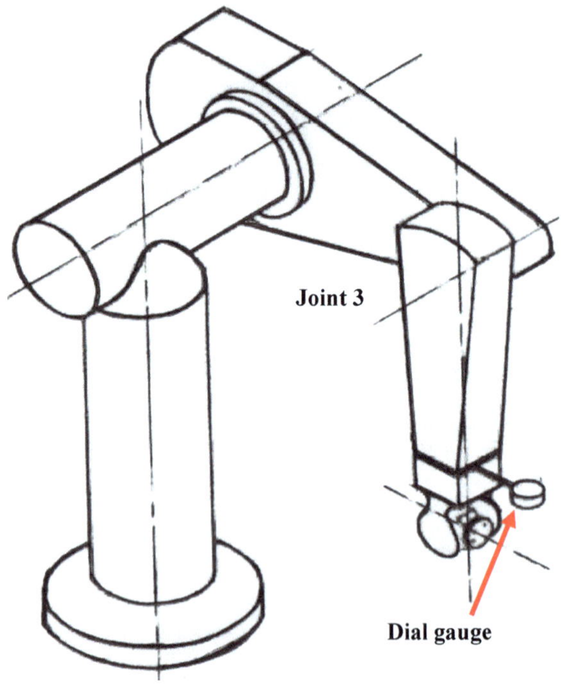

Fig. 5.13 Robot backlash measurement (Bicker et al. 1989)

5.8.2 *Gear Tooth Fault Simulation*

Four different severities of fault are artificially simulated on the pinion bevel of the first stage in the gearbox with only one fault present at a time. The pinion bevel is a very small gear (Fig. 5.14); therefore, the fault simulation was difficult to achieve using specialized machines; so, faults were created manually with the help of a skilful technician using a miniature Dremel drill. The pinion bevel gear was selected for faults simulation and not one from the other gears is explained as follow.

In general, the small size of pinions make their teeth weaker in shape than the gear teeth and also they will be subjected to a larger number of accumulated fatigue cycles (Radzevich 2012). As a result, in practice they are more susceptible to damage than the wheels. Moreover, from experience it has been concluded that dismantling and reassembly of the entire joint 3 of the robot can be accomplished, but the correct mesh between mating gears is not guaranteed. This also was advised in the robot user manual, whereby the manufacturer suggested contacting them if a serious backlash problem has developed. Unfortunately, the manufacturer does no longer support or produce industrial robots. Thus, it was considered better to

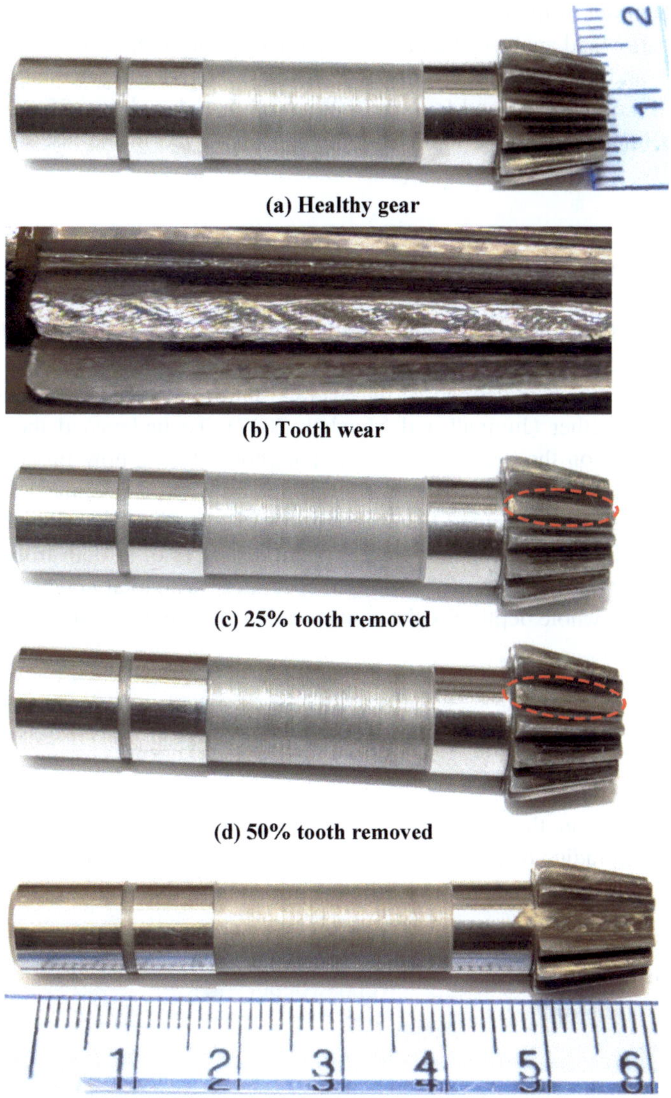

(a) Healthy gear

(b) Tooth wear

(c) 25% tooth removed

(d) 50% tooth removed

(e) Tooth completely removed

Fig. 5.14 Pinion bevel tooth faults

introduce the faults in the bevel pinion as it is easy to dismantle and reassemble with correct meshing with the bevel wheel. All faults have been introduced onto the same gear and on the same tooth, and each time a fault introduced, the bevel pinion must be taken out from the robot, modified and then installed back for robot testing, data collocation and analysis. Firstly, simulation of wear on the entire tooth flanks is introduced, and the other three faults types involve removing a percentage from the tooth depth, i.e. in the second fault the tooth depth is reduced by a 25 %, then by a 50 %, and finally the whole tooth is completely removed in the last stage. These faults can be seen on Fig. 5.14.

5.8.2.1 Effect of Gear Faults on Vibration Severity on the Robot

It is anticipated that the level of vibration will increase when gear backlash or tooth faults are introduced, because this will decrease the contact ratio between the meshed gears, which is defined as the average number of teeth in contact as the gears rotate together (Juvinall and Marshek 2012). To understand the effects of simulated faults on the contact ratio, it is paramount to see how this ratio is calculated and based on which parameters. For this purpose, a schematic diagram of two meshed gears is presented below (Figure 5.15).

The addendum and dedendum are the radial distances of a tooth from the pitch circle to the top and bottom of the tooth, respectively. The total height of the tooth is known as the whole depth. So, the contact ratio can be computed as (Juvinall and Marshek 2012):

$$Contact\ ratio = \frac{\sqrt{r_{ap}^2 - r_{bp}^2} + \sqrt{r_{ag}^2 - r_{bg}^2} - C\sin\emptyset}{P_c\cos\emptyset} \tag{5.8}$$

where r_{ap} and r_{bg} are the addendum radii of the mating pinion and gear respectively (the addendum radius is defined as the distance from the tops of the teeth of a gear to the gear centre); r_{bp} and r_{bg} are the base circle radii of the mating pinion and gear

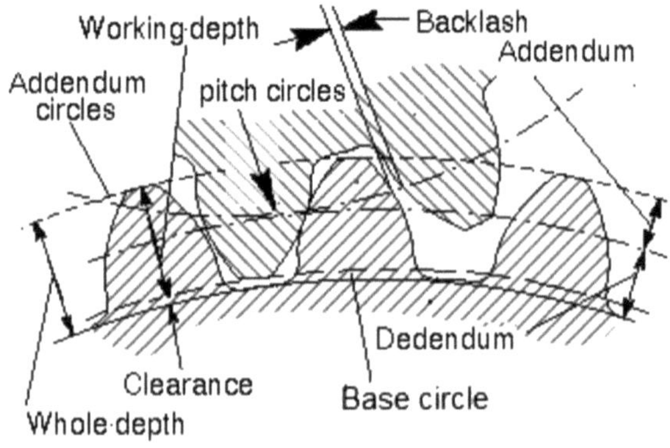

Fig. 5.15 Schematic diagram of two mating (*spur*) gears (Qu et al. 2014)

(the base radius is defined as distance from the base circle to the gear centre); C is the distance between the centres of the mating gears; $\emptyset$ is the pressure angle; P_c is the circular pitch of the pinion gear, known as the distance measured on the circumference of the pitch circle from a point of one tooth to the corresponding point on the next tooth. More detailed calculation and concepts regarding gears can be found in (Juvinall and Marshek 2012).

Generally, the operation of the gears will be smoother and quieter with greater contact ratio, while they are susceptible to interference if it is reduced. As previously discussed if the backlash is increased, the centre distance (C) between the gears is increased too and vice versa, so it can be inferred from Eq. 5.8 that increasing the backlash will decrease the contact ratio and thus high noise and vibration may result from the robot. By contrast, an excessive decrease of backlash, i.e. interference, can also lead to an increase level of vibration. Furthermore, simulating a tooth cut on the pinion gear will cause a decrease in the term $r_{ap}^2 - r_{bp}^2$ until it reaches zero when the tooth is completely removed; and it is obvious from the equation that the deeper the cut the smaller the contact ratio, giving rise to higher the backlash and greater noise and vibration at this point of mesh.

5.8.3 Bearing Fault Simulation

In Fig. 5.1b (see page 90), it can be observed that the input shaft is supported on two bearings of the same type. The one located on the right, adjacent to the bevel pinion, is used for the bearing faults simulation. From the figure it may also be concluded that the applied load on these two bearings is not equally distributed; so, for future studies researchers can investigate the influence of the faulty bearing location on fault detectability. Fortunately, after dismantling the robot joint it was found that identical bearings are still available and a number of NSK bearings type N609 were purchased and utilized for faults seeding.

Two fault types are created in the bearing, inner and outer races faults, using an electrical discharge machining (EDM) technique. However, due to the small size of the bearing, it was not possible to dismantle the bearing, in order to introduce a ball fault, without damaging it, which also affects the fault size controllability of the inner race. The seeded inner race fault has 1 mm width, extending along the bearing width and as deep as the thickness of the inner race. Faults of two degrees of severity are introduced on the outer race of the bearing. Both of them are circular in shape, extended along the outer race thickness and have 1 mm and 2 mm diameters, respectively. The purpose of this variation in the outer race fault size is to test the proposed system in distinguishing the severity of different bearing faults. Figure 5.16 shows the healthy and the three faulty bearings.

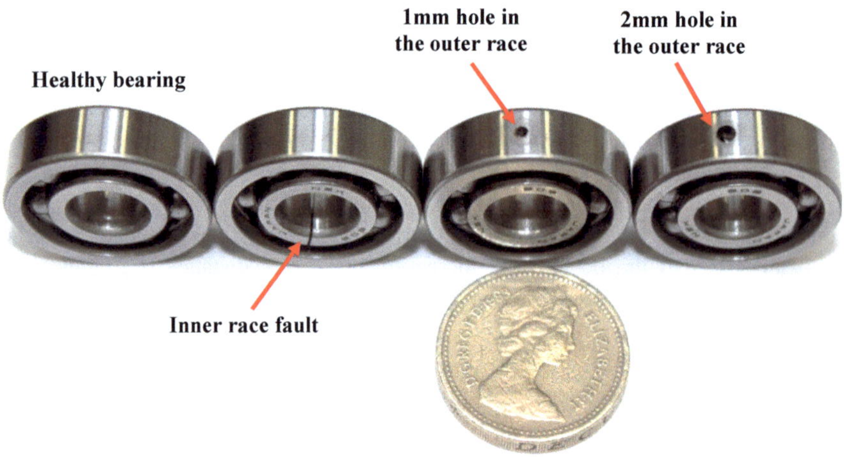

Fig. 5.16 Healthy bearing and three faulty bearings with different fault types

5.9 Data Acquisition Hardware for Preliminary Vibration Analysis

To study the robot vibration behaviour when it is healthy and with different faults simulated, a preliminary vibration signals capturing and analysis set-up was carried out. The vibration signals will be acquired using the selected hardware discussed in the following subsections, and the gathered signals are analysed using software designed based on LabVIEW graphical programing (deliberated in Sect. 5.10). The result of this preliminary analysis will be presented in the following chapter and will also be used in Chap. 7 as an aid to designing the intelligent condition monitoring system.

5.9.1 Accelerometer

Accelerometers are electromechanical devices that convert the mechanical signals, such as vibration and force, to electrical signals, and are an extensively used for fault detection in many machines because of their accuracy, robustness and sensitivity. Selecting an appropriate accelerometer and the way it is mounted on a machine are significant factors in determining the success of any condition monitoring program. Misleading data can be produced if an unsuitable accelerometer is selected for the machine under study or an appropriate one is mounted in an incorrect location on the machine. The major parameters which need to be considered for accelerometer selection are (Sinha 2014): sensitivity, range, bandwidth, frequency resolution, reliability, accuracy, operation environment and cost.

MEMS (Micro-electro-mechanical-sensors) accelerometers compared with the conventional piezoelectric accelerometers, consume less power, are small in size, light weight, low cost and can achieve good performance (Albarbar et al. 2008). Consequently, they are more suitable for an embedded system design, and will be used with the designed embedded system described later. However, a variety of MEMS accelerometers are commercially available. After taking the all-important technical requirements in the consideration, the ADXL001 MEMS accelerometer has been chosen.

– ***The AXDL001 Accelerometer***

The ADXL001 is a MEMS based, single-axis accelerometer developed by Analog Devices. It provides a high performance, wide bandwidth of 22 kHz, and is small in size. The ADXL001 operates on a 3.3 V or 5 V supply, and outputs an analogue voltage, which allows the direct connection of the accelerometer output to the analogue input pins on a data acquisition device (DAQ) or a microcontroller. The accelerometer is available in 3 full-scale dynamic ranges of ±70, ±250, and ±500 g. For this work the ±70 g range was deemed appropriate. An evaluation board is specifically designed by Analog device for this accelerometer making it easier to use. To prevent anti-aliasing, the evaluation board provides a user configurable low-pass filter on the accelerometer output. Moreover, due to stress, over acceleration or fabrication errors MEMS accelerometer can develop flaws; and fortunately, ADXL001 accelerometer like other MEMS accelerometer has a build in self-test pin, which can be used to test both the electrical circuit and the mechanical structure of the accelerometer. Figure 5.17 illustrates the single axis ADXL001.

(a) Schematic diagram of ADXL001 evaluation board **(b) ADXL001 on its evaluation board**

Fig. 5.17 ADXL001 accelerometer and its evaluation board (http://www.analog.com)

5.9.1.1 Optimum Position of the Accelerometer

A review of previous research which has been conducted in the area of industrial robots fault detection, most researchers position the sensors at different locations on the robot without giving a reason. To detect operational abnormalities in a machine, the sensors should be located as close as possible to expected damage locations, although there are instance where the damage is more recognizable at other locations on the machine (Farrar and Worden 2013). Prior to acquiring the data for fault detection, it was crucial to locate the most sensitive positions on the robot. In order to achieve this, a test based on computation of root mean square (RMS) value of the captured vibration signals has been accomplished. Three locations distributed over the robot's structure were selected, as illustrated in Fig. 5.22 (note end chapter). The procedure for this test can be explained as follows.

The accelerometer was attached on the first location on the robot, adjacent to the wrist, and the robot was programmed to move joint 1 in a cyclic movement at different speeds, and the RMS value of the captured signal, which was measured in the direction orthogonal to the axis of rotation, was computed. After that, joint 2 was moved cyclically and also with different speeds and RMS values calculated. This procedure was repeated for all the robot joints. The accelerometer was then relocated to the second (and then the third) position and the same procedure carried out to compute the RMS values. A comparison of the computed RMS values was made and results for joints 2 and 3 are shown in Fig. 5.18. The other joint results are presented in Appendix B.1. The presented results here are when the accelerometer on the three selected locations and joint 2 and 3 moving at three different speeds. After evaluating all the results, it was concluded that the best location to affix the accelerometer is location one, which is consistent with the established best accelerometer location for modal testing in Chap. 4, and thus it will be used throughout this thesis.

5.9.1.2 Mounting of the Accelerometers

After highlighting the most sensitive accelerometer place on the robot it was felt unnecessary to locate an accelerometer on each of the robot's joints. Furthermore, using just one accelerometer was not considered sufficient, since when the robot is performing a general task its joints are rotating around different axes, therefore, the accelerometer may be more sensitive to specific joints than others yielding the monitoring system unreliable. For this reason, it was decided to fix three ADXL001 accelerometers in an orthogonal configuration to measure the vibration in X, Y, and Z directions using a purpose designed aluminium adapter, as shown in Fig. 5.19. This will reduce the length of wiring on the robot and make the embedded system more compact. To fix this adapter correctly on the robot for accurate pick-up of vibration signals, three common mounting techniques were assessed (Girdhar 2004).

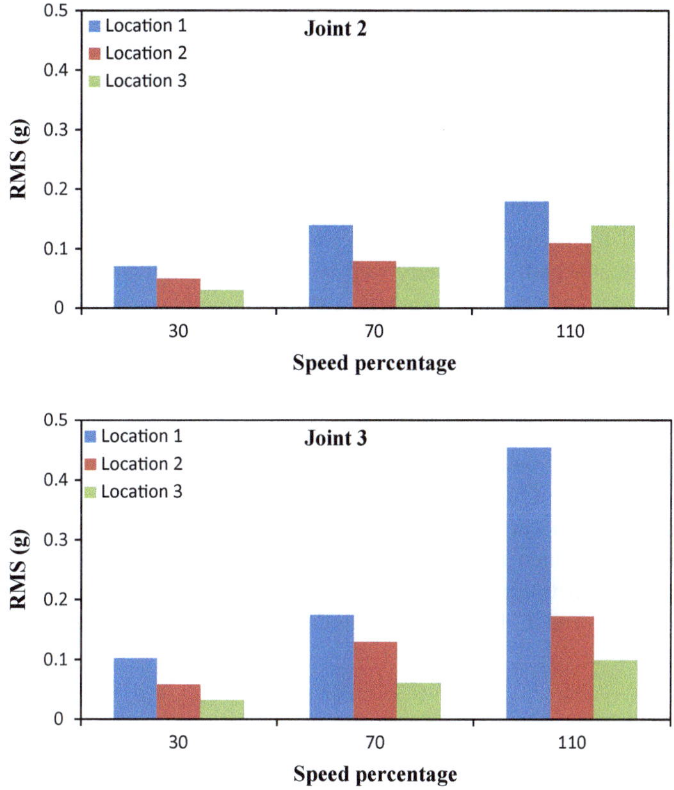

Fig. 5.18 RMS values when the accelerometer on different locations and the robot moving different joints at different speeds

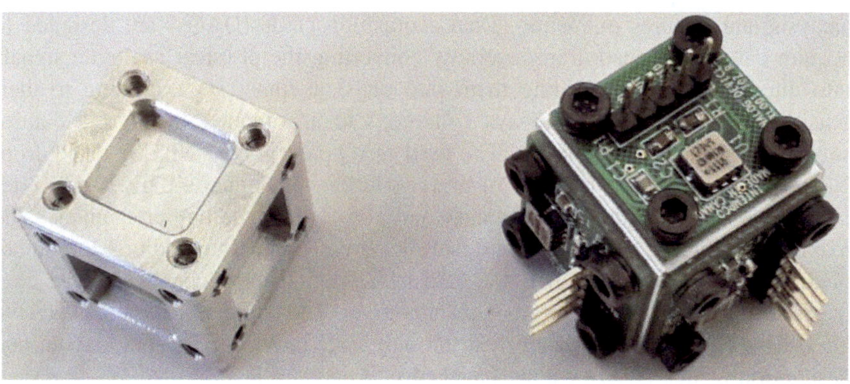

Fig. 5.19 Designed 3 axis accelerometer adapter

Stud mounting—this technique requires a stud to be mounted on the robot surface by drilling and screwing. This is the preferred technique for permanent mounting and gives the best frequency response because the accelerometer and the robot fused together by virtue of the high fastening force of the stud. However, it was not possible to drill into the robot because there is a gear under the established optimum accelerometer location. Magnetic, double sided adhesive tape or beeswax—these methods are conventionally used for temporary measurements, but are not recommended for permanent monitoring because their frequency responses, even when expertly attached, are usually well below that of equivalent stud or adhesive mounted accelerometers. The magnetic mounting method cannot be used in this study as the robot has an aluminium structure. The third adopted technique, which was used here, is adhesive mounting—in this method, the adapter is glued using an appropriate adhesive material such as superglue.

The selected accelerometer was calibrated against the used conventional piezoelectric accelerometer in Chap. 4. The accelerometers are attached to the robot and the Data Physic Quattro analyser along with SignalCalc ACE software was employed for signal capturing and analysis. The selected accelerometer showed very good performance in comparison with the piezoelectric one; the calculated correlation factor between the captured two signals is 0.92 which indicates the two signals are very similar. The results, including time/frequency spectra, are given in Appendix B.2.

5.9.2 Data Acquisition Card (DAQ)

The Data Physics Quattro analyser was used for the preliminary robot vibration analysis; however it cannot be interfaced to Matlab or LabVIEW in order to achieve real-time data acquisition and analysis, which will slow down the process of signals analysis and features extraction. Data acquisition cards (DAQs) are designed to acquire data from sensor/transducers by converting the physical analogue signals into digital form and rescaling them into physical quantities according to their sensitivities. DAQ card is always connected to a computer that has a control software and data storage space. The used DAQ card is type NI USB-6009 from National Instruments. It has 14-bit resolution analogue-to-digital converter (ADC) with 8 analogue input channels and also 2 analogue output channels. The maximum sampling rate of this card is 48 KS/s per channel. In addition to number of digital input/output channels the card has 1 digital input channel can be configured either a digital trigger or an event counter. The communication between the NI USB-6009 and PC is established through USB interface. The pins layout and description of this DAQ card are attached in Appendix B.3.

To synchronize the data acquisition process with the start of the robot movement a triggering signal from the robot controller was used to trigger the DAQ card. The input/output module in the robot controller and in conjunction with a relay

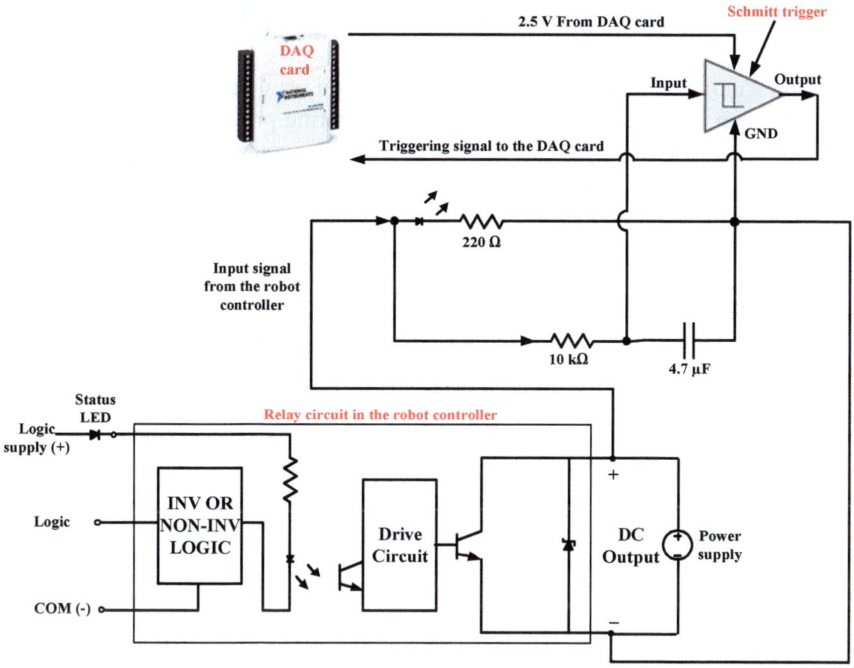

Fig. 5.20 Electronic triggering circuit using Schmitt trigger IC

set allows the user, through the use of various VAL commands, to control external equipment by switching one (or more) of the output lines, or by responding to any one or combination of the outside events that can be represented as on/off states. The control signal from the robot is an analogue, so to convert it to a digital format, either 0 or 1, a Schmitt trigger circuit was built. Schmitt trigger is an integrated circuit (IC) that converts an analog input signal to a digital output signal when it exceeds a certain threshold. Figure 5.20 shows schematic of the output relay connected to the Schmitt trigger circuit; it contains a diode to prevent any damaged that may be caused by back electromagnetic field (EMF). Normally, the output signals are sent via a voltage supply. To put any output signal on/off is simply using these commands inVAL language

Signal	1	for on
Signal	−1	for off

The trigger output was fed into the digital trigger input in the DAQ card (Blum 2013). A Schmitt trigger IC type 74HC14 is used; six seprated Schmitt triggers are included in this chip, however only one is needed in this work (more details about this IC is provided in Appendix B.4). This circuit will also be used for triggering the embedded system later.

5.10 Vibration Data Acquisition and Analysis Software

The data acquisition software sustains the connection between the DAQ card and the host computer, as well as controlling the data gathering process, and includes basic data analysis tools, including time-domain and DWT signal analysis. In terms of data acquisition and analysis in this project, LabVIEW development environment, which is a graphical programming environment from National Instruments used for measurement, test, and control systems development, was selected. It provides built-in libraries for advanced analysis and data visualization and can be integrated with a large number of hardware interface devices. Additional capabilities have been provided by LabVIEW, such as the interface with Matlab and C language, which makes graphical programming in LabVIEW more flexible; Matlab software has many specialized, efficient, and easy to use toolboxes, including the *wavelet analysis toolbox* used in this study and its many options help to accomplish DWT analysis, such as specifying the number of decomposition levels and selecting a suitable wavelet family. Thus by combining the capabilities of Matlab in data processing and the advantages of LabVIEW graphical programming, the developed data acquisition software will have improved capability and greater flexibility.

The LabVIEW programs are called *virtual instruments* or VIs. Each VI has three main components which are the front panel (or user interface), the block diagram (or programming interface) and the icon/connector pane. Programming in LabVIEW consists of placing *express VIs*, specifically designed for measurement and analysis, standard VIs and functions or other programmed elements, such as nodes, terminals, and wires, on the block diagram. A block diagram sample from the prepared LabVIEW code is shown in the Fig. 5.21; the full prepared code is provided in Appendix B.5.

The LabVIEW block diagram begins with a *for* loop, used to control the repetitive execution of the code a predefined number of times (top–left). The DAQ Assistant (centre–left) is an express VI that communicates directly with the 14-bit ADC DAQ card (NI USB-6009) which acquires the analogue measurement signal, in this case is from the accelerometer (s). The DC component of the data signal is subsequently removed, which appears in the signals either because of the electronics associated with the sensor hardware or due to the amplitude of zero frequency component (Chaudhury et al. 2014). This is achieved by computing the median of the time-domain signals using the median finding function, which is then subtracted. The result is then scaled using the accelerometer sensitivity in V/g, to convert them to acceleration values. A DWT analysis is then performed using the Matlab wavelet toolbox (large central block), which is imported into LabView as a Matlab script file (MEX function). For an accurate wavelet analysis a number of parameters in the Matlab code have to be tuned properly, such as which wavelet family, and number of decomposition levels, using **wavedec** command. The

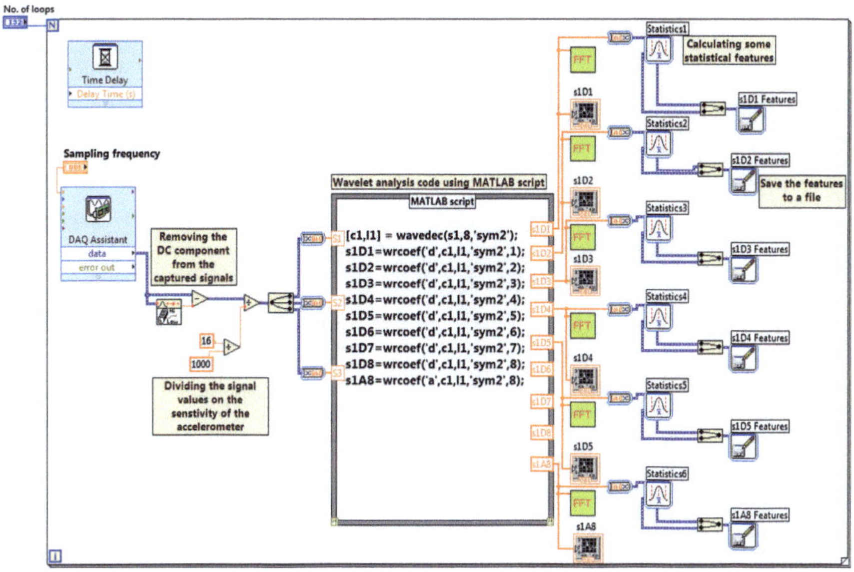

Fig. 5.21 Part from the prepared LabVIEW code

command **waverec** is utilized to return the different components of the decomposed signal from the wavelet coefficients. Researchers who want to accomplish feature extraction using wavelets are advised to read in detail the wavelet toolbox user guide (Misiti et al. 2001). In this research an 8-level wavelet analysis has been applied, as it provided better indication of the robot health (discussed in the following chapter). Following completion of the DWT analysis the statistical features are computed from each signal and saved in text files; these features are utilized in designing the neural network for the fault diagnosis stage (second stage). LabVIEW code for performing the FFT was also incorporated and used concurrently with the wavelet transform code, in order to carry out preliminary evaluation of the bearing and gear frequencies by investigating the frequency spectrum of the sub-signals produced from the wavelet transform.

5.11 The Experimental Set-up

Figure 5.22 shows the experimental set up for preliminary vibration analysis and feature extraction. The selected and tested accelerometer locations are also shown in the figure. The chosen accelerometers are affixed at the optimum accelerometer location (location 1). To power the accelerometers a separate power supply was used, however, power from the microcontroller will be used with the embedded

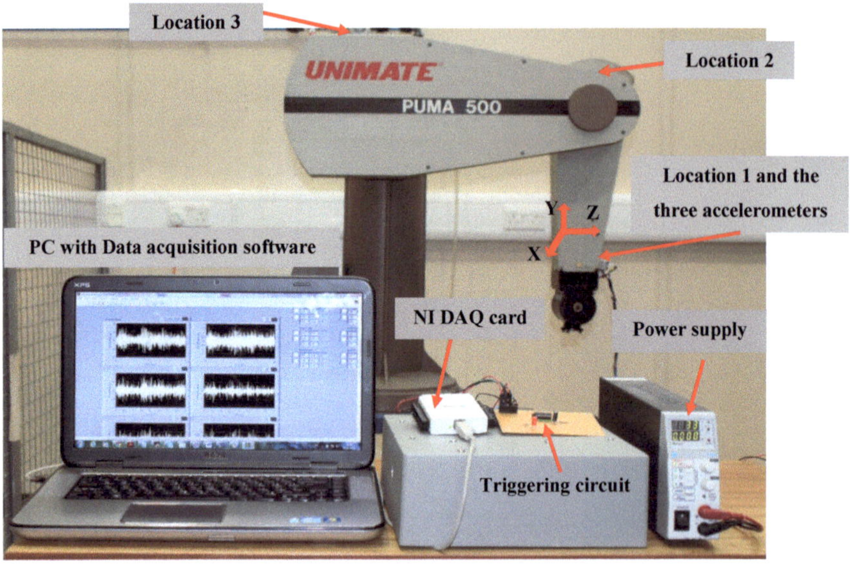

Fig. 5.22 The experimental set-up

system. The outputs of the accelerometers are fed to NI DAQ card. Signal from the robot controller is inputted to the triggering circuit and output of the triggering circuit is connected to the digital trigger pin in the card. The card is interfaced to a PC with the LabVIEW software installed.

5.12 Summary

Joint 3 of the robot was selected in order to simulate different faults in the robot. A detailed description about the mechanical construction of the selected joint along with the different errors that may accompany the power transmission system has been carried out to understand the faults simulation process in the robot. Also, several different possible fault types, such as backlash between gear pair, gear tooth and bearing faults, which may occur in the robot have been assessed, with various severity levels. An appropriate accelerometer was chosen, and the best accelerometer location identified. An aluminium adapter to carry three accelerometers was designed, and accelerometer calibration has been verified. A data acquisition system based on National Instruments (NI) software and hardware has been designed. Hybrid programming, combining LabVIEW graphical programming with Matlab textural programming was effective in developing the signal monitoring and feature extraction system, around which the robot vibration signal capturing and analysis software was constructed.

References

Albarbar, A., Mekid, S., Starr, A., & Pietruszkiewicz, R. (2008). Suitability of MEMS accelerometers for condition monitoring: An experimental study. *Sensors, 8*, 784–799.

Bicker, R., Daadbin, A., & Rosinski, J. (1989). The monitoring of vibration in industrial robots. In *ASME 12th Biennial Conference on Mechanical Vibration and Noise*.

Blum, J. (2013). Exploring Arduino®: *Tools and techniques for engineering wizardry*. New York: Wiley.

British-Standard, (2007). Gears—Cylindrical involute gears and gear pairs—Concepts and geometry.

Chaudhury, S. B., Sengupta, M., & Mukherjee, K. (2014). Vibration monitoring of rotating machines using MEMS accelerometer. *International Journal of Scientific Engineering and Research (IJSER), 2*.

Childs, P. R. N. (2014). *Mechanical design engineering handbook*, Butterworth-Heinemann.

Farrar, C. R., & Worden, K. (2013). *Structural health monitoring: Amachine learning perspective*, New York: Wiley.

Girdhar, P. (2004). Practical machinery vibration analysis and predictive maintenance Oxford Newnes.

Halme, J. (2006) Condition monitoring of a material handling industrial robot. In *19th Internation Congress*, Lulea, Sweden.

Harnoy, A. (2002). *Bearing design in machinery: Engineering tribology and lubrication*. Boca Raton: CRC Press.

Harris, T. A., & Kotzalas, M. N. (2007a). *Advanced concepts of bearing technology*, Taylor & Francis Group.

Harris, T. A., & Kotzalas, M. N. (2007b). *Essential concepts of bearing technology*, Taylor & Francis Group.

Henriksson, M. (2009). *On noise generation and dynamic transmission error of gears*. Ph.D, Royal Institute of Technology.

Hiroaki, E., & Nader, S. (2012). Gearbox simulation models with gear and bearing faults. In: M. Gokcek (Ed.), *Mechanical engineering*.

Ibrahim, G. R. (2011). *Design and implementation of gearboxes vibration based condition monitoring system*. Ph.D Thesis, Manchester Metropolitan University.

Juvinall, R. C., & Marshek, K. M. (2012). *Fundamentals of machine component design*, New York: Wiley.

Kohara-Gear-Industry, (2015). *Gear technical reference*.

Misiti, M., Misiti, Y., Oppenheim, G., & Poggi, J.-M. (2001). Wavelet toolbox for use with matlab, *MathWorks*.

Pettersson, M. (2008). *Design Optimization in Industrial Robotics: Methods and Algorithms for Drive Train Design*. Master Thesis, Linköpings university.

Qu, Y., He, D., Yoon, J., Van Hecke, B., Bechhoefer, E., & Zhu, J. (2014). Gearbox tooth cut fault diagnostics using acoustic emission and vibration sensors—A comparative. *Sensors, 14*, 1372–1393. (Switzerland).

Radzevich, S. P. (2012). *Dudley's handbook of practical gear design and manufacture*. Boca Raton: CRC Press.

Randall, R. B. (2011). *Vibration-based condition monitoring*, New York: Wiley.

Shrivastava, A., & Wadhwani, D. S. (2012). Vibration signature analysis for ball bearing of three phase induction motor. *IOSR Journal of Electrical and Electronics Engineering (IOSRJEEE), 1*.

Sinha, J. K. (2014). *Vibration analysis, instruments, and signal processing*, Boca Raton: CRC Press.

Smith, J. D. (1983). *Gears and their vibration.*

Sommer, A. P. (2011). Vibration-based health monitoring of multiple-stage gear train and differential planetary transmission involving teeth damage and backlash nonlinearity. California Polytechnic State University.

Wasilewski, R. F. (1994). *How to install bevel gears for peak performance.*

Welbourn, D. B. (1979). *Fundamental knowledge of gear noise—A survey.* UK: Cranefield.

Yeýilyurt, I. (1997). *Gearbox fault detection and severity assessment using vibration analysis*. Ph. D, Manchester.

Chapter 6
Robot Vibration Analysis and Feature Extraction

Abstract Using the data acquisition system and experimental set-up shown in Chap. 5, robot vibration with different health conditions are analysed in this chapter. The previously simulated faults in the gears and bearings are introduced in the robot. Based on the proposed robot CM algorithm, which consists of two stages, fault detection and fault diagnosis (as explained in Chap. 3), the vibration analysis presented in this chapter is also performed in two stages. The first stage is when the robot is executing a pick and place task and the second stage involves the robot moving one joint at a time. The effect of different fault types and severities on the robot vibration was investigated and the most effective vibration features are extracted, to be used in the following chapter.

Using the data acquisition system and experimental set-up shown in Chap. 5, robot vibration with different health conditions are analysed in this chapter. The previously simulated faults in the gears and bearings are introduced in the robot. Based on the proposed robot CM algorithm, which consists of two stages, fault detection and fault diagnosis (as explained in Chap. 3), the vibration analysis presented in this chapter is also performed in two stages. The first stage is when the robot is executing a pick and place task and the second stage involves the robot moving one joint at a time. The effect of different fault types and severities on the robot vibration was investigated and the most effective vibration features are extracted, to be used in the following chapter.

6.1 Pick and Place Task

In order to mimic a standard robot task, the robot was programmed to undertake a simple pick and place sequence, as shown in Fig. 6.1. The robot is required to swap the yellow and red ball positions and vice versa. This is achieved by first moving the red ball to the empty position (3) and then moving the yellow ball to position (1) where the red ball was, after which the robot picks up the red ball again and

A.A. Jaber, *Design of an Intelligent Embedded System for Condition Monitoring of an Industrial Robot*, Springer Theses,
DOI 10.1007/978-3-319-44932-6_6

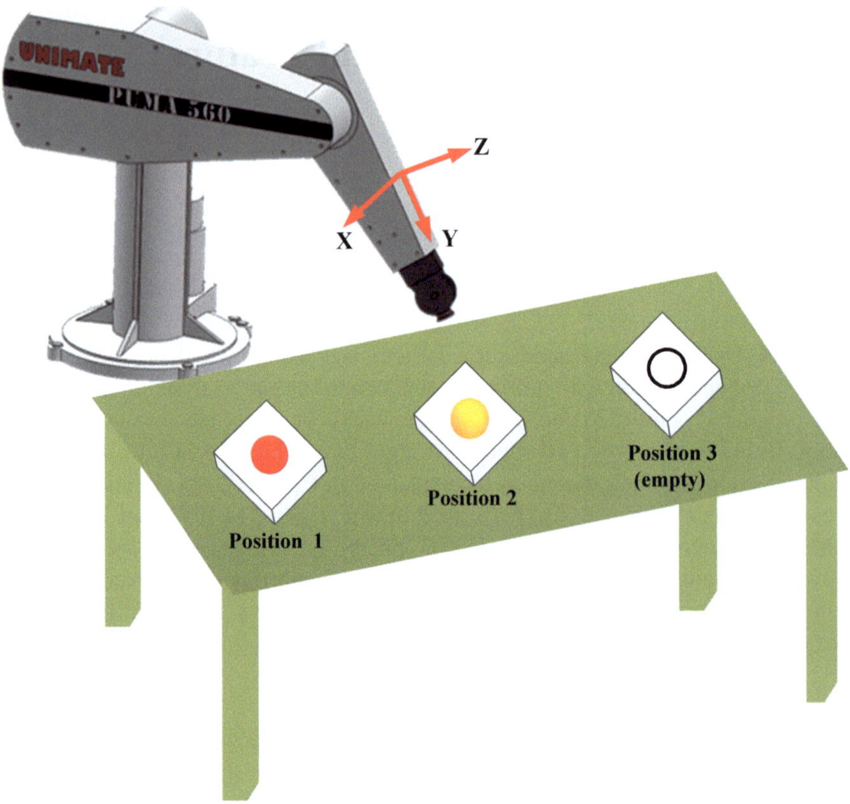

Fig. 6.1 Ball swapping task (Robot CAD drawing from: https://grabcad.com/)

places it at (2). The executed code to perform this task, which is written in VAL programming language, is attached in Appendix C.1.

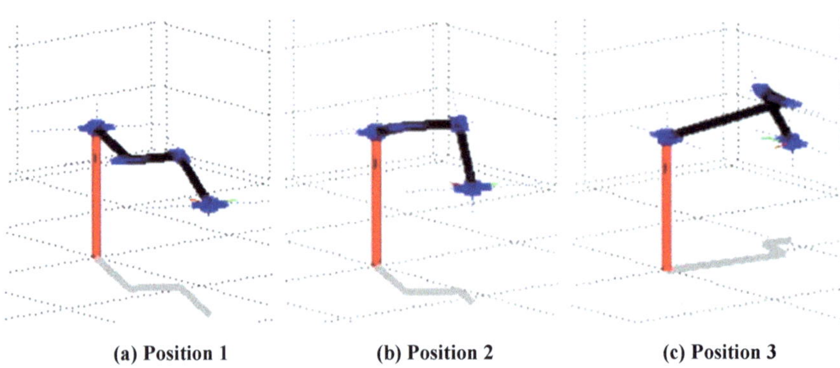

Fig. 6.2 Robot configurations at different balls locations

As the faults are introduced in joint 3 of the robot, the task requires joint 3 to move extensively during the described cycle, to examine the effect of the simulated faults on the resulting vibration excitation. The robot configurations when it reaches these positions are simulated using the robotics toolbox for Matlab and presented in Fig. 6.2. It should be pointed out that all other joints, particularly joints 1 and 2, are involved in this task, and therefore, the captured vibration signature will be made of a combination of signals from different joints motors and gearboxes.

6.2 Robot Fault Detection Whilst Executing Pick and Place Task

For faster robot health evaluation and due to the limitation of the embedded system, this stage has to be simple and reliable, since its purpose is just to report if the robot is healthy or faulty, regardless of the fault type and location, and needs to be generalized to any robot tasks. Thus the warning limits (thresholds), which are discussed and computed in the next chapter, have to be easily calculated and configured in the CM system. In practise this task may need to be conducted by a technician with limited knowledge, and for this reason the signal analysis technique is based on time-domain analysis, as it is simple, straightforward as well as good approach for fault detection if it is combined with statistical control charts (Chap. 3).

As mentioned in Chap. 3 is time-domain signal analysis has some limitations in terms of early fault detection and is not robust in fault identification (diagnosis). However, it has been found that extracting the most sensitive feature from the time-domain signal and utilizing it in computing the threshold values (Chap. 7) using a statistical control chart (SCC) helps in overcoming the early fault detection limit. The previously simulated faults are introduced in the robotic system and it is programmed to perform the ball swapping task repeatedly to complete a predefined number of cycles. Vibration signals from the three axes of measurements are captured at a sampling frequency of 383 Hz with sample size equal to 4096. This frequency is four times higher than the 11th natural frequency of the robot and the highest frequency that may appear in the system, based on the established joint 3 frequencies presented later in this chapter, and fulfils the sampling theory that requests the sampling frequency to be at least two times higher than the highest system frequency (Mohanty 2015). Each time the robot repeat cycle starts, the robot controller sends a trigger signal to the DAQ card through the Schmitt trigger circuit, to synchronize the signal capture with the robot movement.

The raw vibration signals captured under different fault types and severity are extracted from the LabVIEW software front VI panel. Several features from the time-domain signals, such as root mean square (RMS), standard deviation (STD), and kurtosis, are extracted. A comparison of these features, to investigate which is the most faults sensitive one, was accomplished. The sensitive feature must be normally distributed, as this is a condition that has to be met in order to calculate the threshold values using SCC, as will be discussed in Chap. 7. The STD feature, which is normally used as a measure of extent of variation of the processed data and has the same units as the data, was found to be the most faults sensitive, and normally distributed one; and hence it was selected for comparison among fault severities.

6.2.1 Backlash Fault Simulation Results

The vibration signals from the robot with the backlash faults introduced are shown in Fig. 6.3. As mentioned previously, three backlash levels are simulated (high, small and interference). The figure presents the time-domain vibration signals from the robot, when it is healthy and with the three backlash levels, after removing the DC component by computing and subtracting the median from each set of data in LabVIEW code. Limited interpretation of Fig. 6.3 is only possible; however, always vibration of multi-stage gearboxes, as in the robotic system, is very complex and composed of different frequencies, including gear mesh, bearings, and running-speed frequencies. The vibration severity is also connected with the excited resonance frequencies of the robot. The high speed of the robot gears causes cyclic excitation of these resonances, leading to some periodic fluctuations with amplitude proportional to the fault severity. This can clearly be seen in X-axis signals (the first column in Fig. 6.3), which were influenced the most by backlash, while signals in the other two axes were less affected. The high amplitude components present in the signals were observed when the robot changes the rotating direction of the joints. This is leading to developing number of impact between the mating pairs, due to the contact force between teeth. In the interference case the backlash was completely removed, so the impact effect is eliminated between mating gears but the gears are overloaded on the other hand.

It was anticipated that by increasing the backlash the vibration level would increase in the robot, but the opposite has been found. This can clearly be seen by comparing the vibration signals produced with different backlash levels to the healthy robot condition, especially the X-axis signals. It can be seen, for instance, in the case of higher backlash that the amplitudes of vibration signals are lower than those of the healthy case; this also corresponds with the results of research published by Bicker et al. (1989). The amplitudes of vibration increase when the backlash level is reduced, until the increase becomes more significant in the interference case. Although the presence of backlash in the transmission system can cause a transient impact at the reversal of motion, which will lead to an undesirable level of vibration, high backlash between gears will allow more lubricating grease to enter between the mating teeth leading to damping of the vibration. Whereas, the tight gear mesh leads to the lubricant being squeezed out of the mating teeth and the system heats up due to friction between the teeth, resulting in increased vibration.

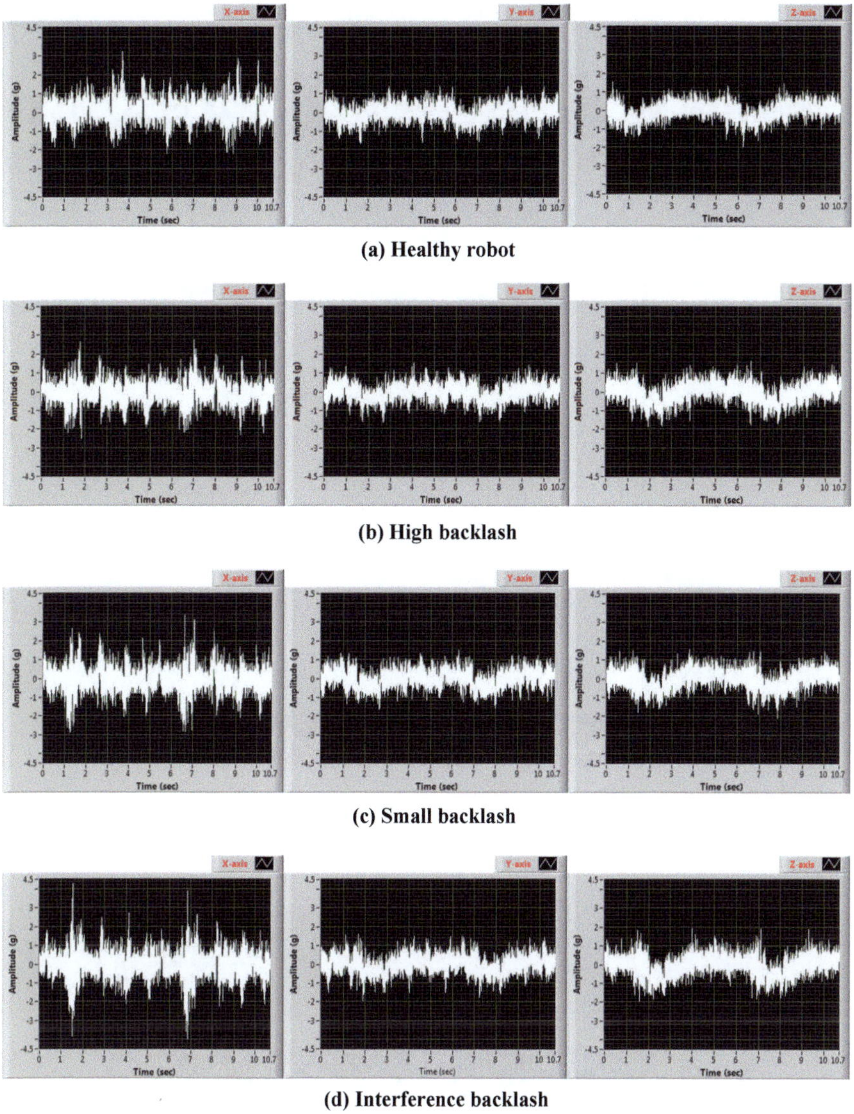

Fig. 6.3 Vibration signals from the accelerometers at different backlash level in the robot gearbox

To further establish the backlash effect and also to compute sets of STD values from the time-domain data to be used in developing and testing of an SCC in Chap. 7, the robot was programmed to execute the pick and place task one hundred times and the STD value for each repetition was calculated; this procedure was conducted at each backlash level. These results are presented in Fig. 6.4 for each of the X-Y-Z axes. The calculated STDs from the high backlash case show the lowest

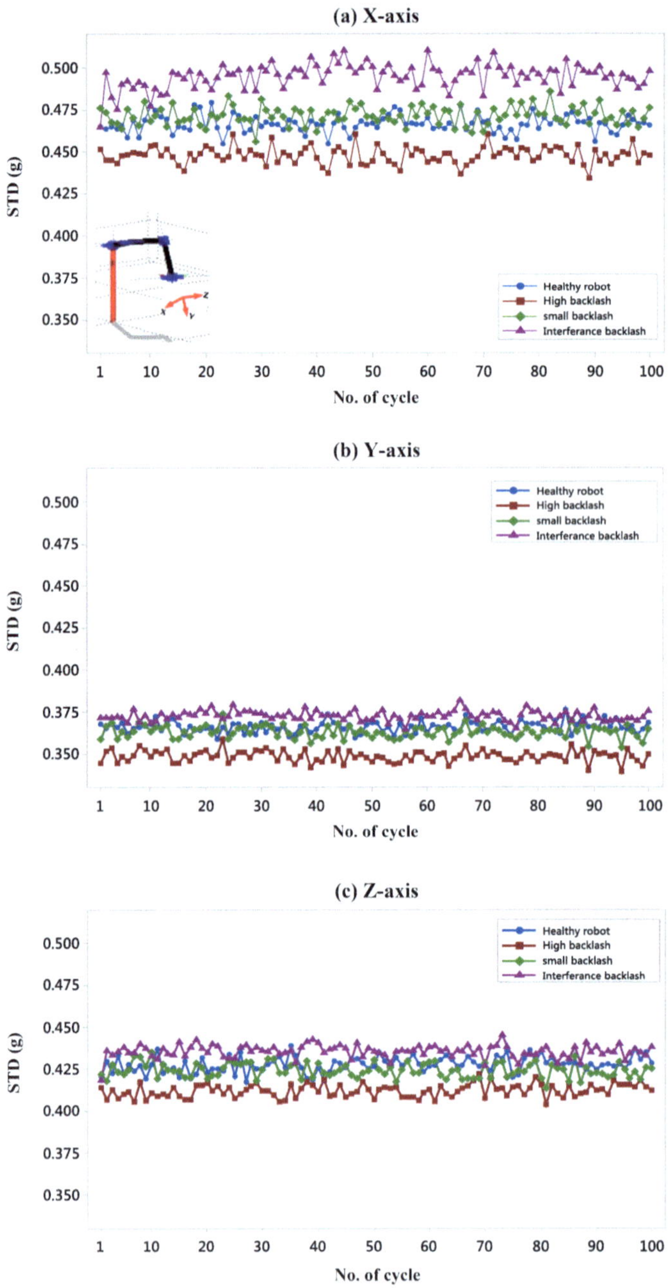

Fig. 6.4 Standard deviation for healthy and with different backlash levels robot

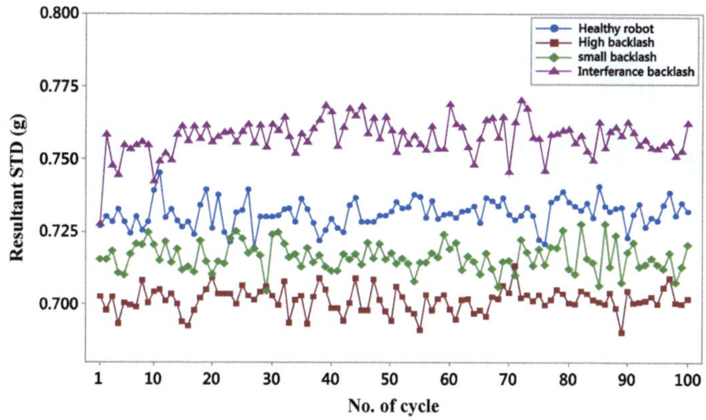

Fig. 6.5 Resultant STD for healthy and with different backlash levels robot

values for all 3 axes; while in the interference case the STD values are the highest. It was observed that the initial STD value in the interference case is always low in comparison to the subsequent values, which is attributed to the lubricant being squeezed out of the gear mesh, reducing the effective damping during the initial cycles on start up. The X-axis shows high STD values than those for both Y and Z, and there is a significant variation between the interference and high backlash conditions when compared with the normal and small backlash conditions which have quite similar magnitudes.

Generally, the STD results when the backlash fault was simulated, and even with the other fault types as shown in the following sub-sections, show that the X-axis vibration is the best to be utilized for monitoring the effect of fault development in the robot, as the different faults can be clearly discriminated. However, the Y- and Z-axis could exhibit higher vibration levels if the robot executes different tasks that involve a different sequence of movements or if a fault is progressed in the other joints. Because of this, the vibration in all three axes has to be considered, in order to achieve reliable fault detection. Thus, the resultant value of the X-Y-Z standard deviations will be computed by taking the square root of the sum of the squares of the three STDs, and this will be used as the fault indicator. Figure 6.5 shows the resultant STD values for the robot when it is healthy and with the three backlash levels. Clear differences can be realized amongst the four trends related to robot health conditions and, as in Fig. 6.4, the high backlash conditions show lower resultant STDs whereas they are the highest in the interference condition.

6.2.2 *Gear Tooth Fault Simulation Results*

The time-domain vibration signals with four different types of gear tooth faults in the pinion bevel, which are tooth wear, 25 % tooth cut, 50 % tooth cut and full

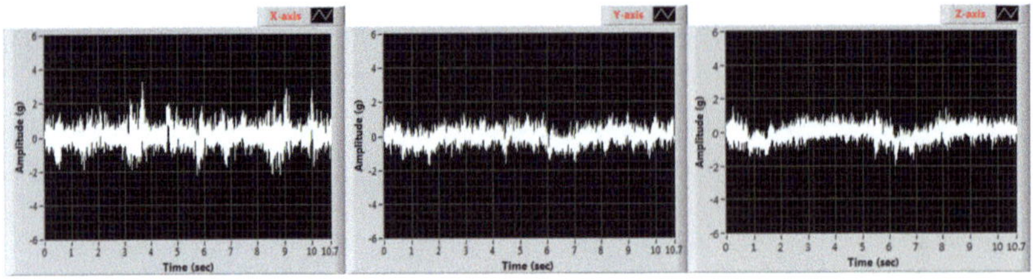

(a) Healthy robot

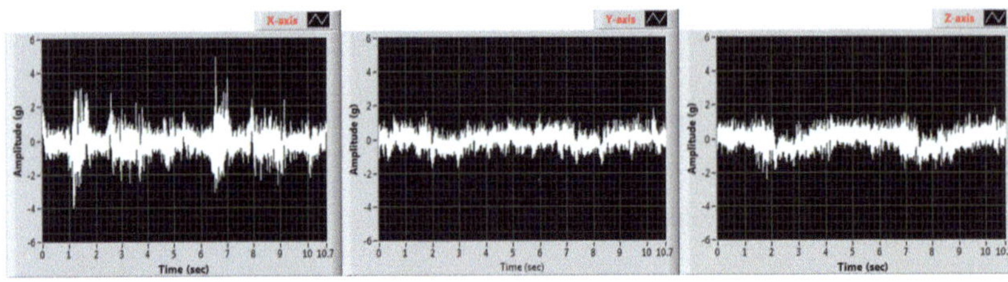

(b) Gear tooth wear

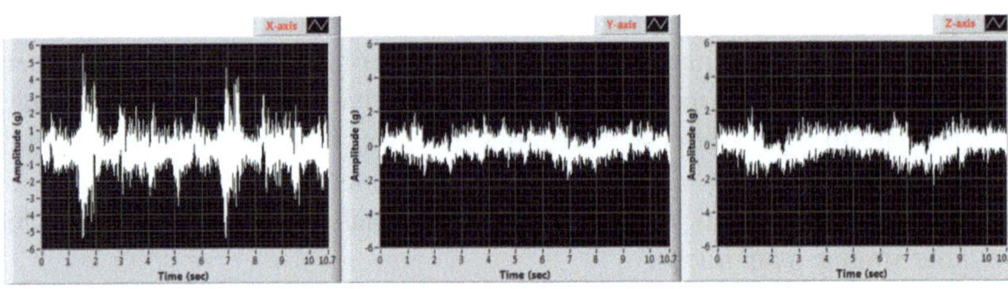

(c) 25% tooth removed

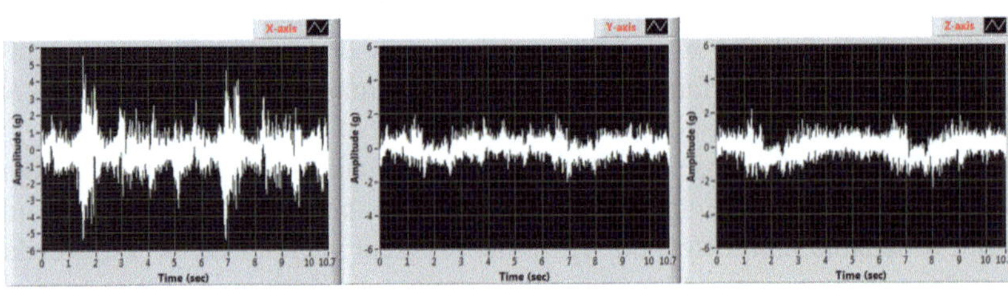

(d) 50% tooth removed

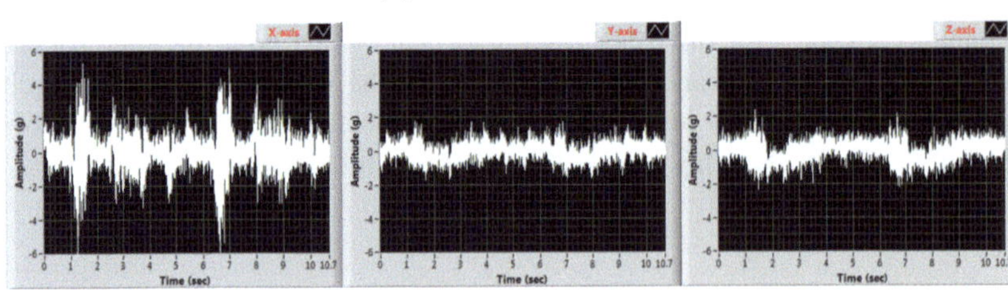

(e) Full tooth removed

Fig. 6.6 Vibration signals from the accelerometers at different gear tooth faults

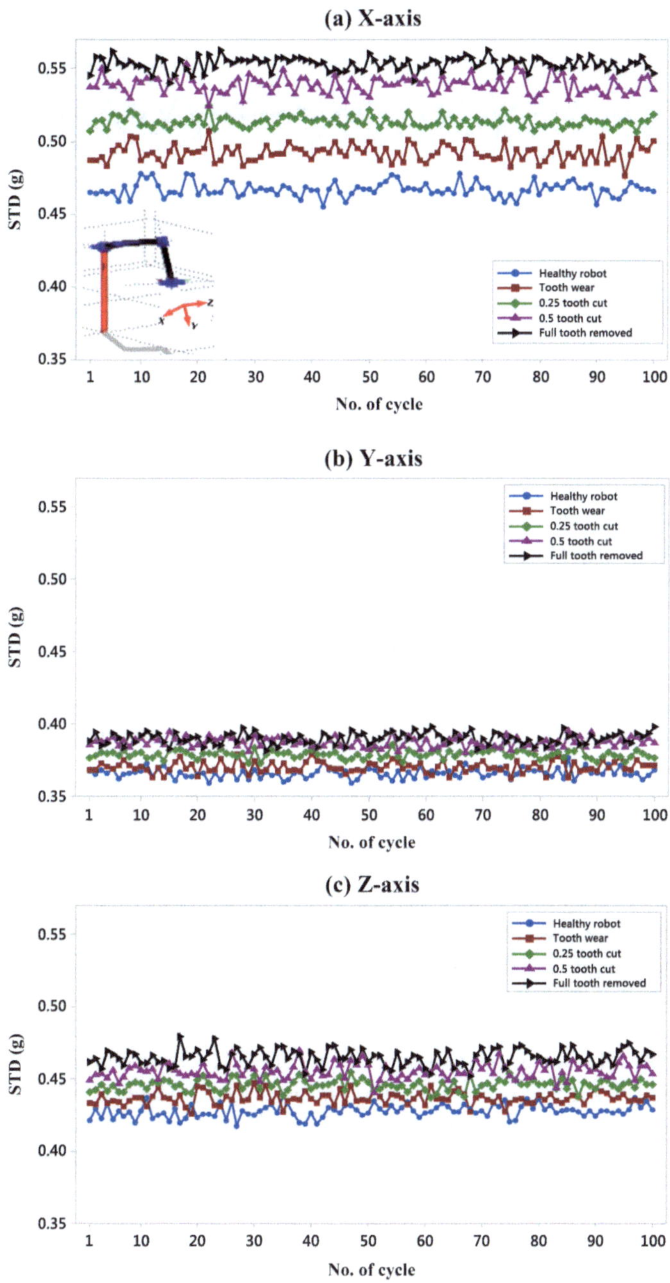

Fig. 6.7 Standard deviation for healthy and with different gear fault robot

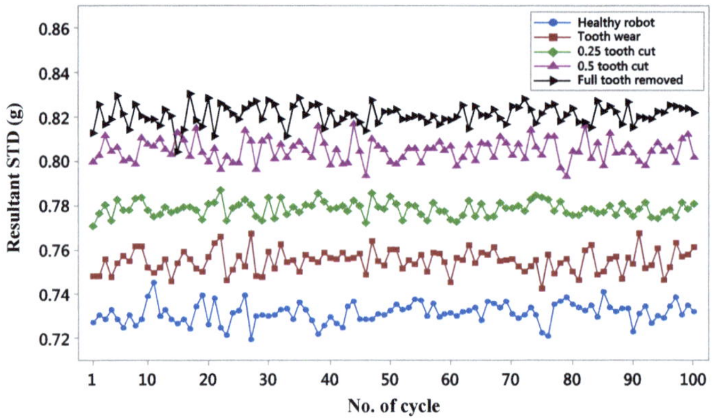

Fig. 6.8 Resultant STD for healthy and with different gear fault robot

tooth removed respectively, are shown in Fig. 6.6. Prior to capturing the data when wear fault was introduced, the robot was run for a short time to allow the gearbox to dynamically settle. As in the previous set of backlash tests the X-axis is the most sensitive to any change in the robot's condition, although the trends are very similar. Looking at the signal from the faulty robot (X-axis) it can be observed that the amplitude is higher than that of the healthy robot. Secondly, the amplitudes increase as the tooth fault becomes severer, as expected; due to the reduction in the contact ratio associated with the tooth cut introduces higher vibration during gear mesh. Furthermore, the impulses in the signals are more prominent when the tooth is completely removed. This can attributed to the adjacent tooth contacting the mating tooth earlier than it would with an ideal geometry.

As mentioned earlier the excessive clearance between gears allows improved lubrication (grease) which will increase damping and reduce vibration. Whilst the same is true here, except that the impact intensity is much higher when there is a tooth cut leading to a high vibration level which increases the computed STD and thus the resultant STD values, as shown in Figs. 6.7 and 6.8. A clear discrimination can be established amongst the four gear tooth faults and the healthy condition of the robot for all three axes of measurement. The STD values are ranging from the lowest trend, when the robot is healthy, to the highest which is when a full tooth removed. These trends are well separated in the X-axis, but a bit compressed in the Y- and Z-axis, and have reduced magnitudes; however, a clear separation among the four cases was got using the resultant STD, Fig. 6.8.

6.2.3 Bearing Fault Simulation Results

An inner race and two levels of outer race bearing faults have been simulated. The purpose of simulation two fault severities in the outer race (as mentioned in Chap. 5)

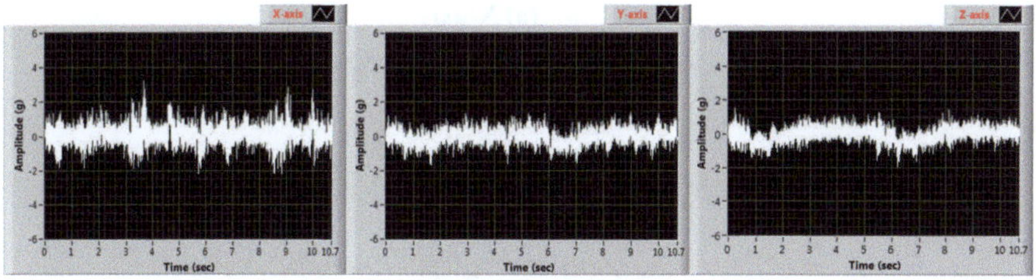

(a) Healthy robot

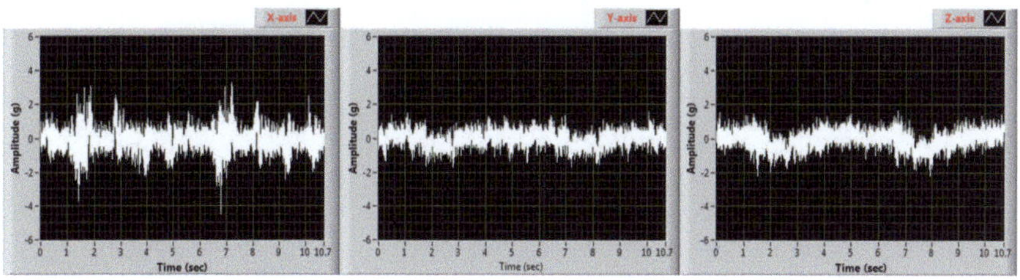

(b) Inner race bearing fault

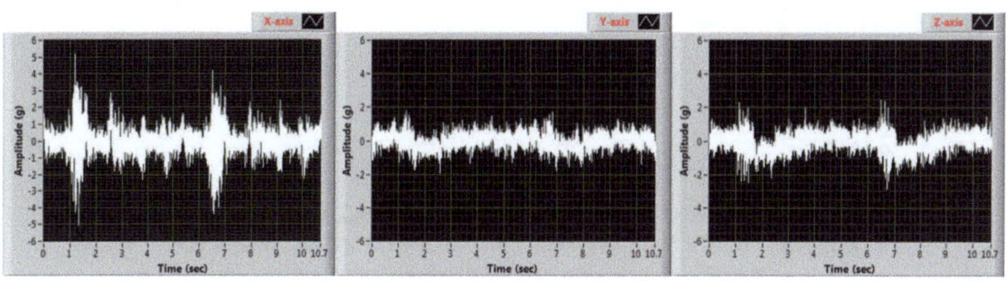

(c) 1 mm hole in the bearing outer race

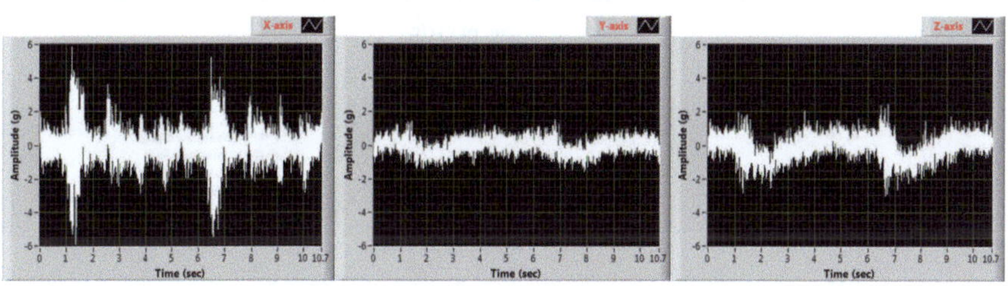

(d) 2 mm hole in the bearing outer race

Fig. 6.9 Vibration signals from the accelerometers at different bearing faults

is to test the effect of the different sizes of bearing faults on the captured vibration signals. The time-domain signals when these faults introduced along with the extracted STD values are presented in Figs. 6.9, 6.10 and 6.11, respectively. Each time a rolling element passes the inner or outer race defect, the collision between the rolling element and fault generate an impulse and excites the vibration resonance of

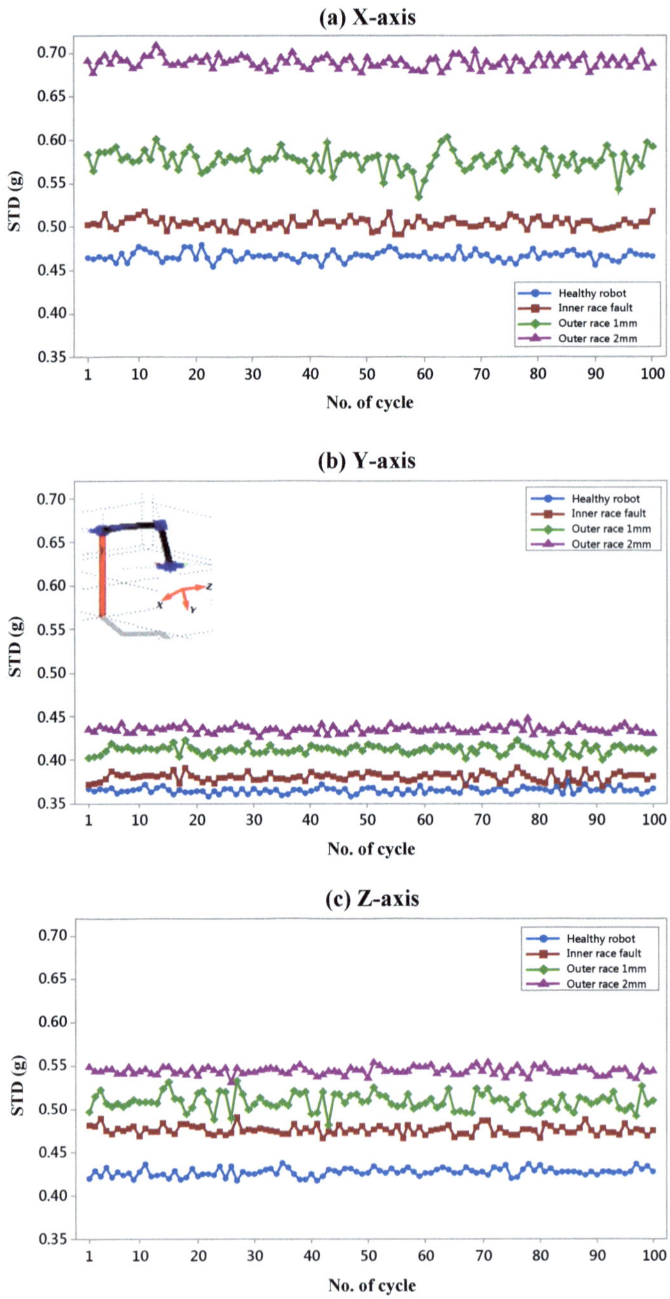

Fig. 6.10 Standard deviation for healthy and with different bearing fault robot

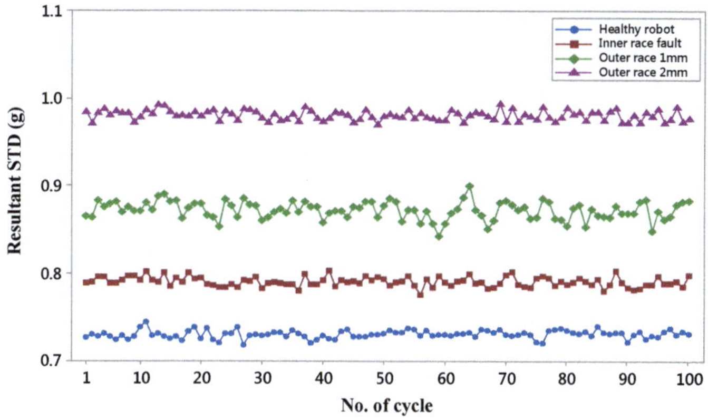

Fig. 6.11 Resultant STD for healthy and with different bearing fault robot

the bearing and the robot structure. This is also true when gear faults are simulated, but since the faulty bearing is supporting the pinion bevel and due to the tangential, radial and axial forces generated by the meshed gears the excited vibration signals have higher amplitudes. This clearly can be seen from the calculated STD values in Figs. 6.10 and 6.11.

When a fault occurs in the inner race or in one of the rolling elements the properties of the induced impulses will change, since the impacts occur at different angular positions as the bearing components rotate. On the other hand, with the outer race defect the excited transient modes do not vary, since the angular position of the defect is fixed at each impact. This is apparent in the signals captured from the two different outer race fault severities in Fig. 6.9; where both faults have the same signal patterns but with different amplitude levels (higher when the fault size is bigger).

6.3 Robot Fault Diagnosis

The previous section was focused on studying the effect of different faults on the vibration level of the robot, with the aim to extract features sensitive to the robot health condition when it is executing a repetitive task using a simple signal analysis technique. This can be utilized with different robot tasks for differentiating between the healthy and faulty conditions irrespective of fault types and locations. However, the next logical step after detecting a fault is to extract features that can be employed to identify which joint has the fault and its type (fault diagnosis), which is the second stage of the proposed CM algorithm and will be the focus of the remainder of this chapter.

To perform this step, each joint should be considered separately by programming the robot to cyclically rotate one joint at a time. For this purpose and since the faults are simulated in joint 3, the robot was programmed to oscillate over its full range of movement (270°) as shown in Fig. 6.12, as some of the robot faults may exhibit direction-dependent symptoms (Datta et al. 2007). This is also in order to relate the produced vibration to the moving joint's transmission components such as gears and bearings. The output from the three accelerometers will be captured and analysed using the DWT signal analysis technique (discussed in Chap. 3) based on the prepared LabVIEW code; from which features are obtained and utilized later (in Chap. 7) in designing the ANN for faults classification. The signals are captured over the entire movement cycle using a sampling frequency of 1031 Hz. The DAQ device needs to be triggered by the robot controller at the start of the movement for synchronization. The following VAL code is used to achieve the cyclic joint movement.

```
1 SIGNAL 1               {To activate output line 1 to trigger the DAQ device}
2 DRIVE 3, −270, 100     {Move joint 3 downward by 270° at 100 % of the motor
                          speed}
3 DRIVE 3, 270, 100      {Move joint 3 upward by 270° at 100 % of the motor
                          speed}
4 SIGNAL −1              {To put the output signal from port 1 off}
5 DELAY. 1E−03           {Delay 1/10,000 s}
```

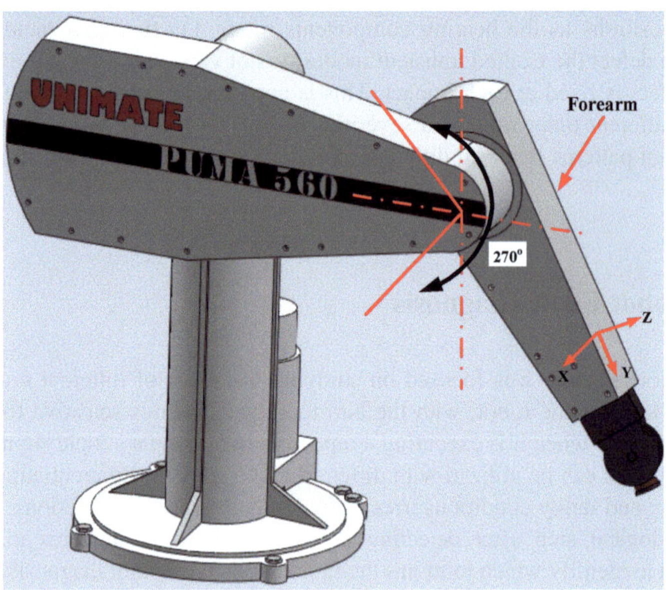

Fig. 6.12 Robot configuration for the fault diagnosis stage (Robot CAD drawing from: https://grabcad.com/)

6.3.1 *Signal Analysis Using DWT*

The multi-resolution DWT analysis was applied using the selected mother wavelet, *db2*. To conduct the DWT analysis, a suitable number of decomposition levels needs to be selected, however, the number is influenced by the lowest frequency bands that needed to be traced as well as the number of samples in the gathered signal; where for a signal of size *N*, the maximum decomposition level is (Abo-Zahhad 2011). Since the number of sample in the captured signal is 4096 so a maximum of 12-level of decomposition can be reached. If the number of decomposition levels of the DWT becomes higher, the extracted frequency band gets lower. Note that the greater the number of decomposition levels the more power and memory is required in the designed embedded system (which is discussed later in Chap. 8). Therefore, it has been decided to depend on the expected lowest fault frequency that may be presented in the signal in selection the decomposition levels.

To calculate the frequencies that are related to joint 3, the mathematical equations and gears and bearing specification presented in the previous chapter are used. The only missing parameter in these equations is the speed of each component. Therefore, the robot's forearm speed of movement has to be known in order to establish the speeds of the other mechanical parts in joint 3. As mentioned, in the experimental work the robot is programmed to move just joint 3 in a cyclic motion within its full range of movement. So, the acceleration signal in the Z-axis has been captured and then by using the numerical integration the forearm speed was obtained. This speed represents the speed of the wheel spur gear and, by using the gear ratios, the other speeds and subsequently the frequencies, are derived. These frequencies are shown in Table 6.1.

From the above table can be seen that the lowest fault frequency that may appear is in the bearing of the second stage (the bearing on the idler shaft), which is indicated by bold in the table. As a result, eight decomposition levels are found to be sufficient for analysing the robot vibration signal; Table 6.2 shows the

Table 6.1 Joint 3 gears and bearings frequencies

Parameter	Gear frequencies	
	Bevel gears (first stage)	Spur gear (second stage)
Mesh frequency (Hz)	95.7	10.96
	Shafts frequencies (Hz)	
Input shaft	7.356	
Idler shaft	1	
	Bearings frequencies	
	Bearing in the first stage (Hz)	Bearing in the second stage (Hz)
FTF	2.94	**0.38**
BPFO	20.62	3.04
BPFI	30.86	4.93
BSF	17.76	2

decomposition levels and their associated frequency ranges (referred to Chap. 3 to see how these frequency bands are computed).

6.3.2 *Result of the DWT Analysis*

Selected results of the multi-resolution signal analysis using DWT extracted from the LabVIEW code are shown in Figs. 6.13, 6.14 and 6.15. These figures represent the analysed signals from the three accelerometers when the severest two fault categorizes, a full gear tooth removed and 2 mm hole in the bearing outer race faults, are introduced in the robot. The remainder of the results when other fault types simulated are included in Appendix C.2. In the above mentioned figures there are 10 sub-figures for each measurement axis; the first sub-figure is the original signal before the DWT is applied, while the rest are the eight constructed sub-signals from the detail coefficients (D1, D2, D3, D4, D5, D6, D7, and D8), and one from the approximation coefficient (A8), where each one contains a specific frequency band presented in the upper left corner of each sub-figure.

Generally, accelerometers provide mixed information about vibration and movement acceleration in the robot (Rodriguez-Donate et al. 2010). The vibration signal is made up of high frequencies, whereas movement acceleration signals are associated with low frequencies (Rodriguez-Donate et al. 2010). In the original signals in Figs. 6.13, 6.14 and 6.15 can be seen that the Y- and Z-axis signals are highly affected by the robot movement, while there is a little (or no) effect in the X-axis signal, due to the main robot movement is in Z-axis and so the angular acceleration, due to the forearm rotation, will have an effect on the captured signal in Z-axis. Also, as a result of the angular movement of the forearm, a centripetal acceleration is generated in Y direction effecting Y-axis signal. However, after conducting the DWT on the original signals and by investigating the produced sub-signals it was concluded that the approximation signals (A8) in the three axes are mostly related to the robot movement and not to its vibration components. Consequently, the approximation signals do not carry any useful information regarding robot faults and are not considered important in the present work.

Table 6.2 Frequency bands associated with each level of decomposition

Level	Signal	Frequency range (Hz)
D1	Detail signals	257.75–515.5
D2		128.8–257.75
D3		64.43–128.87
D4		32.21–64.43
D5		16.1–32.21
D6		8.05–16.1
D7		4.02–8.05
D8		2.01–4.02
A8	Approximation signal	0–2.01

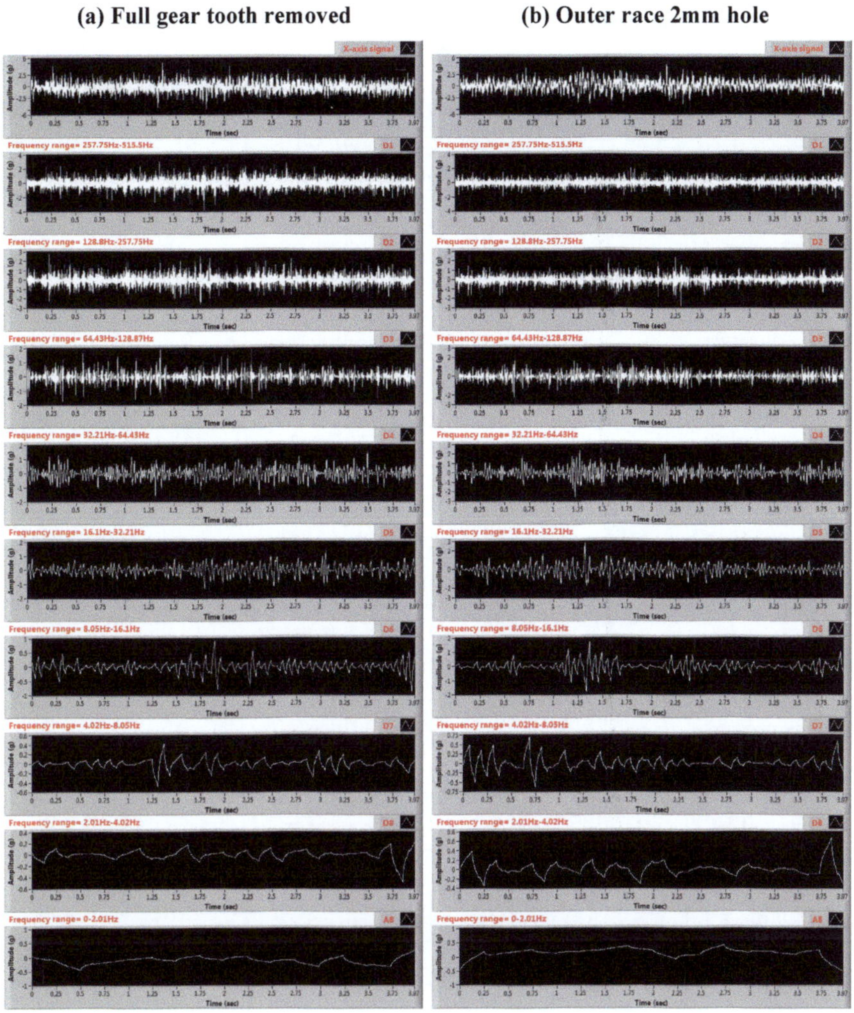

Fig. 6.13 DWT analysis of the X-axis vibration signals from the robot arm when different faults simulated (auto scaled)

However, extracting the robot movement acceleration using DWT could be employed in future work to estimate the kinematics of the industrial robots (Rodriguez-Donate et al. 2010).

Comparing the corresponding time domain signals from the three axes when gear tooth and bearing faults were simulated (or when other different fault types are introduced), it can be concluded that it is not easy to recognize significant variations between them, however, by looking at the generated sub-signals from the three axes can be observed that there are clear differences amongst their amplitudes. It is noticed that the produced sub-signals in Y-axis when different fault conditions are

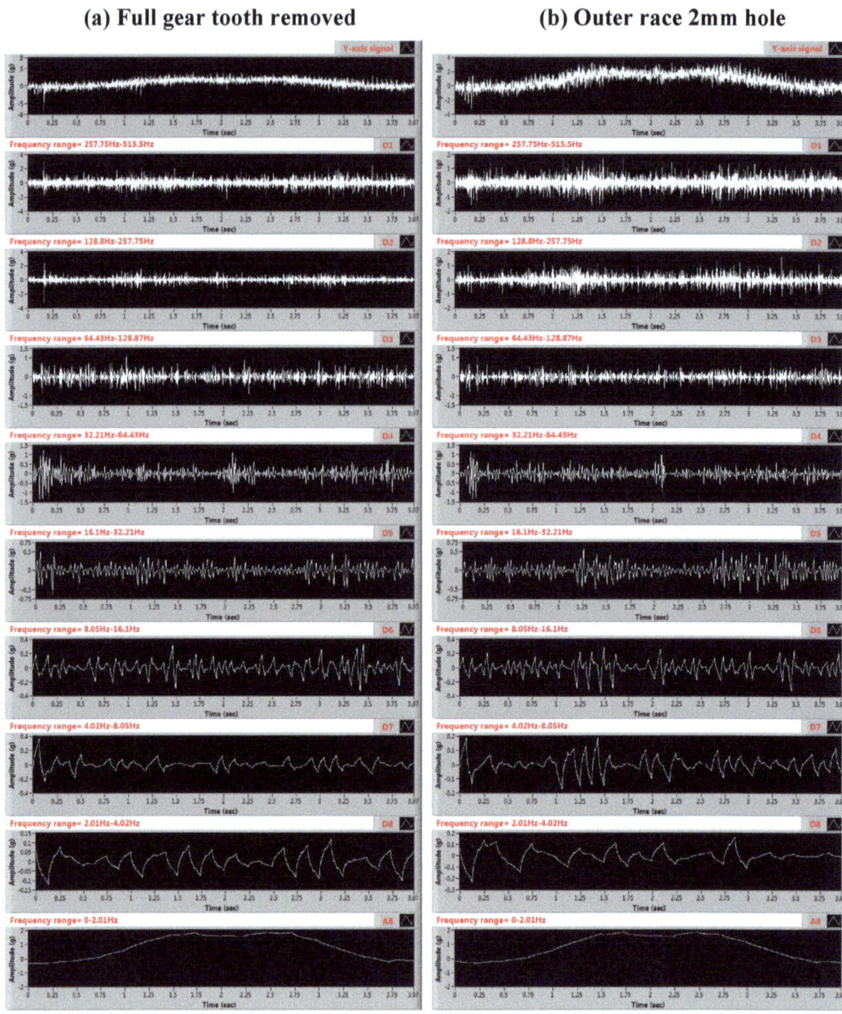

Fig. 6.14 DWT analysis of the Y-axis vibration signals from the robot arm when different faults simulated (auto scaled)

having the lowest amplitudes comparing to sub-signals in X- and Z-axes. Nevertheless, the amplitude in Y-axis signal may get significant if a fault is developed in other parts of joint 3 or in another joint. The reversal of motion can be obviously seen in the Z-axis response (Fig. 6.15) where the higher amplitudes are in the waveform. The high amplitudes at the reversal of motion are due to the impact phenomenon resulting from the backlash, which must be present in the gearbox even when the robot is healthy but within the acceptable range.

When a gear tooth fault introduced was noticed that D3 becomes more sensitive to the fault, since it covers the frequency band corresponding to the bevel gear mesh

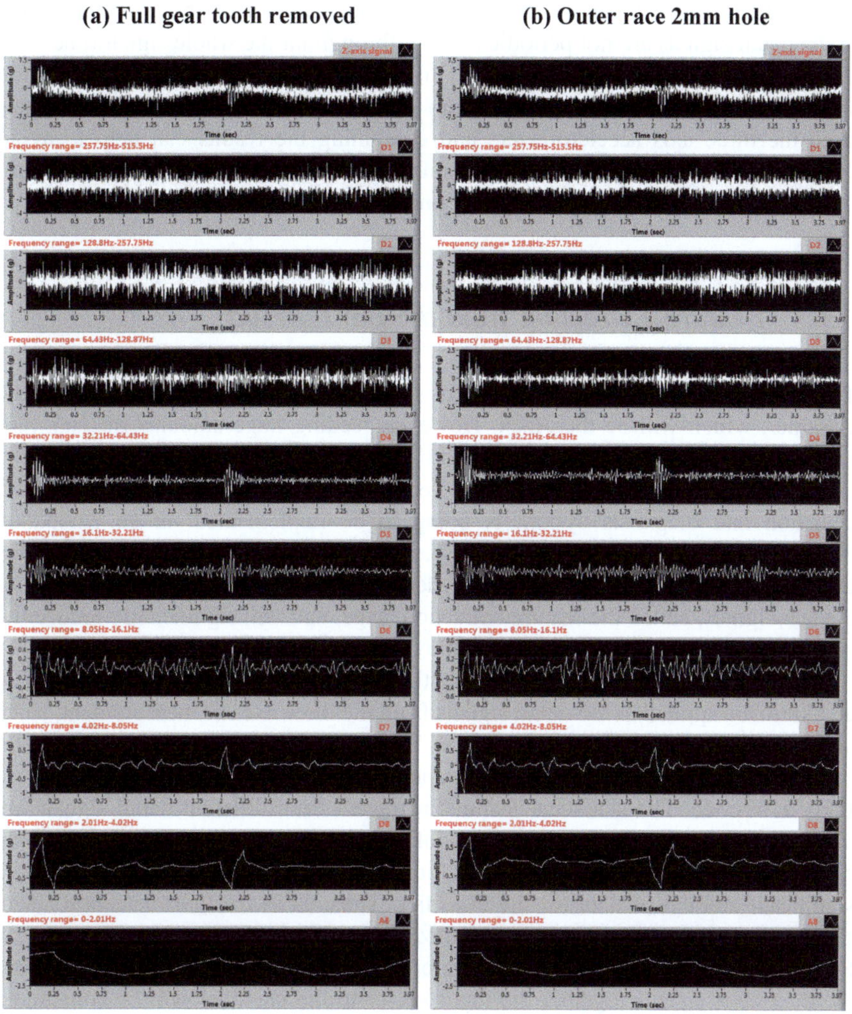

Fig. 6.15 DWT analysis of the Z-axis vibration signals from the robot arm when different faults simulated (auto scaled)

frequency, which is 95.7 Hz, (Fig. 6.13a, 6.14 and 6.15a). Similarly, it is anticipated that D6 will become peaked if a fault is simulated in the spur gear pair, as spur gear mesh frequency is within D6 frequency band. From Table 6.2 can be seen that the ball pass frequency of the outer race (BPFO), which is calculated to be 20.62 Hz, is inside the frequency range of detail signal D5, and is prominent from the signals that contain outer race bearing fault in Figs. 6.13b 6.14 and 6.15b where distinct impulses are evident in comparison to D5 signals when a gear tooth fault introduced. Meanwhile, another detail signal is affected if the inner race fault is simulated, but here will also be D5, since BPFI is within the same frequency band of BPFO. However, it can be observed that the appearance of spikes (or impulses) in the detail signals are not periodic and not throughout the whole signal time as in the rotating machinery. This can be attributed to the variation in joint rotation speed, changing from zero to maximum forth and back. Analysis of the signals from the three axis accelerometers and combining the results can help to avoid uncertainty in fault diagnosis if only one accelerometer is used, since the fault signature can be efficiently detected in one or two axes than others.

6.3.3 Feature Extraction from the DWT Result

The sub-signals from the DWT analysis could be directly used as inputs for the classifier, which will be the ANN. However, feeding huge amounts of data, represented by 4096 samples from each sub-signal, into the classifier will make the classification process more complex; keeping in mind that the proposed fault detection and diagnosis system is going to be implemented on an embedded electronic system with limited memory and processing power. To overcome this limitation, a post processing stage is needed to take out significant parameters from the sub-signals, called the feature extraction stage, and will significantly reduce the data set size required to be passed to the classifier for correct fault diagnosis.

Based on this, several statistical functions are calculated, e.g. the RMS, STD, and kurtosis, applied to the detail signals from the three axes of measurement, since approximation signals are neglected, as discussed earlier. Also, the features are extracted in real-time after the data is captured and DWT analysis is applied using LabVIEW code. The statistical features extracted from different detail signals are always varied, as each signal contains information related to a specific frequency band. If all computed features are used for fault classification, this will also involve significant amount of data for the classifier. Thus, as in the fault detection stage, only the most sensitive feature was utilized, namely the STD value. Therefore, 8 STD features from each axis signal are obtained, and by joining all features from the three axis signals together the input vector to the classifier will be constituted from 24 features.

Figures 6.16, 6.17 and 6.18 show the computed SDTs from the eight detail signals from accelerometer axes X, Y, and Z. Figure 6.16 presents the results from simulating different backlash levels as well as the result extracted from the detail signals of the robot healthy condition, while Figs. 6.17 and 6.18 present the STD values for gear and bearing faults, respectively. First, it can be observed that the STD values when backlash faults are present in the robot are lower in comparison to gear and bearing faults. In particular the high backlash case records the smallest STDs at all detail levels among the healthy and other backlash cases and this is similar in all the three axes. These values are increased by reducing the backlash until they become significantly high when the backlash is completely removed (interference case). This, as discussed in Sect. 6.2.1, is attributed to vibration damping due to entering the lubricant grease between the mating teeth when

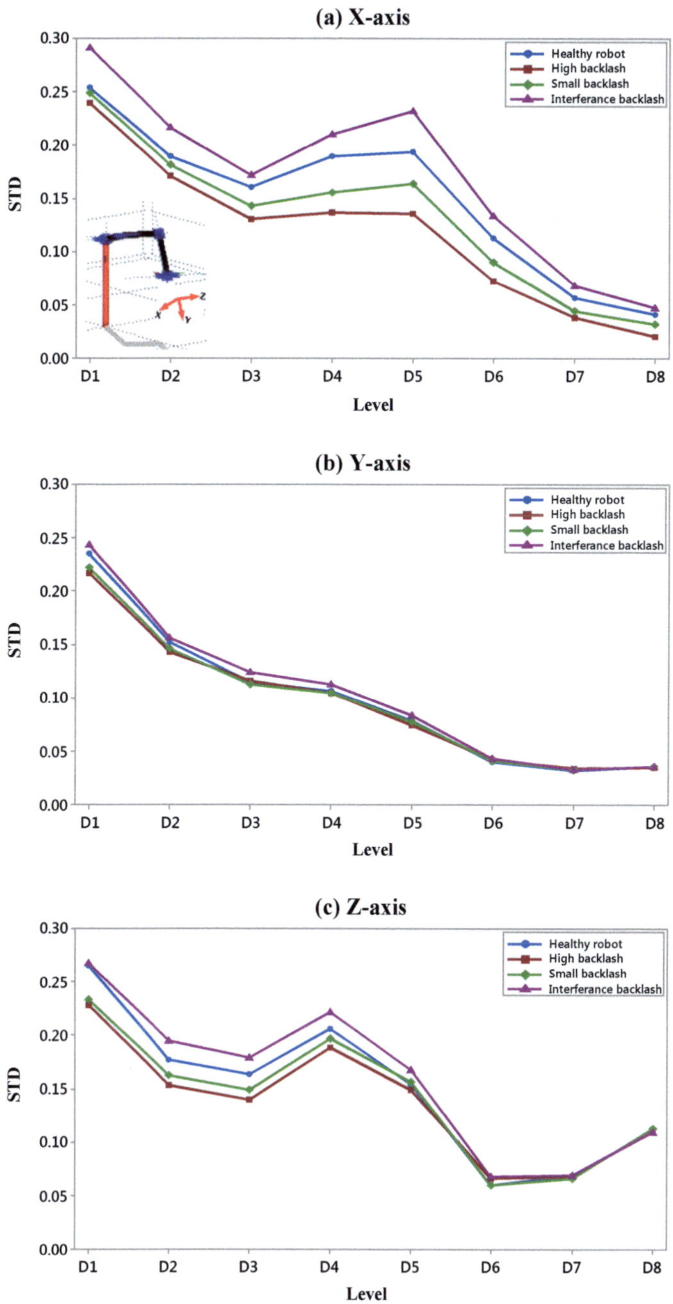

Fig. 6.16 Standard deviation (STD) in all 8 detail signals when backlash faults simulated

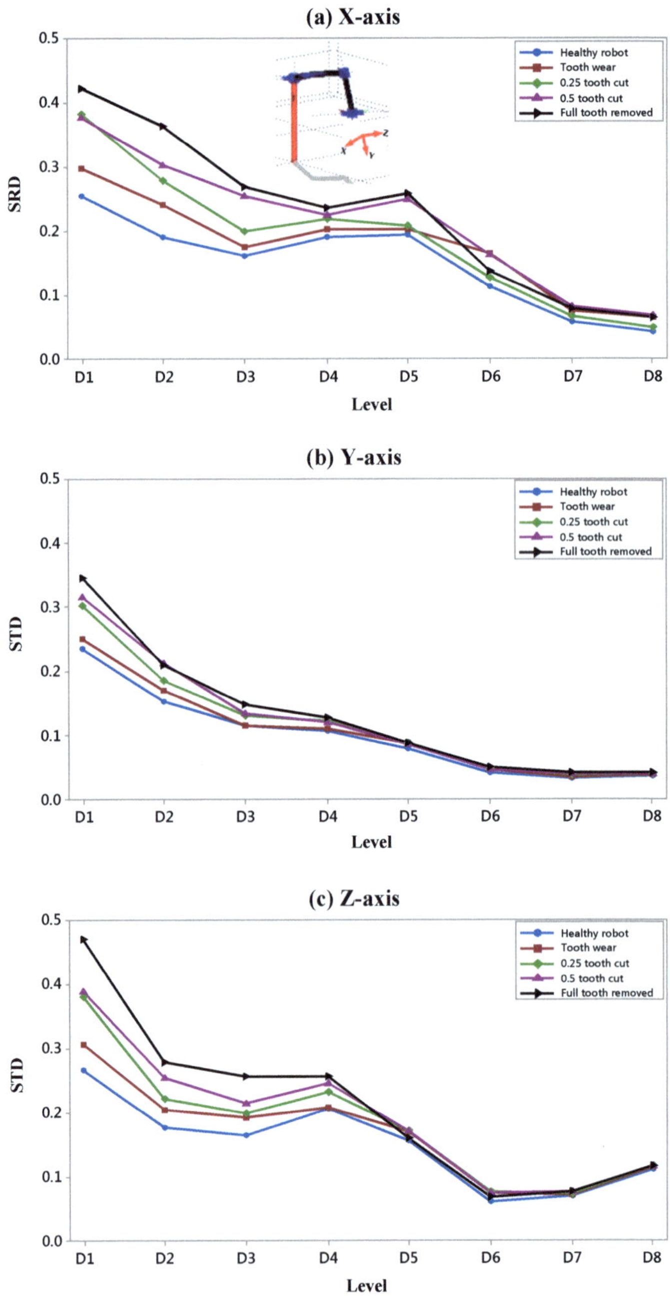

Fig. 6.17 Standard deviation (STD) in all 8 detail signals when gear tooth faults simulated

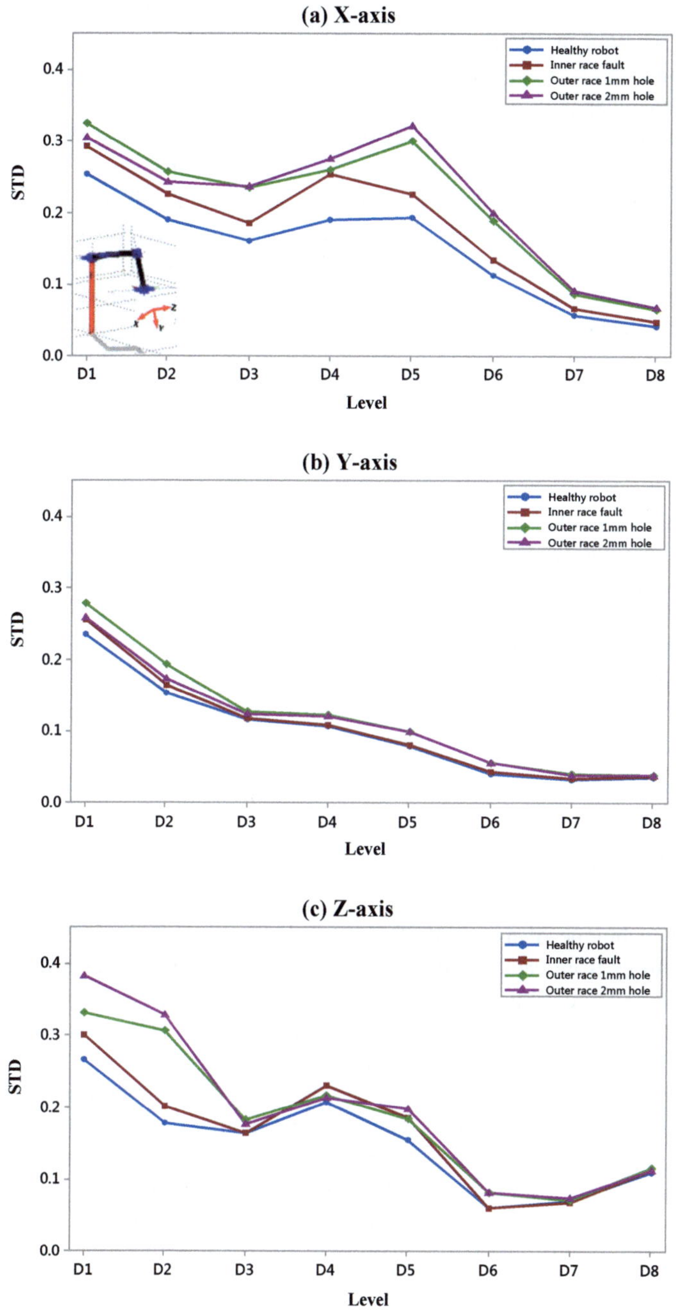

Fig. 6.18 Standard deviation (STD) in all 8 detail signals when bearing faults simulated

excessive clearance between them and squeeze it when they are interfered. The STD vectors at healthy and other backlash levels are clearly separated in X-axis, and also clear separation is shown up to D5 in Z-axis, while they are almost squeezed together in Y-axis.

Mainly, in gear and bearing faults the STD vectors in Y-axis are also squeezed except at higher frequency bands (D1–D3) they are a bit diverged. This is agreed with what have been noticed when at different faults the detail signals in Y-axis are investigated and concluded that their amplitudes are lower than other axes and obvious differentiation cannot be recognized (Fig. 6.14). However, the Y-axis results will not be ignored and will be included in the input vectors to the classifier, as sure the STDs in Y-axis will give a clear indication about a fault if it is developed, for example, in other joints or parts.

When bearing and gear faults are seeded in the robot the STD values in X- and Z-axis are increased as the fault severity increased; this is also identical with the increase of the detail signals amplitudes due to the same reason. Moreover, the STD vectors in these cases are well separated except at the lower frequency bands from D6 to D8, since their frequency ranges are not affect by the simulated faults. However, may the reader is asking why the differences among the STD vectors in X- and Z-axis are very obvious at higher frequency bands; well this can be related to the stimulated gear and bearings higher harmonics due to the simulated faults. Generally speaking, if the STD vectors are very well isolated, the problem of fault classification will be easier when ANN is used, the false diagnosis rate is reduced, and also the designed ANN will be of small size helping in lowering the required processing power and memory of the embedded system.

6.4 Fault Diagnosis Using Combined DWT and FFT

The other technique of using DWT in fault diagnosis is instead of extraction statistical features from each constructed detail signal, the FFT is conducted on each frequency band to establish the excited frequency spectra (Kar and Mohanty 2006), with the idea being to incorporate the advantages of both of these signal processing tools. In DWT specific frequency bands are extracted, and the FFT is used to estimate the frequency spectra within these bands, and in combining DWT-FFT the frequency spectrum of the signal which can be zoomed, without affecting the FFT frequency resolution. Moreover, by using this technique the same frequency resolution with reduced number of data length can be obtained. To extract the frequency spectrum in each decomposition level the DWT is applied first to the time domain signal. Next, an FFT is performed on the generated coefficients, which yields the zoomed frequency spectra where the fault frequency can be identified with significant increase in the fundamental component, leading to enhancing the detectability of the fault frequency component. Then, the frequencies in these spectra are analysed and compared to baseline frequencies; any deviation in the computed frequencies from the baseline ones or appearing some sidebands or harmonics frequencies related to the baseline frequencies are an indication of a fault development in the system. Knowing which frequency is deviating or the generated sidebands or harmonics related to which frequency in the machine components, the faults can be diagnosed.

In this section, the FFT has been taken for each frequency band produced from the three axes of measurement and for different robot faults. This was also done when just joint 3 is moving, in order to be able to link the calculated frequencies to the mechanical components (gears and bearings) in this joint. The excited frequencies at each level are then investigated in more detail to explore their relationship to the joint components, and their relationship to the robot natural frequencies. This can be established by comparing them with the obtained natural frequencies of the robot in Chap. 4 and calculated baseline frequencies described in Sect. 6.3.1. For this purpose, a separate sheet containing the frequencies and harmonics of all gears and bearings in joint 3 as well as the natural frequencies of the robot has been prepared and used in this investigation. LabVIEW code for performing the Fourier transform was also developed and used concurrently with the wavelet transform code. Also, it should be noted that these two codes are run in real-time while the robot is running, which means that online robot health monitoring can be achieved on the robot (or any other machines) by just using the signal acquisition hardware and the prepared code, but with some restrictions due to the long cabling.

The results of the FFT can be observed in the three-dimensional plots of the frequencies produced in all levels and in the three axes using Matlab software, as shown in Fig. 6.19. This figure is for the bearing faults seeded in the robot; the corresponding figures simulating the other faults can be found in Appendix C.3. The top row (a) in all of these figures represents the frequency spectra of the detail signals (frequency bands) for the healthy robot. In these figures each level includes a specific frequency range, i.e. high frequencies are included in level 1 and as the level order get higher the frequency contents of that level becomes lower. It can be noticed that it is not possible to determine any substantial variation across the frequency spectra in the Y-axis, for any of the different fault conditions. This may however not be the case if any of the other robot joints are cycled individually. Significant differences can be observed in the spectra of the X and Z axes for joint 3 in Fig. 6.19, at levels 4–6, and also for the backlash faults, which is shown Appendix C.3, Fig. C.16. By comparing the minimum backlash (interference) case with the high backlash (clearance) case, it can be seen that many frequencies appear within the case of interference whilst there are very few excited frequencies at lower backlash. The tight mesh between bevel gears increases the mechanical stiffness, making components' harmonics appear in the frequency spectrum, and the excited frequencies with high backlash have lower amplitudes in comparison with the healthy and interference cases, most likely due to increased damping.

In the spectra of gear fault simulation case (see Appendix C.3, Fig. C.17) the amplitudes of the higher levels (D1–D3) show significant variation to the healthy condition and get noticeably higher as the gear tooth fault gets more severe. The increased amplitude level is likely due to the increased force of the impact as the

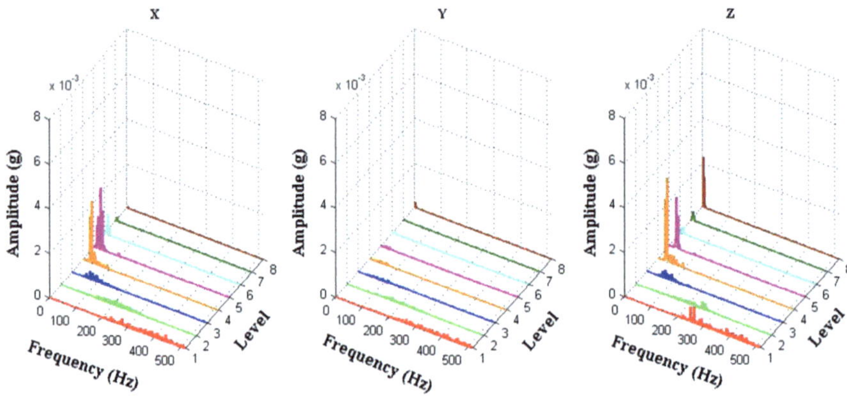

(a) Healthy robot

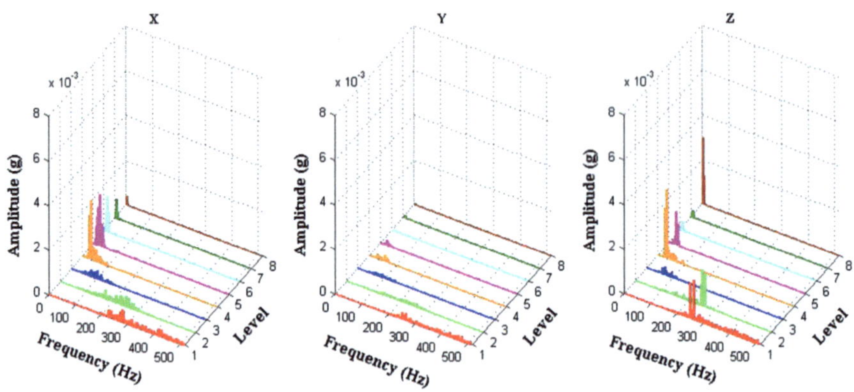

(b) 1mm inner race bearing fault

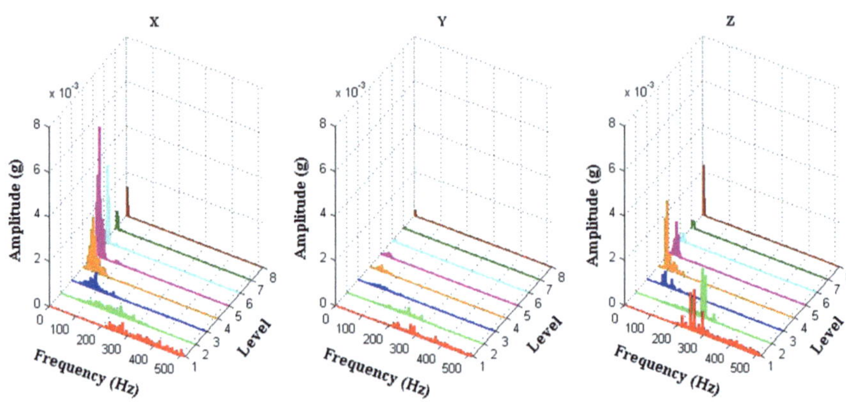

(c) 1mm outer race bearing fault

Fig. 6.19 Frequency spectrum of each detail signal in the X, Y and Z axes in healthy and different bearing faults cases

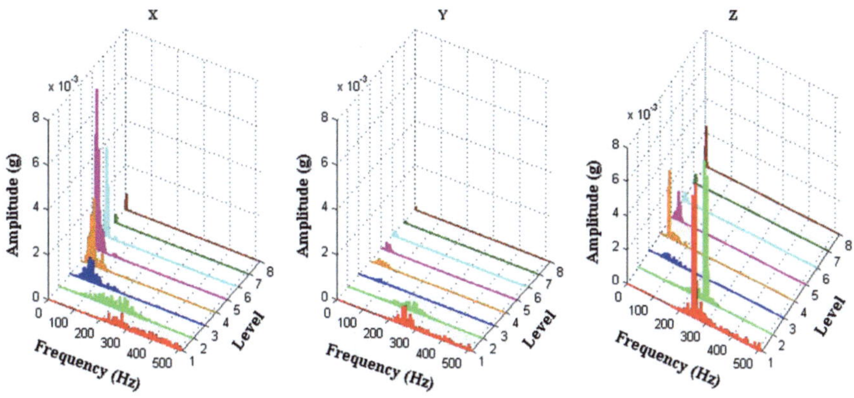

(d) 2mm outer race bearing fault

Fig. 6.19 (continued)

tooth cut gets deeper. These levels are specifically affected at the bevel gear mesh frequency (95.7 Hz), at which the impact is happening, and its harmonics, which are situated within the frequency bands of D1–D3, as illustrated in Table 6.2. In the case of bearing faults (Fig. 6.19), in addition to the variations in the spectra of level 4 and 5 in the X-axis detail signals, a significant difference from the healthy case can be observed in level 6 spectrum. Moreover, detail signals that contain higher frequency bands, such as D1 and D2 show considerable change with the presence of a fault especially in the Z axis. Whereas the variations in the spectra of levels 4, 5 and 6 could be attributed to the BPFI and BPFO of the bearing, which are located in the frequency band of level 5 (16.1–32.21 Hz); whilst the changes in the higher levels, or any other levels, could also be related to either bearings and gears harmonics or to the excited natural frequencies of the robot.

Figure 6.20 shows enlarged images of the X-axis spectra in level 5, extracted from Fig. 6.19. It can be seen that the 4th and 15th harmonics of the input and idler shafts respectively are present in the robot healthy condition and in all bearing fault conditions. The presence of these harmonics can be reasoned to the intended misalignment between the shaft transforming the power from the joint 3 motor and gearbox input shaft. Also, due to the eccentric design of the housings of the input and idler shafts of joint 3 gearbox. Just the mentioned harmonics are detected; however, the frequency itself or its higher or lower harmonics may be recognized in the spectra of higher or lower levels. Interestingly, the 6th natural frequency of the robot (35 Hz) has appeared in all three bearing faults, therefore it can be considered as one of the bearing defect indicators; hence, other natural frequencies presented in other levels could also be used as indicators.

When the inner race bearing fault is simulated in the robot, the BPFI of the faulty bearing (30.86 Hz) is easily recognized in the related spectrum (Fig. 6.20b). Also, in this spectrum the 9th harmonic of the FTF (2.94 Hz) is detected, although it is not detected in the other fault conditions. The 3rd natural frequency of the robot

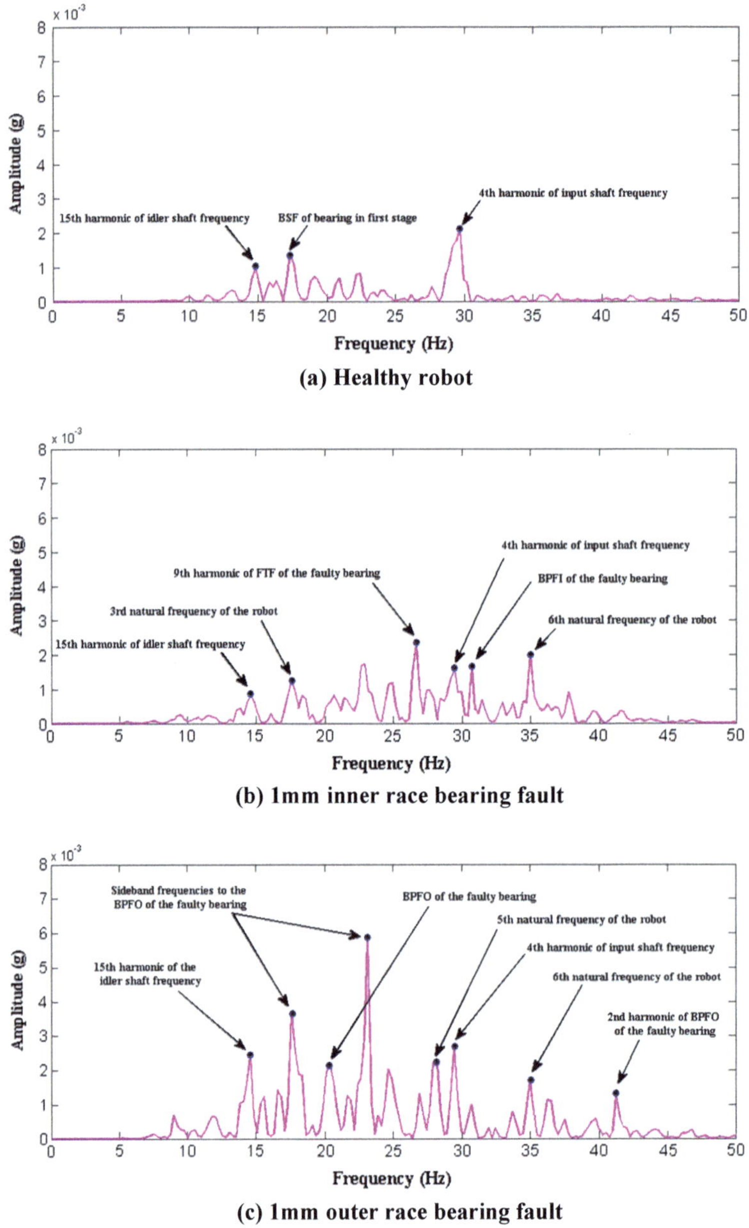

Fig. 6.20 Level 5 and in X-axis frequency spectrum in healthy and different bearing faults cases

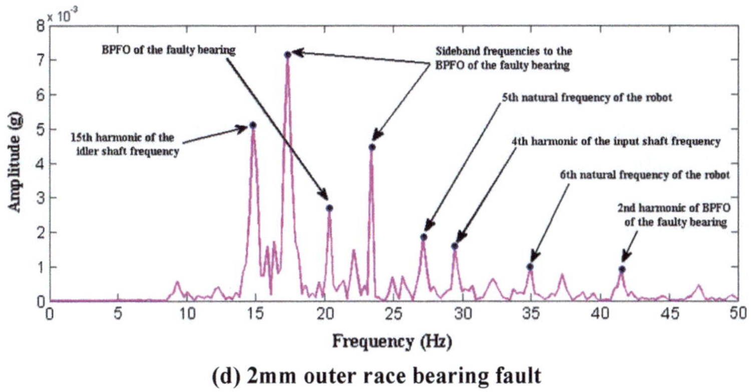

Fig. 6.20 (continued)

(17.19 Hz) has also been identified. Again, in addition to the BPFI of the faulty bearing, the 3rd robot natural frequency can be used as an indicator for inner race fault for the bearing in the input shaft. If the bearing fault is in its outer race, the BPFO will emerge in the spectrum, as observed in Fig. 6.20c, d at 20.62 Hz, with two sideband frequencies. The amplitude of BPFO is increased a little when the hole diameter in the outer race is increased, and the left sideband and the 15th harmonic of the idler shaft are substantially amplified, reflecting the fault harshness. In addition to the detection of the sideband frequencies, the 2nd harmonic of BPFO is also detected in both outer race fault severities. In term of the robot natural frequencies the 5th natural frequency (27.37 Hz), along with the 6th, has appeared with outer race faults; thus, it can be employed as an indicator for outer race fault for the bearing in the input shaft.

Whilst this study not previously been reported on a robot, but on rotating machines, it was initially thought that it would not be possible to detect frequencies related to bearing and gears of the robot, primarily because the non-stationarity nature of the robot vibration signals and secondly the lack of the available information regarding the robot gearboxes (see Chap. 5). However, after repeated practicing of dismantling/assembling the robot joint 3 all required information related to the gears and bearings was acquired. Then, by applying the combined DWT-FFT technique the evidence suggested that this technique can be utilized for monitoring and detecting changes in frequencies related to the different mechanical components of the robot. However, this technique might be not suitable for online fault detection, since the stimulated frequencies in each level need to be carefully examined and may lead to fault identification taking longer time and much effort. As a result, for the purpose of designing an intelligent condition based monitoring (iCBM) system for the robot based on an embedded system, the route of extracting statistical features from the detail signals produced from DWT analysis will be followed; since these features have shown clear changes when faults are developed, making it possible for the ANN classifier to categorize them after training (Chap. 7).

6.5 Summary

The PUMA 560 robot has been programed to execute a handling task that mimics one of its real tasks for number of times and with different faults seeded on joint 3 (elbow). Vibration signals from the three axes accelerometer were captured while the robot was performing this task. These signals were analysed and statistical features were extracted from them. The signals standard deviations (STDs) were used to distinguish between the healthy and faulty robot conditions, as it was found to be the best fault sensitive feature. The obtained STDs from the healthy signals will be utilized in the forthcoming chapter to establish the threshold limits that can be applied in order to realise just the abnormality development in the robot without the fault's location and type.

In the second stage, which is the fault diagnosis stage, the robot has been programed to oscillate just joint 3, where the faults were simulated, over its full range of movement. Also, three axes vibration measurement and analysis was carried out, but this time using the discrete wavelet transform (DWT). DWT represents an efficient time-frequency analysis method. By applying wavelet analysis, the signal can be analysed down to its sub-band frequencies. An eight level wavelet analysis was performed on each axis signal and the STD was computed from each of the produced sub-signals. The extracted STD features from the signals in the three axes will be employed, after concatenating them, as an input vector to the neural network classifier for fault diagnosis in the next chapter.

Another technique for industrial robot fault detection and diagnosis was also investigated in this chapter. It is based on combining the capabilities of the DWT and the fast Fourier transform (FFT). FFT is a useful technique for analysing the spectral content of a stationary signal, but it is inefficient in analysing non-stationary and transitory fault signals, which is the case in industrial robots. Wavelet analysis represents an efficient method of non-stationary signal analysis. Thus, by considering the advantages of these two signal analysis techniques, a wavelet–Fourier signal analysis system has been applied. As a result, the proposed technique has shown its ability in detecting changes in frequencies related to the natural frequencies and different mechanical components of the robot. However, it will not be applied using the embedded system. The technique based on feature extraction from the vibration signals will be invested alongside the embedded system.

References

Abo-Zahhad, M. (2011). ECG signal compression using discrete wavelet transform. *Discrete Wavelet Transforms-Theory and Applications*, 143.

Bicker, R., Daadbin, A., & Rosinski, J. (1989). The monitoring of vibration in industrial robots. In *ASME 12th Biennial Conference on Mechanical Vibration and Noise*.

Datta, A., Mavroidis, C., Krishnasamy, J., & Hosek, M. (2007). Neural network based fault diagnostics of industrial robots using wavelet multi-resolution analysis. In *American Control Conference, USA* (pp. 1858–1863).

Kar, C., & Mohanty, A. R. (2006). Monitoring gear vibrations through motor current signature analysis and wavelet transform. *Mechanical Systems and Signal Processing, 20*, 158–187.

Mohanty, A. R. (2015). Machinery condition monitoring: Principles and practices. Abingdon: Taylor & Francis Group.

Rodriguez-Donate, C., Morales-Velazquez, L., Osornio-Rios, R. A., Herrera-Ruiz, G., & Romero-Troncoso, R. J. (2010). FPGA-based fused smart sensor for dynamic and vibration parameter extraction in industrial robot links. *Sensors, 10*, 4114–4129.

Chapter 7
Intelligent Condition Monitoring System Design

Abstract As explained previously that the condition monitoring system being developed consists of two stages, namely fault detection and fault diagnosis. These two stages will be designed based on statistical (control charts) and artificial intelligence (neural network) techniques. This chapter discusses the main principles of the used techniques, describes the design steps for the proposed intelligent CM system, and presents the results of its performance testing. The extracted standard deviation features, in Chap. 6, that are related to the robot healthy and different fault conditions will be used here for design and testing of these two stages.

7.1 Fault Detection Stage

The fault detection stage will be applied when the industrial robot is carrying out its scheduled task, such as the pick and place task (Chap. 6), and a statistical control chart (SCC) technique is used for fault detection. This stage is used to warn the maintenance operative if any degradation in the robot's health is developing. In this section the concept of SCC and how the control limits can be estimated are first explained, and then the normality requirement in the resultant STD values extracted in Chap. 6, which will be used to construct a control chart for robot fault detection, is elaborated. Selection of an appropriate control chart and its design steps and performance testing are also discussed.

7.1.1 Principle of Control Charts

Control charts are a statistical tool for graphically displaying a *quality* characteristic plotted against the sample number (or time) with a centre line and two (upper and lower) control limits (Montgomery and Runger 2014), as illustrated in Fig. 7.1. Control charts represent one of the most important and widely applied methods for detection of abnormal process operations based on process variables such as

A.A. Jaber, *Design of an Intelligent Embedded System for Condition Monitoring of an Industrial Robot*, Springer Theses,
DOI 10.1007/978-3-319-44932-6_7

dimensions, temperature, vibration and forces etc. They provide a clear differentiation between changes that are a result of unpredictable disturbances in the system and changes that occur as a result of a system fault.

When control charts are being used for health monitoring of any machine, despite how well it is designed and maintained, some amount of inherent or natural variability will always be presented (Kisić et al. 2013). This natural variability, which is usually referenced to as background noise, is the cumulative effect of many small and unavoidable causes. However, from the statistical quality control point of view, a system that has this natural variability is often called a stable system and in-control (Kisić et al. 2013). On the other hand, if the variability in the monitored parameter is large enough when compared to the background noise, this is said due to assignable causes and represents an unacceptable level of system performance. In this case, the sources of variability are not part of the natural causes and commonly entitled *special causes*; a system that is operating in the presence of special causes is said to be an out-of-control.

The conventional control chart is composed of three horizontal lines; a centre line (*CL*) that represents the average value of the quality characteristic corresponding to the in-control state; and two other lines named the upper control limit (*UCL*) and the lower control limit (*LCL*) as shown in Fig. 7.1. These control limits are chosen so that if the process is in-control, nearly all of the sample points will fall between them. It is common practise to connect the sample points on the control chart with straight-line segments since it is easier to visualize how the sequence of points has evolved over time. However, even if all the points fall inside the control limits, but behave in a systematic or non-random manner, then this could be an indication that the process is out of control. If the process is in control, the plotted points should fall in an essentially random pattern.

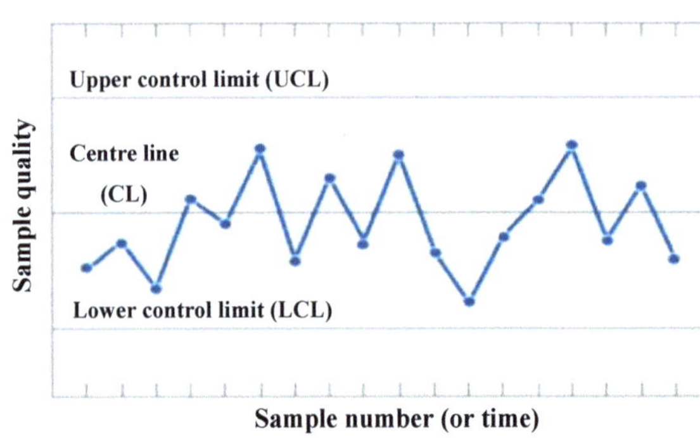

Fig. 7.1 A conventional control chart (Montgomery and Runger 2014)

7.1.2 Control Chart Limits

The choice of control limits is one of the critical decisions to be made when designing a control chart, and is connected to the risk of faulty prediction of being in or out of control. Fundamentally, there is a close similarity between the principles of the control chart and hypothesis testing. By applying the control chart technique, continuous hypothesis testing is carried out through the progress of the process. For a normally distributed process, the null hypothesis (H_0) is that the treated sample mean is equal to the reference (healthy) process mean, while the alternative hypothesis (H_1) assumes that the treated sample mean does not equal the reference mean. The H_0 hypothesis refers to that the monitored parameter is in a good condition and the process should continue, whereas H_1 indicates that there is a problem and actions should be taken. However, two types of error can occur during the fault prediction process using control charts (or hypothesis testing), these are called *type I* and *type II* errors (Montgomery and Runger 2014; Kisić et al. 2013). A type *I* error occurs when a point falls outside the control limits, indicating an out-of-control condition when there is no obvious cause presented (reject the null hypothesis H_0 when it is true). Whereas a type *II* error results if a point falls within the control limits when the process is actually out of control (fail to reject the null hypothesis H_0 when it is false). Therefore, the risk of type *I* error is decreased by moving the control limits further from the centre line. Widening the control limits, however, will also increase the risk of type *II* error. However, the opposite effect occurs if the control limits are shifted closer to the centre line. The general mathematical formulation for a control chart is

$$CL = \mu_w \tag{7.1}$$

$$UCL = \mu_w + k\sigma_w \tag{7.2}$$

$$LCL = \mu_w - k\sigma_w \tag{7.3}$$

where W is a sample statistic that measures some quality characteristic of interest, and μ_w and σ_w are its mean and standard deviation, respectively, while k represents the distance of the control limits from the CL, expressed in standard deviation units. The common practice is to make the UCL and LCL equal three standard deviations (three-sigma) from the CL of the monitored data, i.e. $k = 3$. By using three-sigma control limits it is assumed that the distribution of the quality characteristics is approximately normally distributed. Then, by doing so it is presumed that while the system is in statistical control, nearly 99.7 % of the points will fall within the control limits (Montgomery and Runger 2014; Khan 2013). In this way, a good balance is made between type *I* and *II* errors. In addition to the three-sigma limits, additional warning limits can be utilized. These are named the inner limits, usually placed at two-sigma (El-Din et al. 2006).

7.1.3 Normality Distribution Test

The fundamental approach upon which the control chart algorithms depend assumes that the data are normally distributed; thus, before constructing the control chart, it is required to test the normality of the feature of interest extracted when the system is healthy (Montgomery and Runger 2014; Khan 2013; El-Din et al. 2006). In this project the normality of the healthy features that are extracted from the robot while it executes the repetitive task in Chap. 6 were studied using graphical techniques to assess whether or not the data are consistent with the normal distribution. These are the histogram, which is a bar graph plot of data and represents an approximation to a probability density function, and the normal probability plot. The vertical axis of a histogram represents the data type frequency, while in the normal probability plot it signifies the cumulative probabilities scale. Meanwhile the horizontal axes in both plots are the values of the variable. At the centre of the normal probability plot, a line of normal probability is drawn passing through the mean of the variable and the 50 % cumulative probability. If most of the data points are placed on this line, the data is considered to be normally distributed. However, if the points appear in a curvature shape, then the indication is that the data are not normally distributed. To achieve this and to calculate the upper and lower control limits, as will be explained later, the Minitab 17 statistical package has been used. By carrying out the normality test on all the features extracted from the robot healthy condition, such as the resultant mean and kurtosis, has been found that the resultant standard deviation (STD) feature is the only normally distributed one, as shown in Fig. 7.2.

Generally, the histogram compares the data to the normal probability curve generated from the mean and standard deviation of the data, as shown in Fig. 7.2 (left), which is indicating a bell-shape distribution. The majority of the data points in the normal probability plot, Fig. 7.2 (right), correlate with a straight line except for a few at either end, moreover, the normal distributed line pass through the intersection of the mean and its 50 % cumulative probability. Thus, it can be concluded that the resultant STD feature is a good approximation to a normal distribution, thus it will be used for designing the control chart.

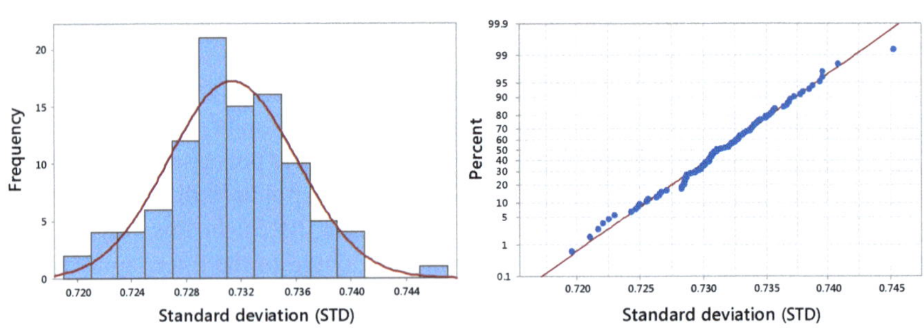

Fig. 7.2 Histogram (*left*) and normal probability (*right*) plots for the resultant standard deviation feature extracted from the robot healthy condition

7.1.4 Control Chart Design

As mentioned in the literature review (Chap. 2) the most commonly used control charts are: Shewhart, cumulative sum (CUSUM), exponentially weighted moving average (EWMA), which are called univariate charts, and T-square control chart, which is one of multivariate control charts. Shewhart control charts are the most popular charts used for process monitoring, and can be easily implemented on microcontrollers since they do not require a lot of computational effort; and for this reason will be used here, based on the following discussion.

It is a standard practice when a variable is monitored using Shewhart charts that both the mean and the variability of the variable are considered. The mean of a variable is monitored with the X-bar ($\overline{X}$) chart and the variability of the variable is monitored using a range chart (R-chart) or a standard deviation chart (S-chart). The X-bar chart informs whether the process is stable with respect to its healthy level, whereas the R- and S-chart provide information regarding the variability of the process and if it is stable over time or not. Significant shifting of the mean and the unusual large variability are indications of special causes or a fault starting to develop in the robot, and which subsequently have to be evaluated using the second stage of the condition monitoring system (Fault diagnosis stage). The R-chart is relatively insensitive to small or moderate shifts for small sample size (Montgomery and Runger 2014), thus, in a situation that demands tight control of process variability, moderately large sample sizes will be required, and the S-chart should be used.

The control limits of the mean and standard deviation charts are derived from the healthy features, using simple calculations, after dividing each data set into subgroups (samples) of ten observations each. The population mean ($\overline{\overline{x}}$) and standard deviation ($\overline{s}$) can then be estimated based on these data, having m preliminary samples, each of size n, with ith sample mean ($\overline{x}_i$) and standard deviation (s_i), as shown:

$$\overline{\overline{x}} = \frac{1}{m}\sum_{i=1}^{m} \overline{x}_i \tag{7.4}$$

$$\overline{s} = \frac{1}{m}\sum_{i=1}^{m} s_i \tag{7.5}$$

The upper and lower control limits and centre line of $\overline{X}$ chart are given by:

$$UCL_x = \overline{\overline{x}} + 3\frac{\overline{s}}{c_4\sqrt{n}} \tag{7.6}$$

$$CL_x = \overline{\overline{x}} \tag{7.7}$$

$$LCL_x = \bar{\bar{x}} - 3\frac{\bar{s}}{c_4\sqrt{n}} \tag{7.8}$$

And the upper and lower control limits and centre line of S-chart are:

$$UCL_s = \bar{s} + 3\frac{\bar{s}}{c_4}\sqrt{1 - c_4^2} \tag{7.9}$$

$$CL_s = \bar{s} \tag{7.10}$$

$$LCL_s = \bar{s} - 3\frac{\bar{s}}{c_4}\sqrt{1 - c_4^2} \tag{7.11}$$

where c_4 is a tabulated constant depends on the sample size and can be found in (Montgomery and Runger 2014). Minitab results for the S-charts, based on the resultant STD data from the healthy robot are presented in Fig. 7.3. It can be seen from the charts that all the points are randomly distributed and within the control limits which means the robot is healthy. The established values of control limits will be incorporated in the designed embedded system in the following chapter for online robot fault detection.

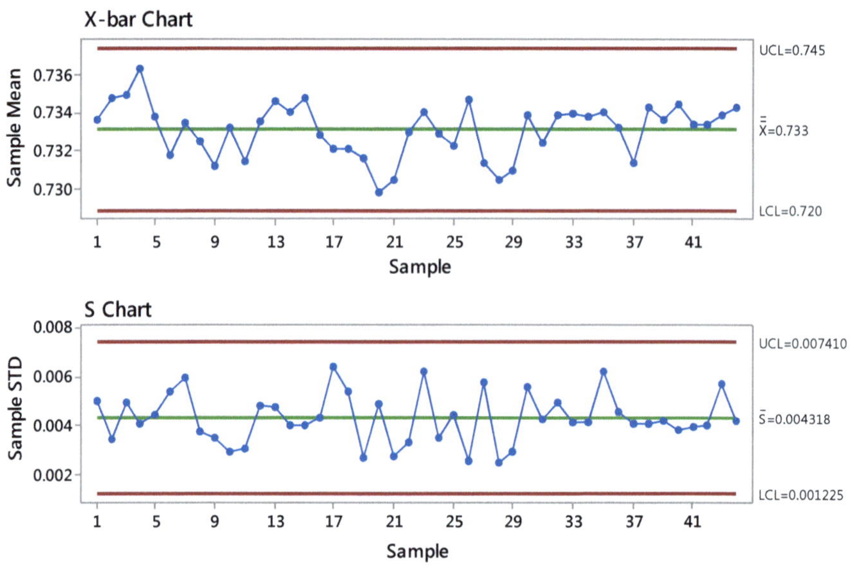

Fig. 7.3 Designed Xbar-S charts

7.1.5 Performance Testing of the Designed Control Charts

Before using the established control charts limits with the embedded system it is important to perform offline testing to confirm that they are able to distinguish between a healthy and a faulty robot. To achieve this first a number of healthy samples have been captured from the accelerometers when the robot executes its programmed sequence. These samples are previously unseen when the control charts first designed. Because it is difficult to introduce a fault in the robot while it is running and also it not recommended allowing the robot working for a long time until it is degraded, different controlled backlash levels have been introduced at periodic intervals when the robot is stopped. First, the backlash is increased (clearance) and then reduced gradually until it is removed totally (interference). With each backlash level the robot is programmed to execute the repetitive task, described in Chap. 6, a number of times and the resultant standard deviation of each cycle signal calculated using LabVIEW programme. The extracted features for all backlash levels (and healthy state) are tested against the control limits and plotted on the same graph, for clarity, using Minitab software as shown in Fig. 7.4.

The red colour markers indicate out-of-control samples. From these figures significant differences amongst the backlash levels compared to the healthy state of the robot can be noticed, particularly on the means of the standard deviation ($\bar{\bar{x}}$) in the X-bar chart, however these differences cannot be distinguished clearly in S-chart graph which monitor the variability of the standard deviation. Accordingly it is concluded that it is justifiable to rely on only the X-bar chart for the robot fault detection, whereas the S-chart will be ignored when the embedded system code is constructed. From these charts it can clearly be seen that most features related to robot's unhealthy state either exceed the control limits, as in the high backlash and interference cases, or on the same side above or below the centre line, as in case of small backlash; while, the healthy features are randomly distributed. Also, as

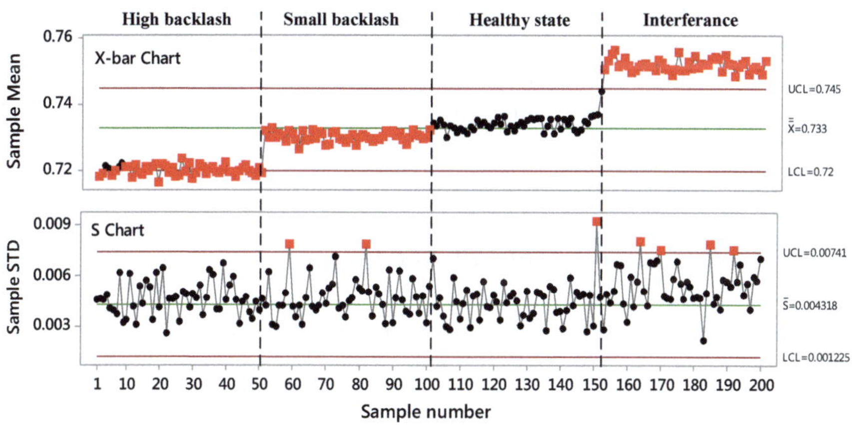

Fig. 7.4 Testing the performance of Xbar-S charts for fault detection under different backlash conditions

indicated by X-bar charts, when the backlash level is increased the means of standard deviation of the samples are decreased. In contrast, the opposite occurs when the interference is introduced to the mating gears. This is attributed to the grease lubricant in the transmission system, as discussed previously in Chap. 6.

There are ten rule-of-thumb associated with control charts used to signal the presence of an abnormality in the process, called the Western Electric Company rules, or run rules (Montgomery and Runger 2014). These rules also have negative effect if all of them are applied, since they greatly increase the number of false alarms (Montgomery and Runger 2014; El-Din et al. 2006). Thus, few of them are going to be utilized here for the robot fault detection. Such as the robot is considered in abnormal condition if three (or more) successive points fall outside the control limits or the points are in a non-random arrangement. The non-random arrangement, for example, is characterised by a run of eight consecutive points on one side of the centre line or six points in a row steadily increasing or decreasing. However, these rules can be applied as warning alarms before serious fault is taken place. This type of test can be used by the robots manufacturers for checking the backlash level before putting the robots in the service. Also, the designed chart has been tested with other types of faults, such as bearings and gears, and very good performance has been shown.

7.2 Fault Diagnosis Stage

After the fault and its severity are detected and estimated in the first stage of the monitoring system, its type and location need to be identified for the operator and maintenance engineers. To achieve this, an artificial neural network (ANN) will be designed using the features extracted in the previous chapter when the robot cyclically runs just joint 3. This section will first present the main idea of the ANN, and then the design steps are discussed. In addition to its accuracy in fault diagnosis, the size of the designed ANN is the most important considered parameter, since the designed ANN in this chapter has to be uploaded onto an embedded system with limited memory size and computational effort. Consequently, features related to different health cases of the robot will be used to evaluate the accuracy of the designed ANN.

7.2.1 Biological and Artificial Neurons

The structure of the artificial neuron is inspired by the concept of a biological neuron shown in Fig. 7.5. Basically, it is a processing element in the nervous system of the brain that receives and combines signals from other similar neurons through thousands of input paths called dendrites.

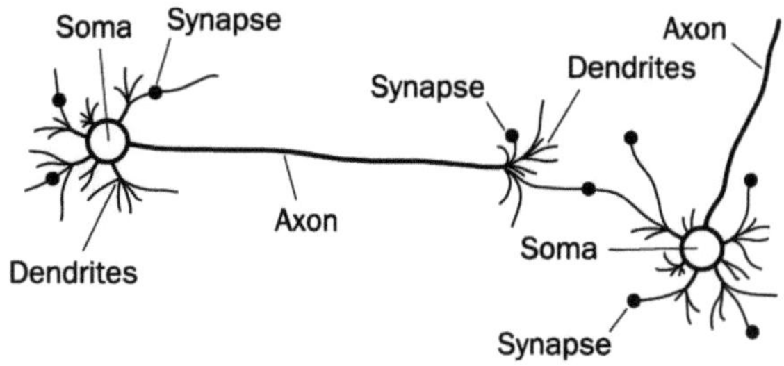

Fig. 7.5 Structure of the biological neuron (Negnevitsky 2005)

Each input signal (electrical in nature), flowing through a dendrite, propagates through a synapse or synaptic junction. The junction is an infinitesimal small gap in the dendrite which is filled with a neurotransmitter fluid that either retards or accelerates the propagation of the signal. Signals from many dendrites are accumulated in the soma (or nucleus), and then nonlinearly modified at the output before flowing to other neurons through the branches of axon as illustrated in Fig. 7.5.

The model of an artificial neuron closely matches the biological neuron and has a summation type of structure that consists of several layers of artificial neurons designed to emulate the biological neurons. Each input signal (continuous variable or discrete pulses) flows through a gain or weight, called a synaptic weight or connection strength and can be positive or negative, integer or non-integer. Figure 7.6 illustrates a typical representation of an artificial neuron, with connection weights.

Where $x_1, x_2, x_3 \ldots x_n$ represent the input vector to the neuron with associated weights $w_1, w_2, w_3 \ldots w_n$. The bias, b, (sometimes called threshold) is often connected to the neuron and it introduces an offset to the transfer function so that even

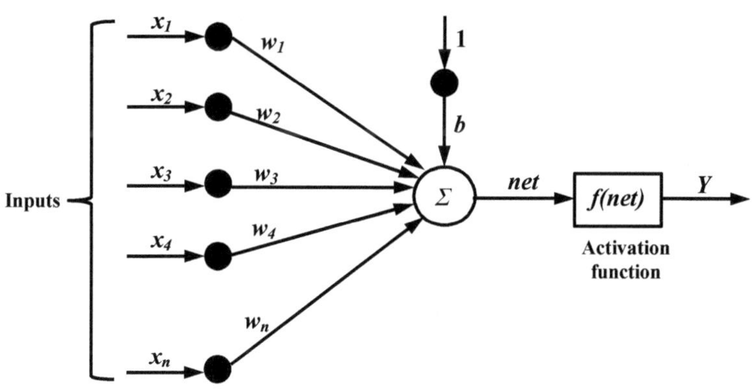

Fig. 7.6 Artificial neuron with connection weights

if the input were zero the neurons would still have an output (Lim 2009). Typically, the bias is set to a value of 1. However, the input to the neuron, which is called *net*, will be the summation of the multiplied inputs with their corresponding weights, and can be written as

$$net = \sum_{i=1}^{n} w_i x_i = w_1 x_1 + w_2 x_2 + w_3 x_3 + \cdots + w_n x_n + b \tag{7.12}$$

where n is the number of inputs to the neuron. The resultant net is then passed through an activation function ($f(net)$) to produce the output which can be represented as

$$Y = f(net) = f\left(\sum_{i=1}^{n} w_i x_i\right) \tag{7.13}$$

7.2.2 Activation Functions

The internal inputs sum of the perceptron has to pass through an activation function, sometimes called the transfer function, which can be linear or nonlinear. If a linear function is used, that will not contribute to a non-linear transformation within a layered structure. However, utilizing the non-linear activation functions allows mapping of non-linear input-output of the neural network. Table 7.1 summarizes and describes the most common types of activation functions.

The *sigmoidal* and *tansigmoidal* functions are commonly used in systems that are nonlinear, such as fault detection systems. Both of these functions are differentiable. As discussed earlier, the nonlinear activation function contributes to the nonlinear transfer characteristics of a neuron which permits nonlinear input-output mapping of ANN. However, with linear activation function, this nonlinearity is lost.

7.2.3 Neural Network Architectures

A neural network is composed of a large number of simple and interconnected computational neurons. These neurons are connected by links that have associated weights. The weights between neurons refer to the significance of the input to each neuron. Normally, either raw data or output from neurons represents the input to other neurons. However, neural networks come in many forms and accordingly there is no generally accepted definition, and the structure of the connections

Table 7.1 Activation functions types

Name	Graphical representation	Mathematical representation	Description
Linear function	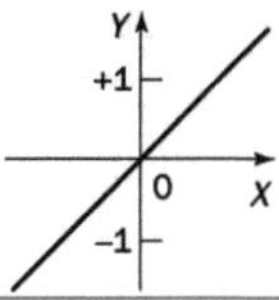	$Y = \text{net}$	The output is equal to the neuron weighted input
Step	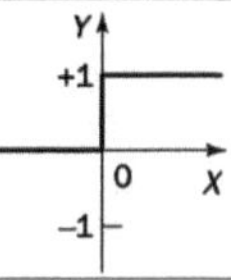	$Y = \begin{cases} 1, \text{If net} \geq 0 \\ 0, \text{If net} < 0 \end{cases}$	In these cases, the output is hard limited depending on the sign of the net
Sign	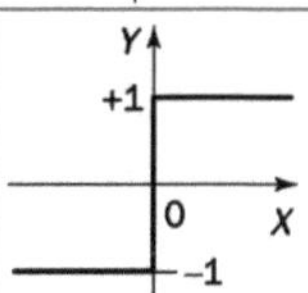	$Y = \begin{cases} +1, \text{If net} \geq 0 \\ -1, \text{If net} < 0 \end{cases}$	
Sigmoid	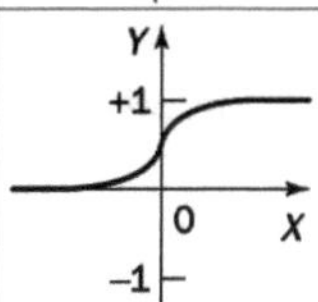	$Y = \frac{1}{1 + e^{-\text{net}}}$	The net neuron will be mapped into values between +1 and 0
Tansigmoidal	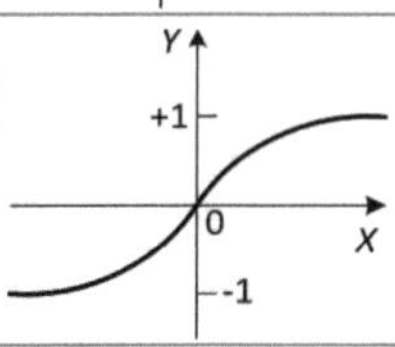	$Y = \tanh(\text{net}) = \frac{1 - e^{(-\text{net})}}{1 + e^{(-\text{net})}}$	The net neuron will be mapped into values between +1 and −1

between neurons and the computation it performs vary between the different neural models. ANNs are generally classified as feed-forward, e.g. the multilayer perceptron neural network (MLPN), and feedback (or recurrent) types, such as Hopfield network, as discussed in the literature review chapter (Chap. 2). In the feed-forward class, the signals travel only in the direction from the input to the output, whereas in recurrent neural network (RNN), the signals can flow in the forward as well as backward or lateral direction. However, the main characteristic of a neural network is that, when the training process is performed well, the mapping formed by the network can show its capability for generalization beyond the training data and not to memorise the training data. The majority of fault diagnosis applications utilizing ANNs utilize a feed-forward architecture, with the MLPN being most popular. A description of this topology is presented in the next section as it adopted for use in this thesis.

7.2.4 Multilayer Perceptron Neural Network

In the multilayer perceptron network (MLPN) or multilayer feed-forward neural network, the input signal is propagated on a layer-by-layer basis through the network in a forward direction, as illustrated in Fig. 7.7, and there are no connections between the neurons in the same layer. The data flows into the network through the input layer, passes through one or more hidden layer(s) and finally flows out of the network through the output layer. Although the network classification will be better when a large number of layers and neurons are used, increased layers and neurons result in a convergence problem and a longer training time (Negnevitsky 2005).

The MLPN is a nonlinear model consisting of number of neurons organized into multiple layers, forming a mapping between the input and the output, adjusted by the weights. The complexity of the MLPN network can be changed from an almost linear model to a highly nonlinear model by varying the number of layers, the number of neurons in each layer, and the values of the weights. The network thus has a simple interpretation as a form of input-output model. Parameter estimation of the model is carried out with a training algorithm, which uses the training data to gradually optimize the parameters (weights and biases) in the MLPN network. The training algorithm is implemented using a search method to estimate the values for the MLPN parameters.

7.2.5 Learning Techniques

The neural network mimics the human brain in learning through training and data storage. Basically, there are three main groups of learning or training algorithms.

Supervised learning: In supervised learning the system will be informed about the exact output vector (Munakata 2008; Negnevitsky 2005), and the difference between the network output and correct (desired) output is calculated as shown in

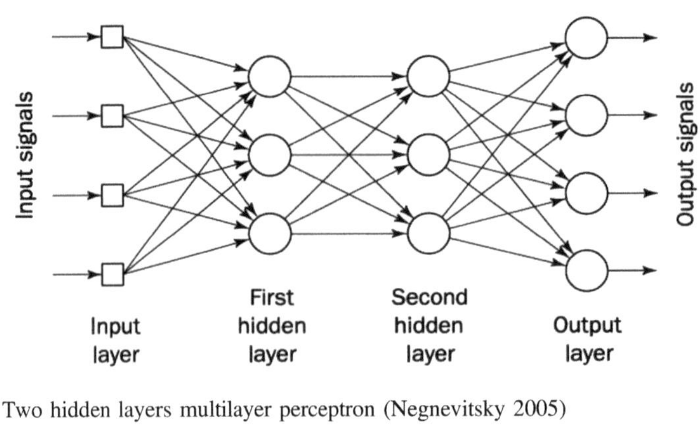

Fig. 7.7 Two hidden layers multilayer perceptron (Negnevitsky 2005)

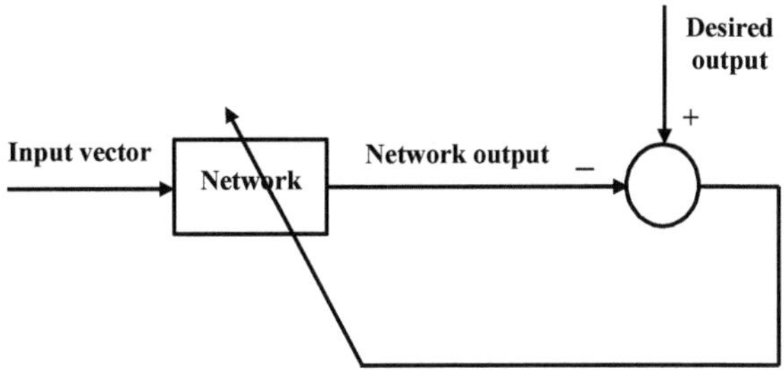

Fig. 7.8 Supervised learning scheme

Fig. 7.8. The weights are changed according to a specific formula. This method can be compared to learning under the supervision of a teacher, who knows what contents have to be learned to get the desired output. In this technique the weights and biases of the network are initially chosen arbitrarily, and then during the training procedure these being updated to reduce the differences between the desired and estimated output. Tasks that fall within the paradigm of supervised learning are pattern recognition, which is also known as classification, and regression (or function approximation).

Unsupervised learning: It is also called self-supervised learning, and in this technique the desired final vector is not identified, but instead the network's weights and biases are updated in response to the network input only (Munakata 2008; Negnevitsky 2005). In the training process, a number of different input patterns are fed to the neural network, and the network discovers significant features in these patterns and learns how to classify them into appropriate categories. Figure 7.9 shows the principle of the unsupervised learning algorithm.

Reinforcement learning: In this technique, a teacher though available, does not present the expected answer but only designates if the computed output is correct or not (Munakata 2008; Negnevitsky 2005). The delivered information helps the network in its learning process, and the computed correct answer will be rewarded and a penalty applied to the incorrect answer.

The supervised learning method is used for the intended application, as both the input (standard deviation features extracted from the wavelet details) and output (desired target) training sets are available.

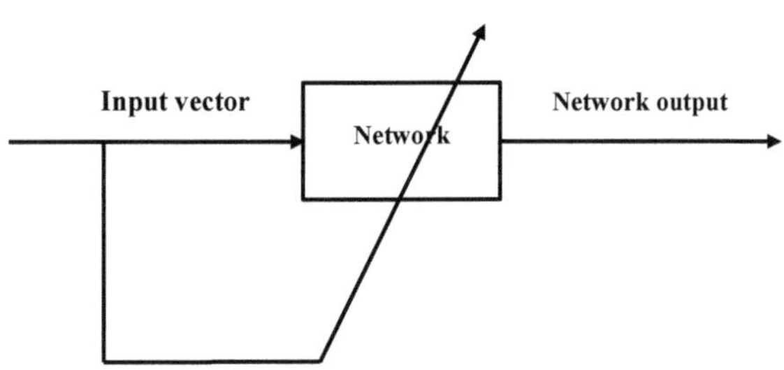

Fig. 7.9 Unsupervised learning scheme

7.2.6 Training the MLPN

The back-propagation training algorithm takes its name from its procedure and represents the most commonly used algorithm. The training can be accomplished when the input and target data pairs are presented to the network which implies a supervised learning process. The computations in the back-propagation training algorithm are passed forward from the input to output layer, and the computed errors are fed backwards through the network, as shown in Fig. 7.10.

The main steps of back-propagation training of a MLPN can be summarised as in the following (Negnevitsky 2005). First, the weights and biases are randomly initialised in order to start the training process. Then, the network takes the input vector of features $x = [x_1, x_2, \ldots x_n]$ and computes the output $y = [y_1, y_2, \ldots y_l]$ and compares it with the target output $t = [t_1, t_2, \ldots t_l]$ to calculate the error (E). This error signal is then back propagated in order to adjust the network's weights and improve its performance, such that the error decreases. Back-propagation training, based on a gradient descent method for numerical optimization, is the most widely used method for feed forward networks. It uses the computed error for updating the weights of the network. There are several forms for error measurements e.g. summed squared error (SSE) (Fernando 2014). The most commonly used one is the mean square error (MSE), which is the SSE of all targets minus outputs, divided by the number of outputs, which is used here:

$$E = MSE = \frac{SSE}{l} = \frac{1}{l}\sum_{k=1}^{l}(t_k - y_k)^2 \tag{7.14}$$

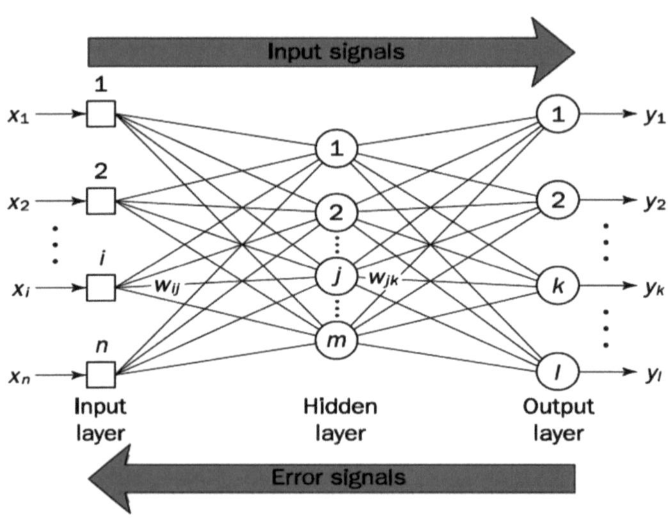

Fig. 7.10 An example of back-propagation training algorithm (Negnevitsky 2005)

where l is the total number of output neurons. After the error being calculated, each weight parameter is updated by an amount proportional to the negative gradient of the error measured with respect to that weight parameter (Eq. 7.15). This process starts at the output layer of the network and works backwards to the hidden layer and then to the first layer.

$$\Delta W_{jk} = -\eta \frac{\partial E}{\partial W_{jk}} \quad (7.15)$$

The factor η is called the learning rate which is a positive constant less than unity and determines the magnitude of the change in weight values (Negnevitsky 2005), and must be selected before starting the training. A very small learning rate leads to a very slow learning process. In contrast, if the learning rate is made larger, the resulting increased changes in the weights may cause instability giving rise to an oscillating behaviour (Negnevitsky 2005). It is worth mentioning that the frequency of weight updating can be performed in incremental mode or batch mode (Fernando 2014). In incremental mode, training sets are presented to the ANN one at a time, and the error is calculated for each input set and back-propagated to adjust and update the weights before the next set is passed. While in the batch mode, all training patterns are presented to the ANN, and all produced errors are accumulated and averaged over the entire training patterns. By using this averaged error, the weights are then updated. This means the updating process of weights occur only after each epoch, and the training process stops when the gradient of the MSE stops decreasing. However, in the training process it is not possible to get the MSE to decrease to exactly zero, so the objective is to get it down below some minimum value, usually ($MSE_{min} < 10^{-2}$) (Mazumdar 2006). If the evaluation criteria satisfy the specifications, the training will be stopped. In contrast, if the error is not low enough, the weights keep modifying until the error satisfies the requirements or the maximum number of iterations are achieved.

Since the inception of back-propagation learning technique there were a number of improvements suggested on improving the speed of convergence (Fernando 2014). Some of these techniques, for example, the conjugate gradient or Levenberg-Marquardt techniques are numerical optimization methods; however, they follow the back-propagation process to update the weights of the network, and thus are still referred to as back-propagation techniques. The used data for training the ANN normally comes in a different range; therefore normalisation of the data is needed to avoid high values from being too dominant and to suppress the influence of the smaller ones (Subbaraj and Kannapiran 2014; Pandya et al. 2012), this also will prevent weights from becoming too large which can lead numerical overflow. So, if all of the input data is normalized to be between 0 and 1 or −1 and 1, then the ANN will give equal priority to all inputs. In this thesis the data will be normalized between 0 and 1 using the following equation:

$$x_n = \frac{x - x_{min}}{x_{max} - x_{min}} \quad (7.16)$$

where, x_n is the normalized value, and x_{min} and x_{max} are the minimum and maximum values among all values of the data. After the training is completed and new input data sets have to be evaluated, which have to be normalised too as the network has been trained in this way. Also, the output results from the network need to be denormalised using the inverse of the normalization function.

7.2.7 Setting Up the MLPN

To model the nonlinear characteristics of the robot fault conditions, as in the work here, there are some issues regarding the number of hidden layers, number of neurons in each layer, and also the activation functions in the hidden and output layers need to be considered. This also because the designed neural network will be programmed on the designed embedded system (Chap. 8). However, the number of input neurons depends on the number of measurements or the number of features extracted from each sensor signal. In this work, the input features to the network are the standard deviation values of the detail signals of the wavelet analysis (D1–D8) extracted in the previous chapter, since the last level of approximation coefficient (A8) does not give useful information regarding robot fault (also as explained in the previous chapter). Therefore, the number of features will be 8 from each sensor (axis) signal with a total of 24 values (from the three axes) input to the network. The number of the robot health conditions to be classified, of which there are ten fault types will be explained in the following section, identify the required number of neurons in the output layer. Consequently, the number of MLPN input and output neurons are fixed.

For any nonlinear system identification problem, at least one hidden layer is required (Negnevitsky 2005). Generally, the computational effort and time of the network increases with increasing the number of layers and neurons in each layers of the network. Likewise, if the network size is too large, it will be difficult to programme on the embedded system since it requires significant amount of memory. Therefore, a trade-off between the computational effort and efficiency of the neural network and the number of the hidden layers is required. Thus, one hidden layer is proposed. Additionally, a nonlinear, differentiable activation function for the hidden layer is needed (Mazumdar 2006). For this purpose, the sigmoidal activation function has been utilized, since it is suitable for applications whose desired output is between 0 and 1, which is the case in this research (Pandya et al. 2012). Also, a linear function, which is normally used in the input and output layers, was used in the output layer. The only variable then remaining is the number of neurons in the hidden layer. The appropriate selection of the number of hidden neurons is based on a balance between output accuracy and network size.

Generally, there is no explicit mathematical or theoretical foundation to determine the best number of hidden layers and neurons in each hidden layer without training several networks and estimating the error of each network. Having too many neurons in the hidden layer may result in a low training error but again the network size and computation time will increase. There are several books and articles that offer a "rule of thumb" to select the optimal number of hidden neurons in the hidden layer to be a starting point, for example:

- The size of the hidden layer should be somewhere between the input layer size and the output layer size (Heaton 2008).
- The number of neurons in the hidden layer should never be more than ×2 the number of neurons in the input layer (Heaton 2008).
- The number of hidden neurons should be 2/3 the size of the input layer, plus the size of the output layer (Heaton 2008).
- Number of neurons in the hidden layer should equal 2× input neurons plus 1 (Sivanandam et al. 2006).

Also, another possibility is by starting with a small number of neurons and gradually increasing them until little or no improvement is observed in the network performance. It is possible to establish the optimal number of neurons in the hidden layer as explained in the following section.

7.2.8 Design of a MLPN for Fault Diagnosis

Matlab's neural network toolbox was utilized to design and test the network, based on the supervised, back-propagation learning technique. Many functions are provided by the toolbox for designing, training, visualizing and simulating the proposed neural networks, and numerous built-in functions for common neural network applications such as pattern recognition and clustering are included. Furthermore, the code for these functions can easily be modified to suit specific requirements. The block diagram in Fig. 7.11 shows the main design steps for the ANN. Some of these steps are adjusted manually before running the prepared Matlab code, such as identifying the network structure, activation functions and number of epoch, whereas others are performed automatically based on the code sequence, i.e. data loading and normalization, and weights and biases initialization.

The neural network input feature vectors consisting of one group representing the healthy state as well as ten groups for different fault types. These groups include three backlash levels: high, small and interference; pinion bevel gear faults: tooth wear, 25 % tooth removed, 50 % tooth removed and full tooth removed; and input shaft bearing faults: inner race fault, and light (1 mm hole) and severe (2 mm hole) faults in the outer race. The data sets sizes of 100 samples for each health condition are used to train the network. After setting the network structure and loading the data, the weights and biases of the network are initialized depending on Matlab's

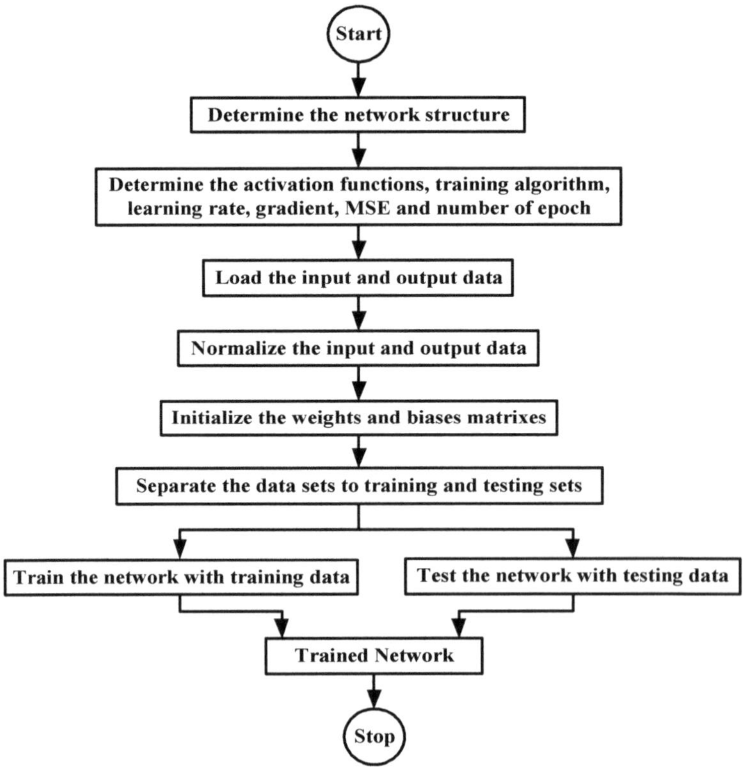

Fig. 7.11 Block diagram of ANN design steps

random number generation function. The function *dividerand* was used to divide the data into three subsets; training, validation and testing sets with ratios of 70, 15 and 15 %, respectively. The parameters used for the network are shown in Table 7.2. Additionally, the training can stop according to any one of the criteria of mean square error (MSE), the number of epochs of training or minimum performance gradient reaches certain value set as shown in Table 7.2.

However, as mentioned in Sect. 7.2.6 the performance of the MLP network can be enhanced if an appropriate back-propagation training function is selected. The neural network toolbox provides many back-propagation training algorithms, with their corresponding Matlab functions are listed in Table 7.3. There are a lot of mathematical equations and discussion behind them and will not be reflected here, since many books and papers have covered them in detail. The literature claims that the Levenberg-Marquardt algorithm is the best training method in the neural network toolbox (Fernando 2014; Gadoue 2009; Ba-Razzouk et al. 1997). Therefore, before finding the optimal number of neurons in the hidden layer, it was decided to training the network on the basis of 24-25-11 neurons architecture against every back-propagation algorithm to identify the best one. In each case, the MSE of the

Table 7.2 The used parameters for designing the neural network

Number of input layer neurons	24
Number of hidden layer neurons	To be identified
Number of output layer neurons	11
Number of hidden layers	1
Hidden layer activation function	Sigmoid
Output layer activation function	Linear
Training algorithm	To be identified
Learning rate	0.05
MSE stopping criteria	10e−4
Minimum performance gradient	10e−5
Maximum number of epoch	50,000

Table 7.3 List of the tested back-propagation training algorithms

Algorithm	Function in matlab
Gradient descent	traingd
One step secant	trainoss
Polak-Ribiére conjugate gradient	traincgp
Fletcher-Powell conjugate gradient	traincgf
Conjugate gradient with Powell/Beale restarts	traincgb
Scaled conjugate gradient	trainscg
BFGS quasi-Newton	trainbfg
Resilient back-propagation	trainrp
Levenberg-Marquardt	trainlm
Bayesian regularization	trainbr
Gradient descent with adaptive learning rate	traingda
Variable learning rate gradient descent	traingdx
And gradient descent with momentum	traingdm

test set has been taken as a performance evaluation criterion as illustrated in Fig. 7.12.

The results show strong evidence in the superiority of *trainbfg* (BFGS quasi-Newton), *trainlm* (Levenberg-Marquardt) algorithms, followed by *traincgp* (Polak-Ribiére Conjugate Gradient), with the *trainbr, traingda, traingdx and traingdm* algorithms performing the worst. As the BFGS Quasi-Newton performs the best amongst all other algorithms, it will be selected for designing the neural network.

Now, to find the optimal number of hidden neurons trial and error testing is performed. However, based on the number of input and output neurons and the rules of thumb presented in Sect. 7.2.7, the starting number of neurons of 11–24 (rule 1), less than 48 (rule 2), 27 (rule 3) and 49 (rule 4) are adopted at the beginning of training. Also, it is tried to avoid the large number of neurons due to

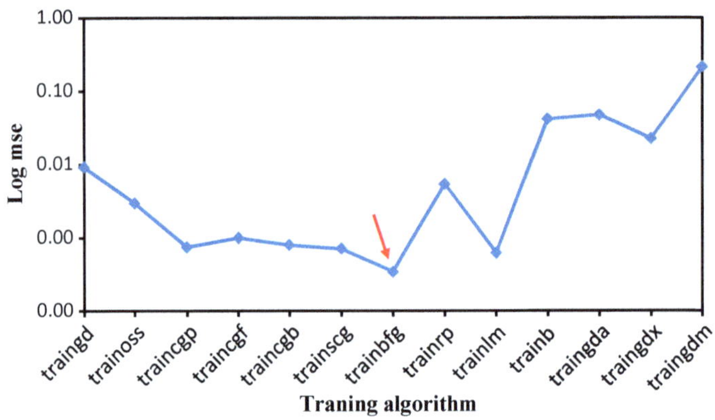

Fig. 7.12 Variation of the MSE for different training algorithms

the limitation that might be faced in the embedded system as discussed earlier in this chapter. Therefore, after many trials it was found the optimum number of neurons that gives a compromise solution for the robot fault diagnosis problem is 17. The ANN learning process and the performance plot is depicted in Fig. 7.13, which shows that the training with 487 epochs met the MSE stopping criteria (MSE less than 10E-4). Furthermore, it was established that the correlation coefficient (R) between the actual and desired (target) outputs has a value above 0.99 for the training, testing and validation data sets, which is indicative of a strong relationship between the outputs and targets of ANN. Thus, this network will be implemented for robot fault diagnosis stage.

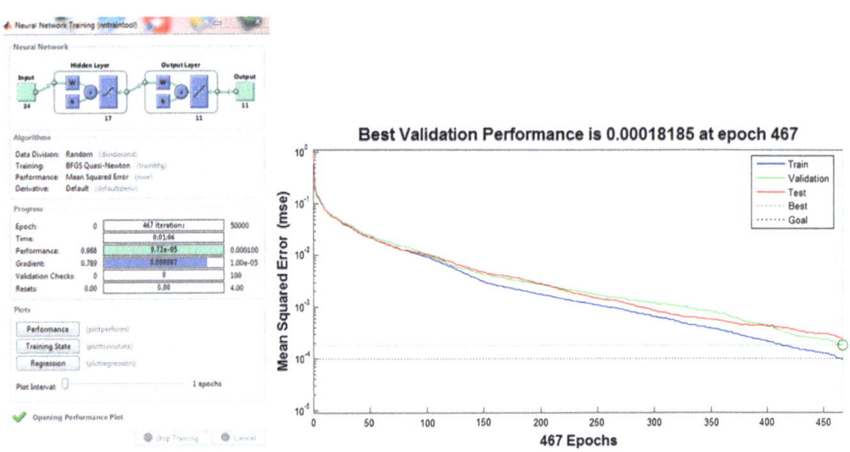

Fig. 7.13 The ANN training process and performance plot

7.2.9 Performance Testing of the Designed ANN

With the ANN's design satisfying the requirements and the training completed, its performance needs to be tested using sets of data not previously presented to the network, and however, the well-trained network must be able to show its capability

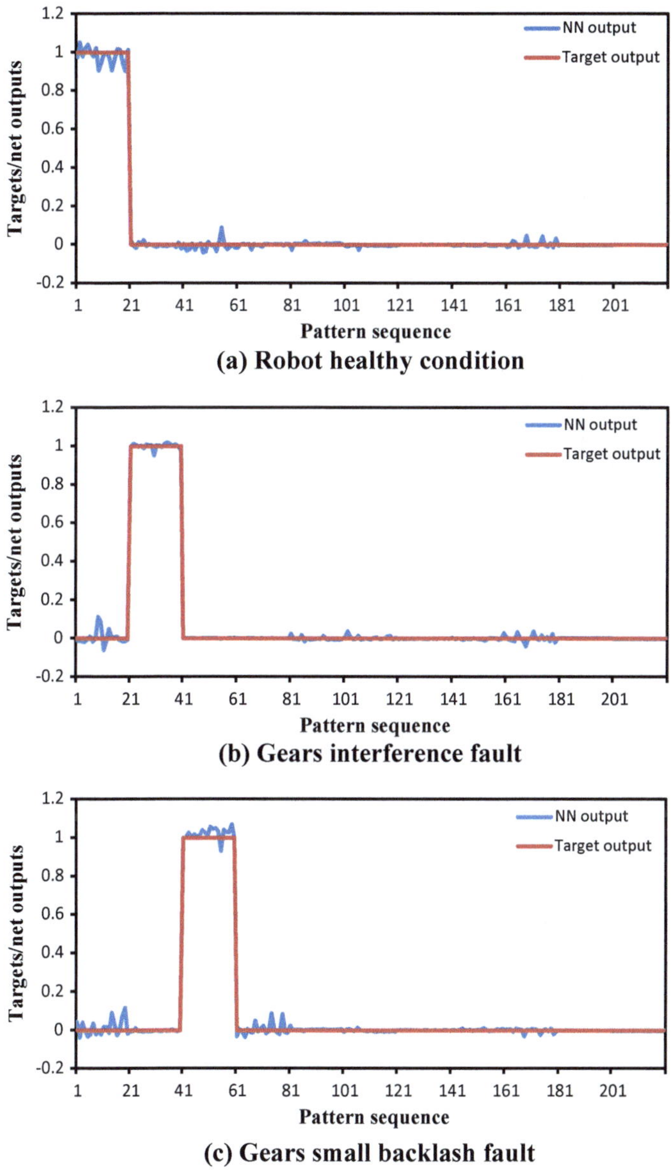

Fig. 7.14 The designed ANN network performance testing with unseen healthy and faulty data

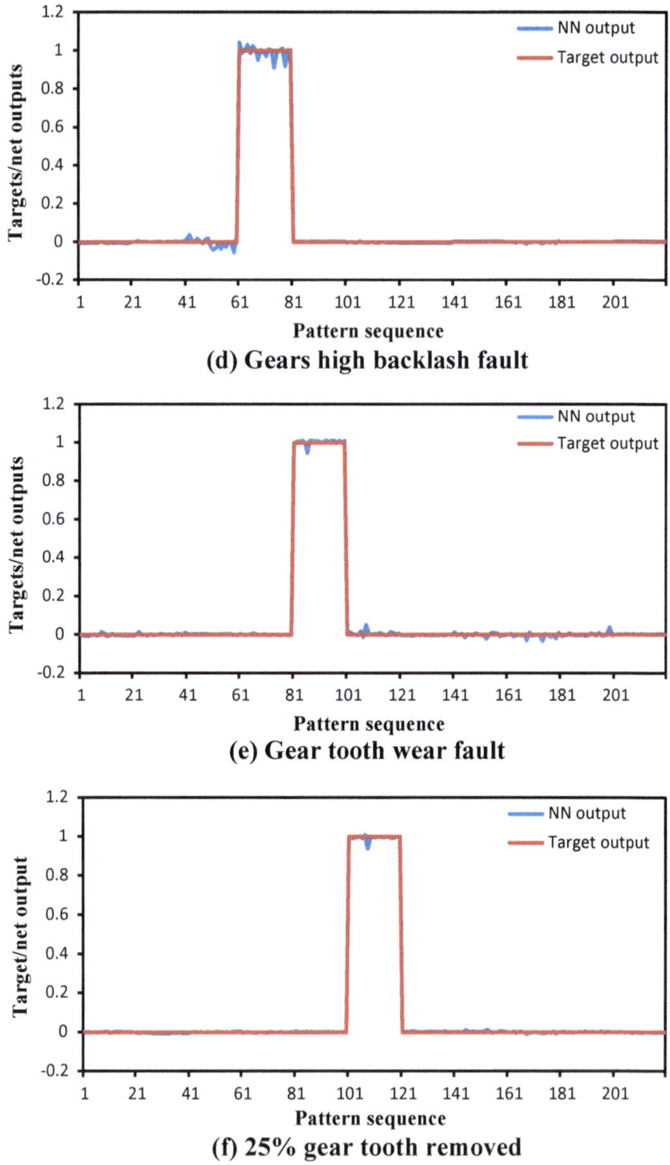

Fig. 7.14 (continued)

in classifying the unseen data samples. The ANN has been trained in a way that produced results from the first neuron in the output layer of the ANN representing the healthy operating condition of the robot, while the remaining ten neurons (of the eleventh) are mapped to represent the ten simulated faults on the robot, and thus a value of 1 for each output neuron is considered the target value. For instance, for the healthy case, the only output neuron that would have a value of 1 is the first one, therefore, it will be [1,0,0,0,0,0,0,0,0,0,0]. The second neuron would take a value of 1 if the first fault type (gear interference) is presented and thus the outputs will be [0,1,0,0,0,0,0,0,0,0,0], and so on with the other fault types.

To test the network, a total of 220 unseen data samples were used. The first 20 samples represent the healthy condition of the robot while the remainder of the samples are divided into ten groups of 20 samples each, corresponding to different robot faults. A selection of the test results are presented in Fig. 7.14, whereas the rest can be found in Appendix D. The tests showed that the network efficiency in term of its capability in classifying the eleven different types of the robot health situation is 100 %, and the ANN design can differentiate the different faults of the robot with very good accuracy when confronted with unseen data.

7.3 Summary

The development of the intelligent CM system has been presented in this chapter. The system is divided into two main stages; the fault detection and fault diagnosis stages. The fault detection stage is based on SCC while the diagnosis stage is on ANN. The extracted features (from Chap. 6), using the designed data acquisition system based on LabVIEW software, from different operating condition of the robot were utilized to design the intelligent CM system. Minitab software has been used to establish the control (threshold) limits in the fault detection stage, and then to test its ability to detect variations in the robot health. The SCC showed very good capability in detecting changes.

To design an appropriate ANN and then evaluate its performance for the fault diagnosis in the second stage, code has been developed based on Matlab neural network toolbox. The ANN has been trained to distinguish among different types of fault in the robot. A significant level of accuracy in fault diagnosis the ANN has been obtained and the percentage of correct classification was approximately 100 %. The designed CM system will then be written in other programming languages, C language for example, to be downloaded on an embedded system. This will be discussed in the coming chapters.

References

Ba-Razzouk, A., Chériti, A., Olivier, G., & Sicard, P. (1997). Field-oriented control of induction motors using neural-network decouplers. *IEEE Transactions on Power Electronics, 12*, 752–763.

El-Din, M. A. S., Rashed, H. I., & El-Khabeery, M. M. (2006). Statistical process control charts applied to steelmaking quality improvement. *International Journal of Quality Technology and Quantitative Management, 3*, 473–491.

Fernando, H. A. (2014). *Artificial neural networks for fault detection and identification on an automated assembly machine*. Masters of Applied Science Master Thesis, Queen's University.

Gadoue, S. M. (2009). *Artificial intelligence applied to speed sensorless induction motor drives*. PhD Thesis, Newcastle University.

Heaton, J. (2008). *The number of hidden layers* [Online]. http://www.heatonresearch.com/node/707

Khan, R. M. (2013). *Problem solving and data analysis using minitab*. New York: Wiley.

Kisić, E., Petrović, V., Jakovljević, M., & Đurović, Ž. (2013). Fault detection in electric power systems based on control charts. *Serbian Journal of Electrical Engineering, 10*, 73–90.

Lim, W. L. (2009). *The application of artificial neural networks for sensor validation in diesel engine condition monitoring and fault diagnosis*. M. Phil., University of Newcastle upon Tyne.

Mazumdar, J. (2006). *System and method for determining harmonic contributions from nonlinear loads in power systems*. Ph.D Thesis, Georgia Institute of Technology.

Montgomery, D. C., & Runger, G. C. (2014). *Applied statistics and probability for engineers*. New York: Wiley.

Munakata, T. (2008). *Fundamentals of the new artificial intelligence*. London: Springer.

Negnevitsky, M. (2005). *Artificial intelligence: A guide to intelligent systems*. Boston: Addison Wesley.

Pandya, D., Upadhyay, S., & Harsha, S. (2012). Ann based fault diagnosis of rolling element bearing using time-frequency domain feature. *International Journal of Engineering Science and Technology, 4*, 2878–2886.

Sivanandam, S. N., Sumathi, S., & Deepa, S. N. (2006). Introduction to neural networks using MATLAB 6.0. New York: McGraw Hill.

Subbaraj, P., & Kannapiran, B. (2014). Fault detection and diagnosis of pneumatic valve using adaptive neuro-fuzzy inference system approach. *Applied Soft Computing Journal, 19*, 362–371.

Chapter 8
Embedded System Design

Abstract The previous chapter concentrated on the design of the fault detection and diagnosis system based on SCC and ANN respectively; based on the extracted features described in Chap. 6. In this chapter the implementation of the proposed robot health monitoring system using an embedded system is discussed. A short overview of the currently available embedded systems for condition monitoring applications is provided, to evaluate their appropriateness for the work here. Then, a systematic methodology has been followed for the electronic hardware, which makes up the embedded monitoring system, selection process by investigating the available options in the market and suitability for this project. The configuration of several parameters of the developed embedded system is discussed and results presented, and the final system enclosure and installation described.

8.1 Rationale

As discussed in the previous chapters the proposed intelligent condition monitoring algorithm is made up of two stages, namely fault detection and fault diagnosis. The first stage incorporates SCC for fault detection, while diagnosis stage utilizes the DWT for signal analysis followed by statistical feature extraction; the extracted features are then passed to the ANN classifier for fault classification. A wired data acquisition system together with signal analysis and feature extraction software was initially developed based on National Instrument hardware and software, constructed and tested on a PC. With some modification the designed system could be easily adopted for health monitoring of different robots or machines. However, the main aim of this work is to develop an embedded system for real-time implementation of the proposed CM algorithm.

By thoroughly examining the proposed CM system it can be seen that the key steps are: signal acquisition, feature extraction for fault detection stage, fault detection by applying the SCC technique, conducting the signal analysis using DWT and also feature capture for the fault diagnosis stage, and finally performing the fault classification using ANN. Each one of these steps requires a substantial

A.A. Jaber, *Design of an Intelligent Embedded System for Condition Monitoring of an Industrial Robot*, Springer Theses,
DOI 10.1007/978-3-319-44932-6_8

computational effort, particularly the DWT and ANN, since they involve a large numbers of mathematical operations. For this reason and also due to the non-availability of high performance, low-cost, compact size processor boards it is not feasible to implement the full two-stage system on a single processor board. Therefore, it was decided to apply the first stage data acquisition and fault detection utilizing a small, low-cost board that can be attached to and move with the robot arm, while implementation of the fault diagnosis stage (DWT and ANN) on another high-performance board remote from the robot arm and interfaced to a PC for reporting the robot health condition. The communication between these two boards was established wirelessly; this has raised the concept of wireless sensor network (WSN) in this project. The board on the robot together with the three accelerometers is called the end device (or the sensor node), whereas the board connected to the PC is called the base station. This chapter discusses the requirements and design for building a wireless embedded system, and component selection.

8.2 Available Wireless Sensor Network Modules

Wireless sensor networks (WSNs) are comprised of a scalable number of distributable, lightweight wireless modules termed nodes. Each node is equipped with sensing, computing, and communication capability. The nodes are organized in the network with one configured as a coordinator (or base station) (Faludi 2011). The coordinator represents the network controller and is responsible for setting up and maintaining the network, which should be connected to a reliable, uninterrupted power source, and requires to be interfaced to a PC for further data processing and visualization. Another configuration of the wireless devices is a router which can send, receive and route the information; networks may have multiple router nodes, which act as a messenger for communications between other widely separated devices. An end device is the third possible configuration of the nodes, which always needs a router or coordinator to communicate with. It cannot work as a messenger between any other devices and can only transmit data to the devices that is connected to. Generally, wireless network can be composed of one coordinator, multiple end devices, and no routers.

Different wireless modules serve different purposes are available in the market, including the TelosB and Waspmote modules, shown in Fig. 8.1. These modules are required to be low cost and low power consumption, and are designed with limited memory size and restricted computational capabilities (Feng et al. 2015), i.e. the Waspmote module has only 8 KB (kilo bytes) random access memory (RAM), 128 KB program flash memory and an 8-bit processor running at 8 MHz. These modules are not powerful enough for real-time implementation of digital signal processing or intelligent algorithms, such as DWT and ANN; thus, it was decided to develop wireless modules capable of handling the tasks of the proposed condition monitoring system (fault detection and diagnosis).

(a) TelosB module [http://www.willow.co.uk] (b) Waspmote module [http://www.libelium.com]

Fig. 8.1 Some examples of the available wireless sensor modules

8.3 Wireless Sensor Network Topologies

A wireless network topology indicates how the wireless sensor nodes are linked to each other and to the base station/coordinator. There are four main types of WSN topologies, as shown in Fig. 8.2 (Faludi 2011). The simplest one is the *Pair* topology, which basically has two nodes where one node must be a coordinator while the other one is configured to work as an *end device or a router*. The second network configuration is *Star* topology, in this network many end nodes communicate with a central coordinator and do not communicate with each other directly. The coordinator routes the received messages between the nodes as requested.

A *Mesh* network arrangement consists of one coordinator node, and in addition to end nodes there are number of router nodes. Also, the end nodes can be connected to the coordinator or routers and the routers can communicate with each other as well as with the coordinator. The last topology is *Cluster tree* network which is similar to mesh network. Here, the end nodes are clustered around the coordinator or each router and the routers do not communicate with each other; instead, they pass the messages along to the coordinator and the coordinator sends them as required. In this research, the topology best suited is Pair network, since only one sensor node and base station are used for robot monitoring; for future work may be more than one sensor node can be employed for health monitoring of more than a robot simultaneously and in that case the Star, or other networks should be utilized.

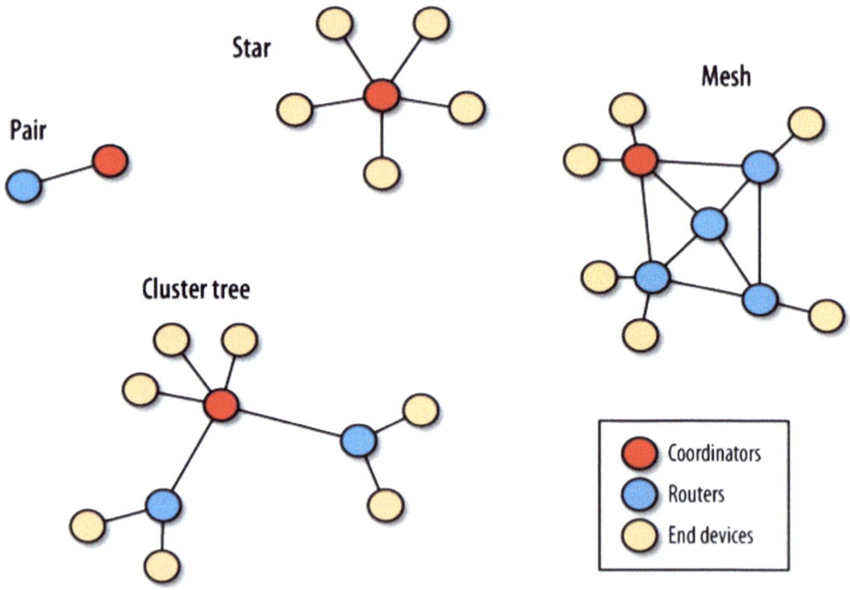

Fig. 8.2 Wireless network topologies (Faludi 2011)

8.4 Design of the Sensor Node (End Device)

The sensor node, as depicted in Fig. 8.3, is an embedded system usually outfitted with a microcontroller unit (MCU) [or digital signal processor (DSP)], radio frequency (RF) transceiver, power supply, and memory along with various sensors and actuator depending on the application. The microcontroller with memory to store the data acts as the central computing and controlling device of the sensor node. The radio transceiver module, which allows two-way radio communication between several nodes in order to distribute the information, represents the communication subsystem of the node. The power supply subsystem is in charge of powering the whole sensor node, and is normally composed of batteries, which offer an easily available, low cost, and high capacity source of power, and have become companion with sensor nodes.

Fig. 8.3 Wireless sensor node hardware architecture (Yang 2014)

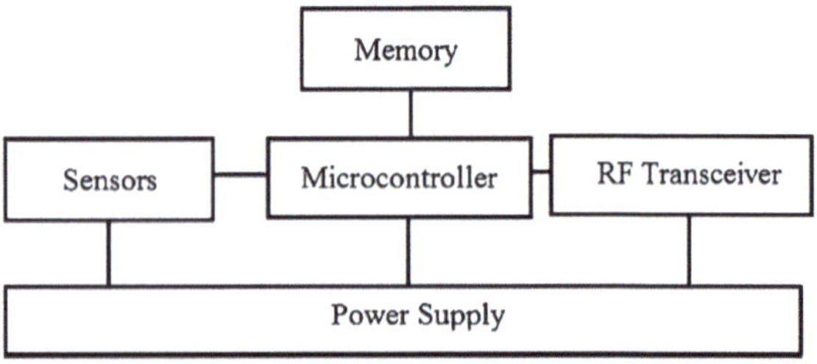

In this work, one node will be fixed on the robot arm and will be in charge of performing the first stage of the proposed monitoring system, which is collecting vibration signals, execute preliminary evaluation of the robot health (fault detection), and then sending the information to the base station. So, in addition to good computational capabilities, the node has to be of light weight and small size, easily attached and does not add extra loading on the robot arm. Powering of the node is one of the most critical points in building any wireless node, since from practical point of view it is not convenient to replace batteries on the nodes of a WSN at short intervals. The node has to be of a reasonable price and easy to produce, for future expansion of the embedded wireless CM system. The previously selected accelerometers in Chap. 5 represent the sensor part of the sensor node, and in this section the selection of the microcontroller, wireless protocol, transceiver and the power supply source of the node is discussed. However, the proposed design methodology is not limited for industrial robots and could be utilized for health monitoring of different machines.

8.4.1 Microcontroller Selection

A microcontroller unit (MCU) is a programmable device that combines electronics components, include central processing unit (CPU), memory and peripheral devices, integrated into a single chip (Emilio 2015). There are two families of microcontrollers: CISC (Complex Instruction Set Computer) and RISC (Reduced Instruction Set Computer). CISC is characterized by a single bus between the CPU and memory, which means a single memory for both data and instructions, and is commonly known as Von Neuman architecture. Because of this, it involves more than one machine cycle for the execution of each instruction; this constitutes the disadvantage of this family. RISC microcontrollers, on the other hand, are the Harvard architecture, which provides the program and data memory physically separated from each other, two separate buses between the CPU and data-memory, and between the CPU and program memory.

The microcontroller to be utilized in a sensor node is responsible for the acquisition, processing, compression, recording and storage of data. There are a considerable number of microcontroller boards produced by different companies available, and key requirements that need to be taken into a count when selecting a microcontroller for constructing a sensing node are energy consumption, cost, processing speed, memory size, physical size and support for peripherals. The selected microcontroller must have a fast, high resolution analogue to digital converter (ADC) with at least three input channels, one for each of the three single axis MEMS accelerometers. The sampling frequency of the ADC has to be fast enough to cover the highest sampling frequency used for capturing the vibration signals from the accelerometers in the fault diagnosis stage, which was 1031 Hz. So, in order to select an appropriate microcontroller to be used in the sensor node, a comprehensive evaluation of the available products has been performed. Several microcontroller families from, for example, Microchip, Arduino, and Texas Instruments companies have been assessed. The Arduino was found to be well suited for this project, as it meets the requirements and also has a large community support which makes the development process simpler.

Arduino offer several microcontroller models with different characteristics. The main differences between these models are the type of processor, the number of inputs and outputs ports, and the capacity of memory. Arduino controllers are relatively efficient, consume less power, cheap, and suitable for use in a harsh environment. To reduce the number of available Arduino options it was decided to look at Arduino boards that have high specification and meet the desired requirements, such as DUE and Intel Galileo, which is a new board designed to be compatible with Arduino hardware and software and is based on Intel architecture. The Intel Galileo board was thoroughly tested, but despite the high functionality it was established that the signal capturing process takes very long time in order to capture 4096 samples from one analogue input channel; as a result was rejected. The Arduino DUE board was tested and found to be more than capable for this work in terms of signal capturing speed, processing speed, memory size, power consumption and, of course, cost.

The Arduino DUE (Fig. 8.4) is an open-source, single-board microcontroller based on a 32-bit, RISC, Atmel SAM3X8E ARM Cortex-M3 processor. It offers a relatively small size form, measuring 101.6 × 53.3 mm, and compatibility with most of the standard Arduino shields. The Arduino DUE board has an 84 MHz clock frequency, USB connection, four UARTs (Universal asynchronous receiver/transmitter) serial ports, a power jack, a reset button, and an erase button. There is a 12-bit resolution analogue to digital convertor (ADC) built in inside the processor with 1 MSPS (mega samples per second) sampling frequency and 12 input channels. The board comes with 512 KB flash memory, and 96 KB of

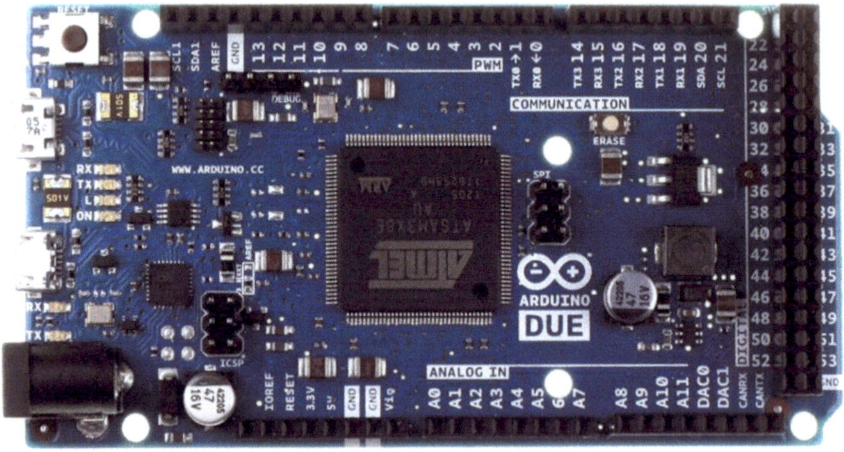

Fig. 8.4 Arduino DUE microcontroller board (https://www.arduino.cc)

SRAM. Whilst other Arduino boards accept up to 5 V on I/O pins, the Arduino Due board is based on 3.3 V on the I/O pins, fortunately, the selected accelerometer can work using 3.3 or 5 V which makes the connection of the accelerometer to the board does not require any conditioning circuit.

8.4.2 Wireless Technology Selection

Communication devices are used to exchange data between the nodes of wireless networks. The advances in wireless applications have led companies to develop different types of wireless standard. These standards are usually classified by their capabilities and properties, and designed to suit different applications, such as fault detection or human health monitoring. In this section, three widely used wireless network technologies are discussed, in order to investigate their pros and cons and also to select the appropriate technology for the work here. These technologies are (Giannoulis et al. 2013):

IEEE802.11x

The IEEE802.11x family of standards are meant for wireless local area network (WLAN), which is also known as Wireless Fidelity (Wi-Fi). There are four generations of Wi-Fi products available which are IEEE 802.11a/b/g/n, operate in high frequency, unlicensed Industrial, Scientific, and Medical (ISM) radio bands ranging from 2.4 to 5 GHz. Typically, it is adapted for relatively high bandwidth and high data transfer rate, ranges from as low as 1 MBPS (megabyte per second) to over 50 MBPS, and commonly used for mobile computing devices, such as laptops. With the use of a standard antenna the transmission range can be up to 300 ft, and it can be significantly improved by utilizing a directional high gain antenna. Although the data transmission and rate ranges are enough for wireless sensor network application, the power requirement generally limited its usage in wireless sensor application.

Bluetooth (IEEE802.15.1)

Bluetooth is lower power than IEEE802.11, has an operating frequency 2.4 GHz within the ISM band, and represents a personal area network (PAN) standard. It is specifically aimed to serve applications that require short range communication, such as data transfer between computers and other peripheral devices like keyboards or cell phone to replace wire connectivity. Bluetooth supports star network topology, and can enable up to seven remote nodes to communicate with a single base station. However, in addition to its short range application and scalability problem (low number of nodes per network), the other disadvantages of Bluetooth are high power consumption and nodes that need a long time to be synchronized with the network when returning from sleep mode, and which increases the average system power.

IEEE802.15.4/ZigBee

The IEEE802.15.4 standard was designed for low data rate transmissions, low cost, and low power consumption wireless personal area networks (WPAN). In terms of communication range, this standard can be considered as a middle ground solution between IEE802.11 (Wi-Fi) and IEEE802.15.1 (Bluetooth), and supports multiple transmission frequencies, multiple data rates, and two topologies, star and point-to-point (pair topology), which makes it a flexible standard. It operates in the unlicensed ISM bands at 868 MHz in Europe, 915 MHz in the USA and 2.4 GHz worldwide, with data rates 20, 40 and 250 Kbps (kilo bit per second), respectively.

ZigBee is a standard designed by the ZigBee Alliance, which is an association of companies working together to enable reliable, cost-effective, and low-power wireless network. It is based on IEEE802.15.4 standard which means that ZigBee can take full advantage of this standard. In addition it can accommodate multiple networks topologies like star, point-to-point, and mesh networks. Also, a ZigBee network can have at most 65,000 nodes, making it a very scalable standard. Because of the aforementioned features, the ZigBee standard has been adopted in many wireless sensor network applications, and will be the best candidate for the work described in this thesis. Additionally, ZigBee modules are nowadays available in the market in small size with very affordable prices.

8.4.3 ZigBee Module Selection

There are many parameters which need to be considered when selecting a ZigBee module, these include power consumption, operating frequency, flexibility, coverage range, the module and testing costs, and the compatibility with the micro-controller unit. However, many certified semiconductor companies, such as Texas Instruments, Freescal, Digi International, are providing successful design for ZigBee products. After revising the available products the XBee module from Digi International has been selected, as it meets the above mentioned requirements, is compatible with the selected Arduino board, and includes other factor, i.e. previous successful experience, popularity, and available development resources.

XBee is a trade name from Digi international, and it is a wireless module designed for applications that require reduced data communication while having long range capabilities with less power consumption. XBee modules come in different formats for different kinds of application. There are two types of XBee modules, series 1 (s1) and series 2 (s2). As shown in Table 8.1, XBee s2 consumes slightly less power and has a better range than the s1. The XBee s1 and s2 are pin-for-pin compatible, but based on different chipsets and running different protocols. The s1 module uses the IEEE 802.15.4 standard protocol, while the s2 module relies on ZigBee protocol. Furthermore, both series come in two different transmission powers, regular and Pro. The regular version is simply called XBee, and is less expensive than Pro version; the Pro version uses more power and is

Table 8.1 Comparison between XBee s2 and s2

Specifications	XBee s1	XBee s2
Indoor range (m)	30	40
Outdoor range (m)	100	120
Frequency band (GHz)	2.4	2.4
Transmit power (mW)	1	2
Supply voltage (V)	2.8–3.4	2.8–3.6
Data rate (Kbps) (kilo bit per second)	250	250
Transmit current (mA)	45	40
Receive current (mA)	50	40
Receiver sensitivity (dbm)	−92	−96

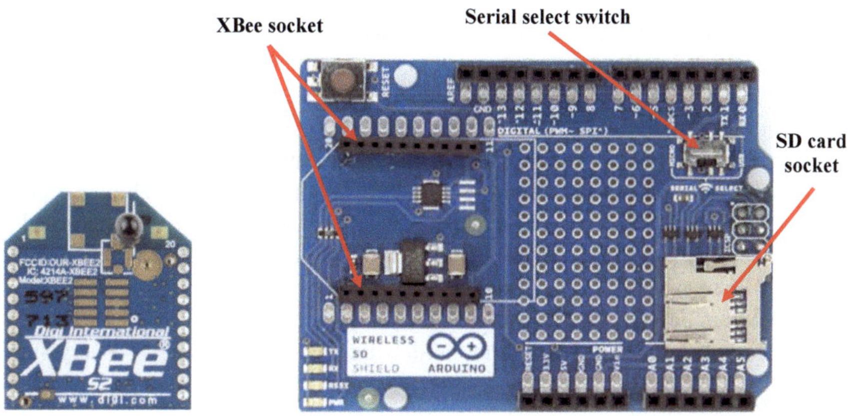

Fig. 8.5 XBee module and the wireless shield for Arduino (https://www.sparkfun.com; https://www.arduino.cc)

slightly larger than the regular version, but, on the other hand, has a longer communication range. The XBee s2 (Fig. 8.5) has been selected for this research, as it provides good in-door range, good data rate, low-energy consumption, better receiver sensitivity and it supports mesh and tree networks. This makes the system scalable and reliable for future development.

XBee modules can be attached to DUE using a wireless shield (Fig. 8.5) designed for easy connection to the XBee. The shield acts as a daughter board and is attached on top of the DUE. It has an optional on-board micro SD-card connection capability for serving data on a card over the network. In the shield there is an on-board switch labelled ‘serial select’, which determines how the wireless shield’s serial communication connects to the serial communication between the microcontroller and the USB-to-serial chip on the Arduino board. The switch allows two settings which are Micro and USB; in Micro mode, the XBee module will communicate with the microcontroller and the sent data from the microcontroller will be transmitted to the computer through USB as well as being sent wirelessly by the wireless module, but the microcontroller will not be programmable via USB in this mode. In USB mode, the microcontroller on the board is bypassed and the module can communicate directly with the computer, and helps in utilizing the Arduino's USB-to-Serial connection to configure the XBee modules.

8.4.4 Antenna Type

XBee modules must have an antenna to send and receive signals, of which there are four different types offered on XBee s2, namely whip, chip, U.FL, and RPSMA, as illustrated Fig. 8.6 (Faludi 2011). The wire (or whip) and chip antennas come pre-connected to the XBee modules, while the U.FL and RPSMA are derivatives of the connector types, offering chip with connectors on the board. The wire antenna is a single piece of wire that protrudes from the XBee and provides an omnidirectional radiation, which implies the maximum transmission distance in all directions is the same if the antenna is oriented in upright direction perpendicular on the module.

The chip antenna is a flat ceramic chip that is mounted on the XBee. The form of signal radiation by this antenna is heart-shaped or cardioid, which means that the signal will be attenuated in many directions. However, because the chip antenna is nearly flush, that makes it a suitable choice for any sensor that needs to be located in a small space. It is also robust compared to the whip antenna which is subjected to mechanical stress.

In some application the XBee module needs to be fitted inside a metal box; in which case a U.FL antenna has to be used. This type of antennas has a very small

Fig. 8.6 Types of antenna (Faludi 2011)

connector on the module and to achieve the physical connection to the external antenna an adapter cable is used. The advantage of this option is the XBee module can be enclosed in a casing and the antenna mounted on the outside of the case. Similarly, the RPSMA connector is a different type of socket from the U.FL connector. It is bigger and mounted directly to the XBee without a connecting cable. These last two options are more expensive.

Generally, an external antenna allows better signal transmission and reception, and a larger range. The indoor range of the selected XBee module is only 40 m and sometimes noise in the medium distort the signals and shortening the range. Therefore, it was decided to utilize an external antenna type RPSMA on the coordinator module (Base station), and either chip or whip or RPSMA antenna on the sensor node, depending on the transmission range tests which will be performed later in this chapter.

8.4.5 Power Source Selection

Rechargeable batteries are commonly employed as a power source for the wireless sensor nodes. Several characteristics should be considered when choosing a battery for a wireless sensor node. The most important of these attributes are energy density, charge/discharge cycle, size, self-discharge rate and cost. There is no any battery technology that meets all these criteria, so a compromise must be made. Each battery type has its advantages and disadvantages with a wireless sensor node. The most common rechargeable batteries types are: lead acid, nickel-cadmium (Ni-Cd), nickel-metal hydride (Ni-Mh), lithium-ion (Li-Ion), and lithium-ion polymer (Li-Po). The comparison of the different types of batteries can be seen in Table 8.2 (Buchmann 2003).

Lithium-based batteries are technically more advanced, most widely used and fastest growing energy sources. Comparing to the other three battery types, Li-Ion and Li-Po have higher specific energy densities and offer a lower self-discharge rate than both Ni-Cd and Ni-Mh, with only lead acid having a lower rate. Lithium-Po batteries are similar to Lithium-Ion however with a different type of electrolyte used, and furthermore offer the advantage of being very thin and light weight, thus allowing them to be easily included in sensor nodes while occupying very little space. In this regard, a Turnigy Li-Po battery (Fig. 8.7) has been chosen as a power

Table 8.2 Comparison among rechargeable batteries

Specifications	Lead acid	Nickel-cadmium	Nickel-metal hydride	Lithium-ion	Lithium-ion polymer
Energy density (Wh/kg)	30–50	45–80	60–120	110–160	100–130
Charge/discharge cycle	200–300	1500	300–500	500–1000	300–500
Self-discharge/month	5 %	20 %	30 %	10 %	~10 %

Fig. 8.7 Lithium-ion polymer (Li-Po) battery

source for the wireless node. The battery has an energy capacity at 7.4 V/1000 mAh (1 h continuously working at 7.4 V with 1 A discharge current) with weight 62 g and size 74 × 35 × 13 mm, which can fit easily in the node enclosure as shown later.

8.5 Design of the Base Station Node

The base station architecture is the same as of the wireless sensor node with some differentiation in the selected components, configurations and the tasks entrusted to it, and it includes the network coordinator connected to the central processing unit which in turn is interfaced to a PC. This unit is responsible for implementing the second stage of the condition monitoring algorithm. The received signals from the sensing units will be analysed in this station using the DWT based on the selected wavelet family (*db2*) and decomposition levels. Then, the features are extracted and utilized as input vector to the designed ANN, which responsible for robot health diagnosis depending on feature classification. The result of analysis will be transmitted to the PC to report the robot's health situation to the operator. However, wavelet transform and neural network are computationally demanding algorithms and require a processor with quite advanced processing characteristics. Therefore, it was decided to apply theses computational processes on a digital signal processor (DSP), as it has features designed to support high-performance, repetitive, and numerically intensive tasks. The radio transceiver and its antenna type have been previously discussed.

8.5.1 Digital Signal Processor (DSP) Selection

A DSP is a microprocessor characterized by high computational power, specifically designed and optimized to efficiently handle a variety of complex mathematical operations such as digital signal processing tasks (Dargie and Poellabauer 2010). Selecting a suitable digital signal processor (DSP) is an important issue in real-time signal processing, as a processor may be well suitable to some applications, but it is an inappropriate choice for others. The main aspects of selecting a DSP are as follows: arithmetic format, memory size, speed, data bus as well as the support for peripherals, such as ADC, depending on the requirements of the case study (Emilio 2015).

In terms of arithmetic format, DSPs are divided into two categories, fixed point and floating point, which refer to the format used to store and manipulate numbers within the devices. The numbers in the fixed point DSPs are represented as positive or negative integers in a fixed range; and they are generally cheaper, faster, consume less power, but produce higher quantization error or noise (Kehtarnavaz 2008; Emilio 2015). This noise reduces the signal to noise ratio of the system, as it will be added to the signal. Additionally, with fixed point DSPs the signal has to be scaled to avoid the overflow problem, which is a case where the result of an arithmetic operation exceeds the capacity of the register used to hold that result (Kuo et al. 2006), and therefore extra code has to be written. In contrast, floating point DSPs have the capability to represent the numbers by the combination of a mantissa (or a fractional part) and an exponent part; thus are easier to program, have a higher dynamic range, a shorter development cycle, and a better precision. Because floating point DSPs have high dynamic range for numeric calculation the overflow problem does not need to be considered, but, on the other hand, they are more expensive and consume higher power than fixed-point ones (Smith 1999).

The RISC architecture is adapted in most DSPs and they can execute multiple instructions in parallel, resulting in fast operations. On- and off-chip (external) memory sizes are considered another key factor when evaluating a DSP processor. However, most DSPs provide several kilobytes of fast on-chip SRAM, which allows the core processor fast access without wait states. The on-chip storage capacity is not always sufficient to hold the program's code and data; in such a case, an external memory interface can be used to connect the DSP to an external memory.

A key measure of the appropriateness of the DSP is its execution speed and data width. The most common way to evaluate the DSP speed is to measure the required amount of time to execute the fastest instruction on the processor. Therefore, floating point devices can be specified by MFLOPS (millions of floating point operations per second), and MIPS (millions of instructions per second) to specify fixed point devices (Kuo et al. 2006). The size of data bus has a significant effect on the chip size, the pin numbers required for the packages and the size of external memory device that can be connected to the DSP. A DSP with larger data word size is the best choice for the application that needs more precision. Most floating point DSPs use 32-bits/word data bus, while in fixed-point DSP 16 bits/word data size is the most common.

Depending on the above specifications, the search task is concentrated on the large semiconductor companies such as Texas Instruments, Analog devices. Many different types of chips and evaluation boards can be found listed on these companies' webpages. However, it is time consuming looking through every chip's data sheet; fortunately, the companies have grouped their products by different application fields, such as imaging, video, and industrial. In the industrial field, which is related to this research, the semiconductor companies have delivered some application notes for signal processing, where some microprocessors are recommended. As a first step in the selection procedure, all the suitable microprocessors in the list are checked to establish if there is a corresponding evaluation board, because of the limited time available. Eight evaluation boards are nominated, and listed in Table 8.3. The power consumption and physical size of the DSPs are not considered here, since the base station will be interfaced to the PC. Also, in this work an ADC is not required on the base station, as it will receive the signals directly from the sensor node via the wireless link; however the resolution and number of channels of the ADC are included in the table for comparison purposes and for the benefit of other researchers.

Table 8.3 provides a succinct summary of the key details of the investigated boards. If the price criterion is used, the second (STM32F103RB) or third (TMS320F28335) options in the table would be selected. However, the issue with the second (and first) one is fixed-point DSP, while TMS320F28335 is mainly designed for control application. The last option in the table is the fastest and more powerful one, but in contrast is the most expensive. Therefore, TMS320C6713 Development Starter Kit (DSK), which has been in production for approximately ten years but is still available and useful (Wright et al. 2013), was selected for its very good specifications and rich set of peripherals (as discussed in the following subsection). It is very prevalent in laboratories and extensive information and support are available from other users on the Internet, helping in speeding of the development process. For brevity, this board will be referred as C6713 DSK or just DSK throughout the thesis where is needed.

8.5.2 TMS320C6713 DSK

The C6713 DSK, which is manufactured by Spectrum Digital for Texas Instruments (Wright et al. 2013), is a powerful, relatively cost-effective development board with the necessary hardware and software for real-time applications. The DSK board, with an approximate size of 127 × 203 mm is shown in Fig. 8.8. It is provided with a 225 MHz (4.4 ns cycle time) TMS320C6713 floating point DSP chip based on the VLIW (very long instruction word) architecture, which is very well suited for numerically intensive and multifunction applications as it uses multiple functional units to execute multiple parallel instructions (Kehtarnavaz 2008). The DSP chip is capable of delivering up to 1800 million instructions per second (MIPS) and 1350 million floating-point operations per second (MFLOPS).

Table 8.3 Investigated DSP boards

DSP evaluation board	CPU type	CPU speed	ADC No. of channels Max. resolution Max. sampling freq.	Memory capacity	Price
Explorer 16 Starter Kit Microchip	16-bit Fixed-point	40 MHz	32 channels configurable as 10-bit, 1.1 MSPS or 12bit, 0.5 MSPS	256 KB Flash 30 KB SRAM	$130
	http://www.microchip.com/stellent/idcplg?IdcService=SS_GET_PAGE&nodeId=1406&dDocName=en024858				
STM32F103RB Development Board STMicroelectronics	32-bit Fixed-point	72 MHz	16 Channels 12-bit 1 MSPS	32– 128 KB Flash 6–20 KB SRAM	$70
	http://www.st.com/web/catalog/mmc/FM141/SC1169/SS1031/LN1565/PF164487				
TMS320F28335 Experimenter Kit Texas Instruments	32-bit Floating-point	150 MHz	16 Channels 12-bit 12.5 MSPS	256 KB Flash 34 KB SARAM	$99
	http://www.ti.com/tool/tmdsdock28335				
TMS320C6713 Development Kit Texas Instruments	32-bit Floating-point	255 MHz	TLV320AIC23 Audio codec	256 KB Flash 16 MB SDRAM	$395
	http://www.ti.com/tool/tmdsdsk6713				
56F8037EVM Freescale	16-bit Fixed-point	32 MHz	16 Channels 12-bit 2.67 MSPS	64 KB Flash 4 KB RAM	$199
	http://www.freescale.com/webapp/sps/site/prod_summary.jsp?code=MC56F8037EVM&fsrch=1				
Zoom™ OMAP-L138 eXperimenter Kit	32-bit Fixed-point	375 MHz	TLV320AIC3106 Audio codec	8 MB Flash 128 KB SDRAM	$495
	http://www.logicpd.com/products/system-on-modules/zoom-omap-l138-experimenter-kit/				

The C6713 DSK includes 16 MB of synchronous dynamic random access memory (SDRAM) and 256 KB of flash memory. It also has interfaces to analogue audio signals through an on-board AIC23 codec and four 3.5 mm audio, but will not be used in this project. The board also incorporated a 32-bit external memory interface (EMIF), on which external memories and other devices can be connected, and a 16-bit host port interface (HPI) through which a host processor can directly

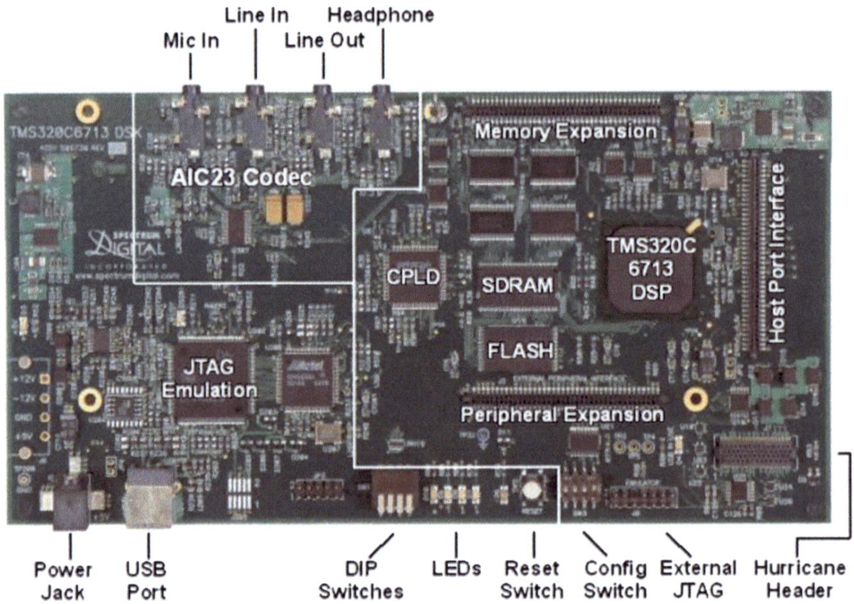

Fig. 8.8 Texas instrument C6713 DSK (http://www.dsprelated.com)

access the DSP memory. There are four LEDs (light-emitting diodes) and DIP (dual in-line package) switches on the board that can be read from a program and provides the user with an interactive feedback interface. A complex programmable logic device (CPLD) is employed to implement logic that ties the board components together. The CPLD has a register based user interface that lets the user configure the board by reading and writing to its registers.

The C6713 DSK processor contains an enhanced direct memory access (EDMA) controller, which is a highly efficient data transfer engine that serves the purpose of releasing the DSP from data transfer between the DSP and its external peripherals or between the on-chip peripherals and the memory. The C6713 DSK has two 32-bit timers that can serve a variety of purposes, such as timing and counting events, generating pulses, interrupting the CPU and sending synchronization events to the EDMA controller. A joint test action group (JTAG) emulator that can directly access the register and memory state of the DSP chip through a standardised JTAG interface port is included on the C6713 DSK, allowing on-board programming and debugging.

8.5.3 *Daughter Card Interface*

The three expansion connectors, peripherals, memory, and HPI, can be used for integrating plug-in daughter cards with the DSK. The daughter cards can obtain

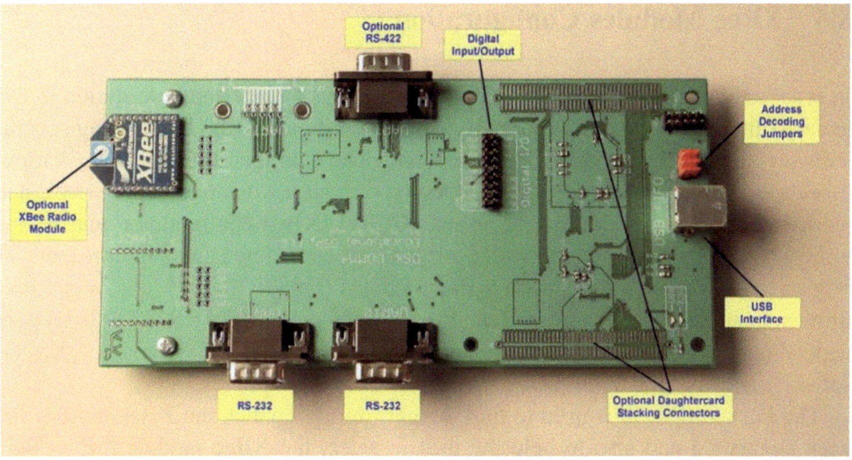

Fig. 8.9 DSK_COMM4 communication daughter card (http://www.educationaldsp.com/stockproduct_dsk_comm4.htm)

power directly from the DSK, as each connector includes multiple ground, +5, and +3.3 V power signals. The peripheral expansion connector additionally provides both +12 and −12 V to the daughter cards. To expand the capability of the C6713 DSK to achieve the suitability for this project, the DSK_COMM4 daughter card (Fig. 8.9) was used. The DSK_COMM4 is a multichannel communication daughter card designed to operate with some of the Texas Instruments DSKs from C6x family, such as C6713 DSK. The basic configuration of this card from the vendor includes a USB interface capability of up to 1 MB/s, two RS-232 UART channels, eight digital outputs, and eight digital inputs. Also, there are two optional UART channels on the board can be inhabited with either RS-232 or RS-422 or XBee radio transceivers.

Two of the DSK peripheral connectors (external memory interface and external peripheral interface) must be utilized for a memory mapped integration of the DSK_COMM4 board to the C6713 DSK platform. Moreover, the daughter card itself can come with pass-through connectors, if requested, to allow stacking of other daughter cards if needed. One XBee module (the coordinator) is installed on this card and only one digital input channel is needed, to receive the triggering signal from the robot controller by which the DSP is triggered to start processing the received data. This card is delivered with XBee s1 wireless module soldered on it and as discussed earlier the XBee s2 is going to be used in this project, therefore, a minor modification was achieved on the card by de-soldering the on-board wireless module and putting a socket that allows installing and de-installing the XBee s2 module without soldering/de-soldering.

8.6 XBee Modules Configuration

In order to allow the transceiver modules on the sensor node and base station to talk to each other, they have to be correctly configured individually, by utilizing specific software, before using them in a network. The configuration includes various aspects such as classification of the node (coordinator, router or end device), network ID, destination address and so on. X-CTU (XBee configuration and testing utility) is windows-based application software developed by Digi and represents the official configuration software for XBee modules (Faludi 2011). Many versions of firmware can be selected and written into a ZigBee module via RS232 or USB port, depending on the used interface board. Figure 8.10 shows the layout of X-CTU software and its four main tabs.

The PC settings tab is used to find the port through which the XBee module has been connected and also to help the user to select the XBee module for configuration from a range of plugged in devices. The range test tab is utilized to perform a wireless communication coverage range test, as will be explained later in this chapter. Terminal tab is used to open the X-CTU terminal window, which can be used to read the data being received by the connected module. The modem configuration tab allows changing the firmware version, writing the firmware setting to

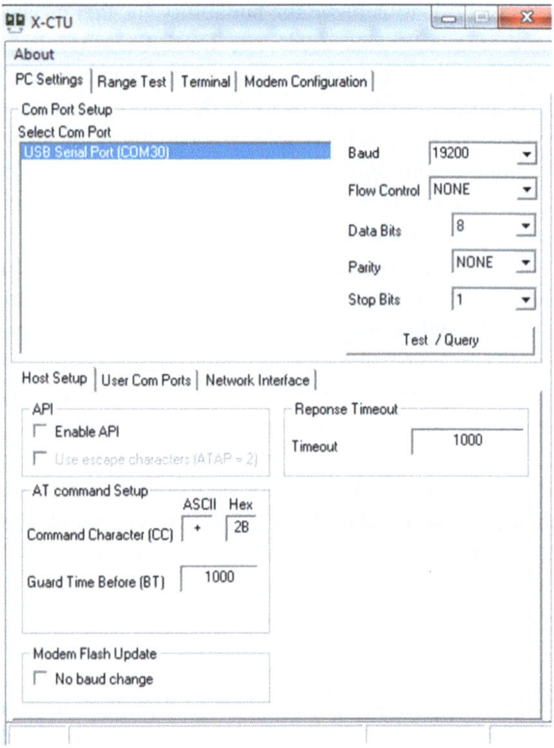

Fig. 8.10 Layout of the X-CTU configuration software

Fig. 8.11 XBIB-U-DEV adapter for XBee configuration

the module, as well as setting the module as a coordinator, router or an end device by using its three sub-tabs drop-down menus named modem XBee and function set, respectively, as illustrated in Fig. 8.12. The Firmware is instructions programmed in the module's memory which controls the device and provides several instructions on how the devices can communicate with other hardware.

However, to configure the XBee modules using this software, it is necessary to interface the XBee to the computer. For this purpose, the XBIB-U-DEV development kit (Fig. 8.11) from Digi International was employed, and will be utilized later for the coverage range test. The configuration progress includes two stages, coordinator configuration and end device configuration. In practise, when these modules are configured, the coordinator automatically scans to select a communication channel and it always listening to the end device and receive the incoming data which it sends to the DSP board for processing.

The firmware type XB24-B, which supports the full functionality of ZigBee protocol, is selected from modem sub-tab in the modem configuration tab for both coordinator and end device, and the latest version of this firmware is required to be downloaded on the modules. The firmware supports coordinator, router and end device configurations. The configured parameters are listed in Table 8.4 and the screenshots showing the configuration of both modules are as shown in Fig. 8.12. The function set sub-tab was used to configure the functions that the XBee modules provide, which are as a coordinator and an end device. However, all used XBee modules (here just two) must have the same personal area network (PAN) ID and baud rate e.g. 1234 and 19,200. The destination address high (DH) and low (DL) for the coordinator should be the same as the serial high (SH) and serial low

Table 8.4 The configured parameters in the XBee modules

Parameters to be configured	Setting	
	Coordinator	End device
Firmware	ZNET 2.5 COORDINATOR AT	ZNET 2.5 ROUTER/END DEVICE AT
Firmware version	1047	1220
PAN ID	1234	1234
Destination address high (DH)	13A200	13A200
Destination address low (DL)	409C8E69	40AC1682
Node identifier	Coordinator	End device
Baud rate	19,200	19,200
Other parameters	Default	Default

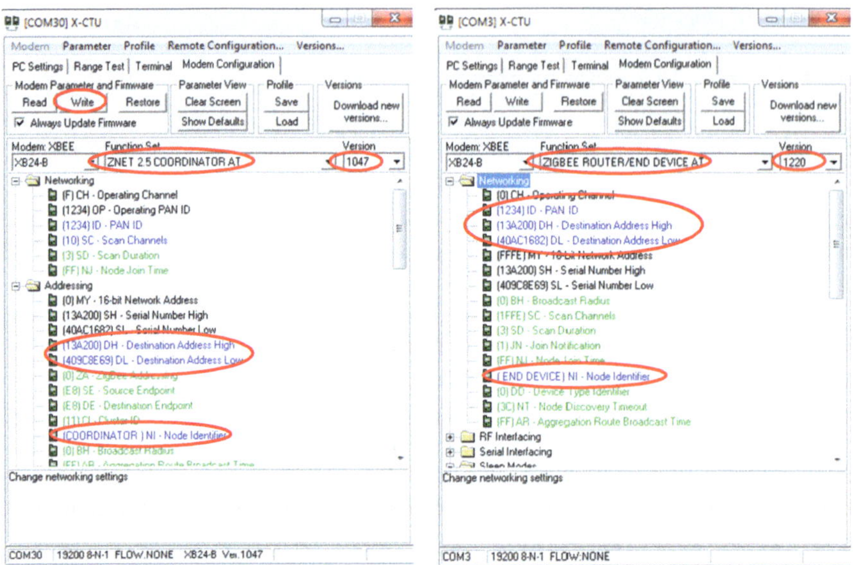

(a) Coordinator configuration **(b) End device configuration**

Fig. 8.12 XBee modules configuration using X-CTU

(SL) of the end device and vice versa, and which can be found written on the back side of the XBee modules. The node identifier is a user configurable text name that can be set to easily identify a module. Accordingly, by clicking the write button in the modem configuration tab, these settings will be downloaded on the connected module.

8.7 Coverage Range Test

The purpose of this test is to study the effects of antenna type and XBee transmit power on the actual coverage range and communication quality between the base station and the sensor node under real working conditions. The optimum configuration parameters will be the outcome of this study, and the received signal strength indicator (RSSI), which is defined as the signal strength level of a wireless device measured in (dBm) of the last received packet (Dargie and Poellabauer 2010), is used. A loop-back test using X-CTU software is performed to investigate the relationship between RSSI and the distance for a point-to-point communication when different configurations are introduced. Figure 8.13 shows the screenshots of range test in X-CTU, it provides the RSSI indication bar in dBm where −40 dBm represents the strongest signal received by the module and −104 dBm is the weakest. The range test procedure was carried out as follows.

The coordinator module is connected to PC and sends a packet of data to a remote module. Each of the coordinator and remote module are installed on a XBIB-U-DEV kit, as it has a feature to perform a loop-back test as shown beforehand in Fig. 8.11. The remote module will send the received packet directly back to the coordinator module and the X-CTU estimates the value of the RSSI based on the last received packet by the coordinator module, and shows it in dBm (Piyare and Lee 2013). “Packet delay”, “Data packets number” and “Data received

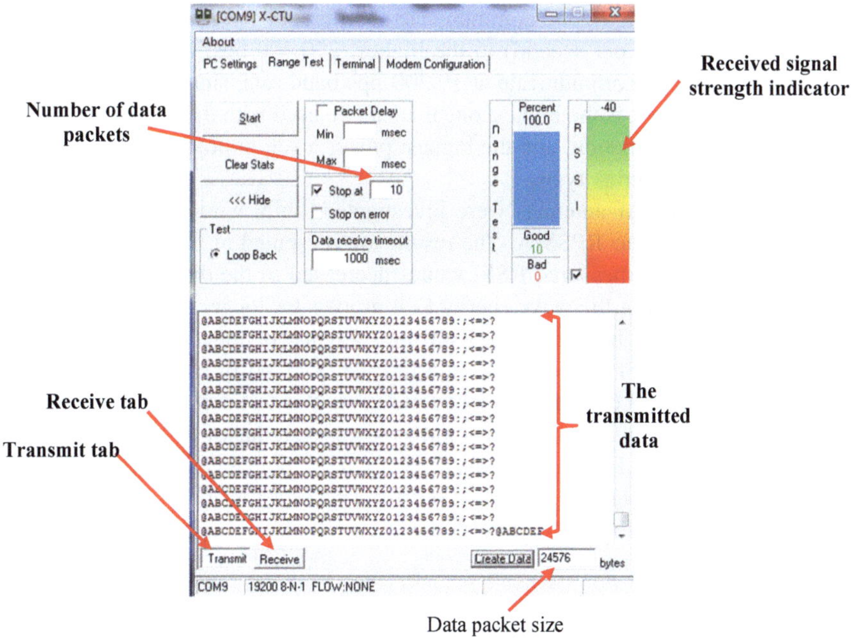

Fig. 8.13 Range test screenshot

timeout" are parameters by which a wide range of scenarios can be simulated, and also, the data size that is needed to be sent during the experiments can be controlled using "Create data" tab.

8.7.1 Test's Scenarios and Result

Two scenarios were accomplished in this study and in both the base station node is equipped with RPSMA antenna type, and powered from the PC. The distance between the two modules starts at 5 m and is then increased in 5 m steps until the wireless connectivity was lost. Throughout the experiments, a 24 KB data size (typical size of the measured vibration signals) was created by the X-CTU software and transmitted by the coordinator to the sensor node 10 times; every time the RSSI was measured and then the averaged RSSI value computed and used for comparison. All the experiments were carried out in an indoor environment, in the Robotics Lab in the Stephenson Building at Newcastle University.

The first test was to establish the optimum antenna type for the sensor node. In this test, chip, wire and RPSMA antenna types were equipped with the sensor node alternately and each time the RSSI is measured at different distances. However, there are five power levels at which the XBee module transmits the information, and they correspond to −8 (the lowest), −4, −2, 0 and 2 dBm (the highest) respectively. The power level can be set through the configuration modem tab of the X-CTU, and the nodes can be configured to communicate at different baud rates ranging from the lowest 1200 bps (bit per second) to the highest 230,400 bps. In this scenario the modules were set to communicate at 19,200 bps baud rate, since experimentally it was established that it is the highest one that can be used with the specified data size without loss of information, and the highest power level, in order to investigate the effect of just the antenna type.

Three categories of antennas were investigated, chip, wire and RSPM, as the U.FL is very similar to RPSMA. The results are presented in Fig. 8.14, and it can be observed that the measured RSSI values decreased as the distance is increased, due to the depletion of the wave energy as it propagates longer. The fluctuations in the RSSI values can be correlated with the presence of reflection and multipath phenomena because of the walls and interference from Wi-Fi routers located in the building (Piyare and Lee 2013). Before conducting this test it was expected that the RSSI of the chip and wire antennas would be lower than that of RPSMA antenna, as can clearly be seen in Fig. 8.14. In the case of wire or chip antenna (when the RSSI is low) the full number of data packet were not received back by the coordinator; this is illustrated in Fig. 8.14, where just 7 packets were captured (indicated by

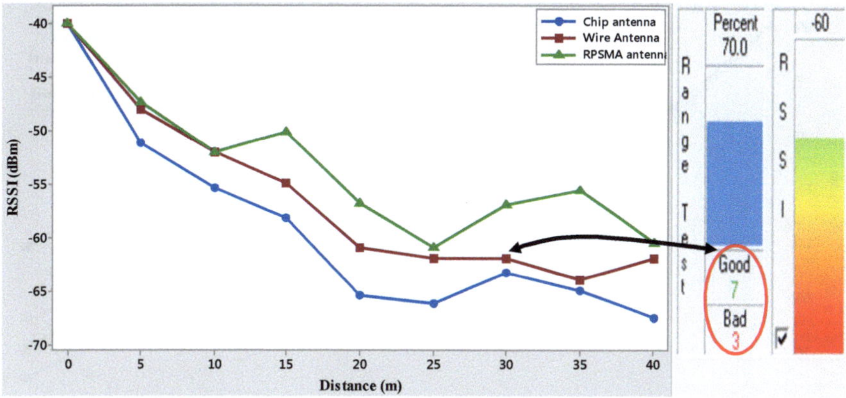

Fig. 8.14 Measured RSSI values versus distance with different antenna types

good in the figure) and 3 lost (indicated by bad) for the wire antenna. Based on these results, the RPSMA antenna will be used, since it records the highest RSSI particularly at longer distances.

Next, after establishing the optimum antenna, the second scenario was performed to investigate the effect of the power transmitting levels. There are many parameters that affect the power consumed by the XBee modules. Operation modes, such as transmitting and receiving, draw different amount of currents, and therefore their power consumption is different. Furthermore, as mentioned earlier, the modules can work at different transmitting power levels and by increasing the power level the consumed current will increase. There is no problem with power consumption in the base station node as it will be powered from the main power supply plug in the practical conditions, but the consumed power in sensor node needs to be optimised. The point now is to carry out a test to measure the RSSI considering the effect of transmit power levels with existing RPSMA antenna and at the same baud rate as previously.

The findings of this test are illustrated in Fig. 8.15; it is apparent that increasing the transmitting power has improved the transmit performance, but that will increase the power consumption level. Fortunately, it was indicated that all transmitted packets of data were received back by the coordinator when the sensor node transmit power was in the medium level; and consequently the device was set to send the data at this power level. As stated in the XBee s2 datasheet and shown previously in Table 8.1 that the indoor range is approximately 40 m, which was found to be the case according to the results obtained from both experiments. The transmitted data packets were fully received up to 40 m with RSPMA antenna and medium power level, whereas the RSSI and data packets drop significantly when the distance is increased beyond this limit.

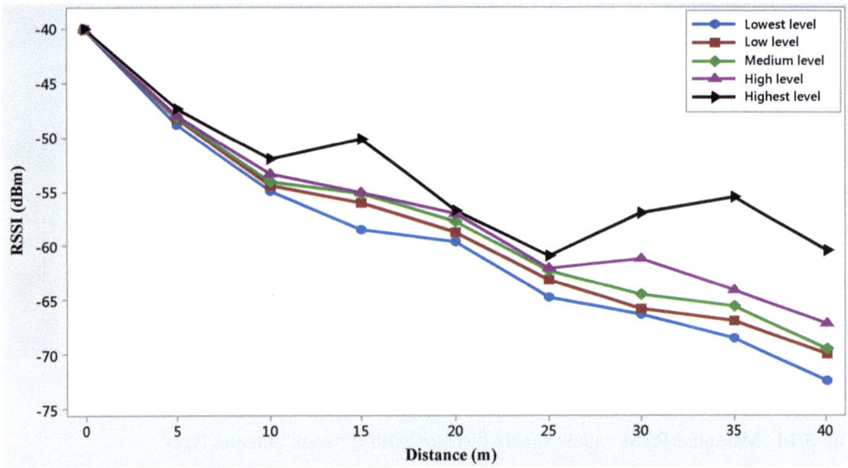

Fig. 8.15 Measured RSSI values versus distance at different transmit power levels

8.8 Final System Hardware

To protect the wireless node and base station from dust and other contaminants, enclosures have to be used. Available enclosures for the Arduino DUE and C6713 DSK, are designed to enclose just the boards (with the daughter card in the case of the DSK) without the other connected peripherals such as the sensors, triggering circuit, battery, XBee shield and the antennas; thus, it was necessary to look for alternatives.

Metal or plastic boxes are commonly used for creating custom-built enclosures for embedded equipment. For this purpose, a 179 × 86 × 51 mm plastic box, for the wireless node, and a 250 × 250 × 100 mm aluminium box, for the base station, were purchased and modified. Bolts were used to fix the Arduino microcontroller and the C6713 DSK to the base of their boxes. Small rectangle and circular cuts were made on the base and right side of the sensor node box, to allow attaching the sensors adapter to the robot using super glue, and the antenna, via the wireless shield, and power supply port, from the battery, to the microcontroller, as shown in Fig. 8.16. A double sided adhesive tape at the bottom of the battery was utilized to affix it on the box.

The base station should be located next to a host PC, for reporting the analysis results and also to permit control of the entire monitoring system through the user interface. Therefore, holes were machined into the base station box walls for the

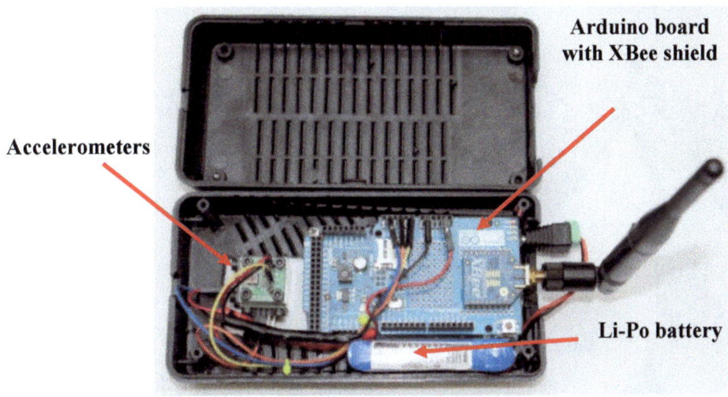

Fig. 8.16 Wireless sensor node hardware

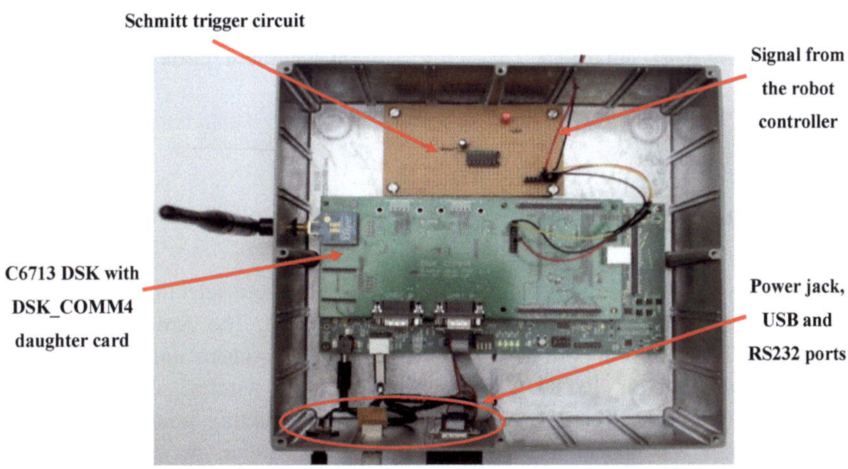

Fig. 8.17 Base station hardware

USB, RS232 and power ports as well as for the antenna and the wire that transmitting the signal from the robot controller to the Schmitt trigger circuit (Fig. 8.17). The sensor node was attached to the robot using two cable ties; this along with the final embedded system layout is depicted in Fig. 8.18.

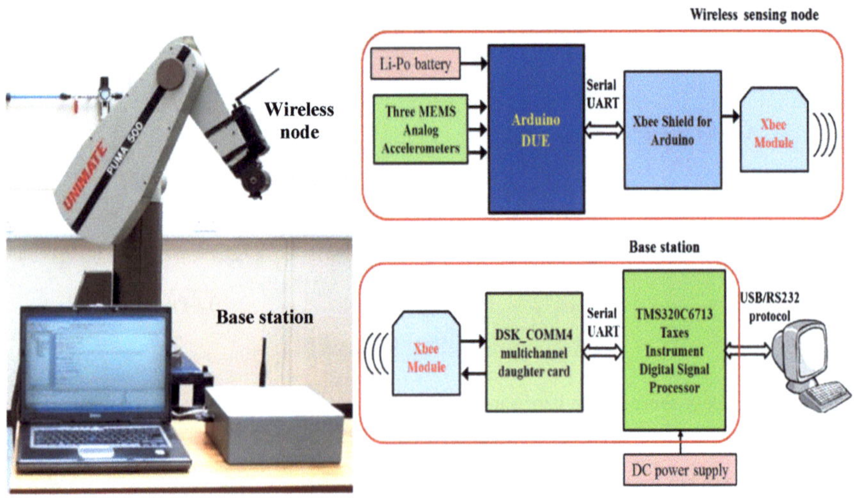

(a) Embedded system installation (b) Embedded system layout

Fig. 8.18 Designed embedded system installation and layout

8.9 Summary

This chapter has detailed the design approach and hardware selection requirements based on the proposed condition monitoring algorithm. The Arduino DUE was selected as being appropriate as a core of the wireless sensor unit. The wireless capability was put on the DUE board using Arduino wireless shield. A Texas Instrument C6713 DSK board, which is based on a floating point digital signal processor chip, was chosen as the base station of the wireless network, because of its advantageous processing features. To enable the wireless communication on the base station a daughter card, entitled DSK_COMM4 has been utilized.

A comparison between the existing wireless technologies like Wi-Fi, Bluetooth and ZigBee was presented and it was established that the ZigBee wireless transmission consumes less power and is very reliable when it operates within the recommended range. The XBee wireless module, which is based on ZigBee protocol, was employed to establish the communication between the sensor node and the base station. Configuration of the XBee transceivers for both sensor node and base station was crucial in the implementation of wireless networks. The configuration method of the XBee transceivers in addition to several experiments to test the transmission range and to select the optimum antenna type and transmitting power on the sensor node were carried out. Enclosures for the sensor node and base station hardware were developed relying on commercially available boxes, and then the final system installation illustrated. In the next chapter, development of the embedded software, description of the system operation and the implementation result will be discussed.

References

Buchmann, I. (2003). What's the best battery? [Online]. http://batteryuniversity.com/learn/article/whats_the_best_battery

Dargie, W. W., & Poellabauer, C. (2010). *Fundamentals of wireless sensor networks: Theory and practice*. New York: Wiley.

Emilio, M. D. P. (2015). *Embedded systems design for high-speed data acquisition and control*. Berlin: Springer.

Faludi, R. (2011). *Building wireless sensor networks*. Sebastopol: O'Reilly Media, Inc.

Feng, G. J., Gu, J., Zhen, D., Aliwan, M., Gu, F. S., & Ball, A. D. (2015). Implementation of envelope analysis on a wireless condition monitoring system for bearing fault diagnosis. *International Journal of Automation and Computing, 12*, 14–24.

Giannoulis, S., Koulamas, C., Emmanouilidis, C., Pistofidis, P., & Karampatzakis, D. (2013). Wireless sensor network technologies for condition monitoring of industrial assets. *IFIP Advances in Information and Communication Technology*.

Kehtarnavaz, N. (2008). *Digital signal processing system design: LabVIEW-based hybrid programming*. Amsterdam: Elsevier.

Kuo, S. M., Lee, B. H., & Tian, W. (2006). Real-time digital signal processing: Implementations and applications (2nd ed.). New York: Wiley.

Piyare, R., & Lee, S.-R. (2013). Performance analysis of XBee ZB module based wireless sensor networks. *International Journal of Scientific & Engineering Research, 4*, 1615–1621.

Smith, S. W. (1999). *The scientist and engineer's guide to digital signal processing*.

Wright, C. H. G., Morrow, M. G., & Welch, T. B. (2013). Comparison of DSP boards for classroom use. In *Proceedings of 2013 IEEE Digital Signal Processing and Signal Processing Education Meeting, DSP/SPE* (pp. 273–277).

Yang, S.-H. (2014). *Wireless sensor networks: Principles, design and applications*. Berlin: Springer.

Chapter 9
Embedded Software Design, System Testing and Validation

Abstract In general, the embedded system design is broken down into two main parts, hardware and software. In Chap. 8 the selection of the required components for developing the wireless embedded system has been discussed, this chapter describes the programming of the proposed robot condition monitoring algorithm on the developed hardware. The software tools used to build the embedded system code are first discussed, and the flow chart of the developed code for implementing the proposed robot condition monitoring algorithm described; the main steps of this code are also explained. To validate the embedded system functionalities in signal capturing, analysis, fault detection and diagnosis the system were tested and the results presented.

9.1 Wireless Node Software Development Tool

Arduino microcontrollers are programmed using an open source Arduino programming language, which is based on C/C++ and contains a specific set of structures created to teach core programming and computing concepts through electronics to nonprogrammers (Faludi 2011). An integrated development environment (IDE) makes it possible to write code for the boards which can be downloaded directly from the Arduino website. The Arduino IDE is split into three major parts (as shown in Fig. 9.1): a toolbar across the top of the window contains seven buttons that control the program behaviour; the white area in the centre is where the code can be entered and modified; the black section (text console) in the bottom of the window is where the code status messages and used memory appear, which helps in code debugging (Fig. 9.2).

The code is commonly known as a sketch in the Arduino programming community, and includes two blocks: setup and loop. The setup block runs only once each time the Arduino is started or restarted, and allows configuration of Arduino parameters such as pin modes, communication setup, etc. The core of most Arduino programs is the loop function; it runs continuously as long as the Arduino board is powered. The development environment also includes a serial monitor to monitor

A.A. Jaber, *Design of an Intelligent Embedded System for Condition Monitoring of an Industrial Robot*, Springer Theses,
DOI 10.1007/978-3-319-44932-6_9

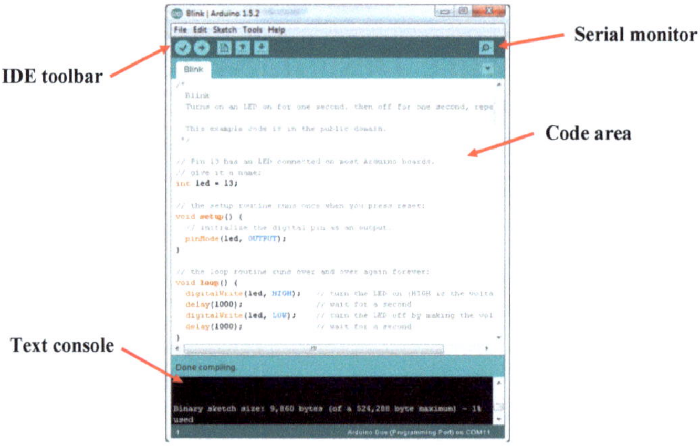

Fig. 9.1 Arduino IDE

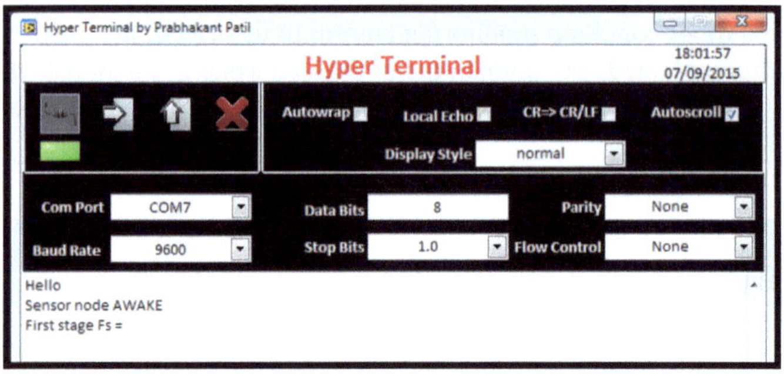

Fig. 9.2 A screenshot from the serial communication software

the serial communication of the board when a program is run, making testing of the program easier, and there is no need to use the wireless communication each time the code is tested. However, to create a simple program that would run on an Arduino board implies writing and saving a sketch within the IDE. The written sketch is then converted to C and compiled to produce a binary code which the microcontroller on the Arduino board will be able to execute. The binary code is uploaded to the Arduino board via a USB connection.

9.2 Base Station Software Development Tools

A set of code development tools are accompanied with the C6713 DSK for application development, including Code Composer Studio (CCS or CCStudio) integrated development environment (IDE), a highly optimising C compiler and the DSP/BIOS kernel (Kehtarnavaz 2008). CCS is an IDE package made by Texas

Instruments (TI) for developing embedded applications on their processor families such as TMS320C6x. It has a user friendly graphical environment for building and debugging C and assembly codes on a variety of TI processors. A number of debugging features are available in CCS IDE, including setting break and probe points to stream data to and from the target DSP, watching variables using watch window tool, viewing memory and registers, and graphing capabilities.

In traditional code debugging processes, the target processor is normally stopped and a snapshot of the processor state is examined. This is, however, not an efficient method to test for real-time glitches. The CCS complements the traditional techniques using the so-called DSP/BIOS feature, which provides mechanisms to analyse an application program as it is running on the target DSP without stopping the processor. The TMS320C6x DSP family has also a feature of real-time communication with the host PC; this can be performed using real-time data exchange (RTDX). The RTDX property allows users to transfer the data from PC down to a DSK board and vice versa, as well as data analysis in real time again without stopping the program execution on the target. This data exchange occurs via the universal serial bus (USB) port emulating the joint test action group (JTAG) connection.

TI provides several optimized software libraries to implement algorithms on the C6713 DSPs. The chip support library (CSL) has been used in this project, which can be configured within the DSP/BIOS configuration tool. This library provides a set of C-language interface functions that simplify the configuration and initialization of on-chip peripherals such as the EMIF, DMA, serial ports and timers, which reduces time-consuming manual configurations. This also helps in configuring the DSP peripherals for the daughter card interface. A free version of the CCS is available, but it is code-size limited. Fortunately, the DSK comes with CCS version 3.1 accompanied with a free license. In the CCS, code is written in C language (with extension.c) which is compiled by the C compiler to produce an assembly source file (with extension.asm) (Kehtarnavaz 2008). The assembler is utilized to convert an assembly file into a machine language object file (with extension.obj). The linker combines object files and object libraries as input to produce an executable file (with extension.out), which can then be uploaded on the DSP board.

9.3 Designed Embedded Software

The embedded system code is divided to work on the base station and wireless node. The code on the wireless node (written using Arduino programming language) is responsible for conducting the first stage of condition monitoring (fault detection), while the base station code was written using CCS environment and responsible for fault diagnosis stage. The flowcharts of these two codes are shown in Figs. 9.3, 9.4 and 9.5, respectively. Both codes start by initialization and setting up the needed hardware, such as input/output pins, timers and other communication peripherals. In addition to the fault diagnosis stage implementation, the base station is also in charge of controlling the whole embedded system. However, in each of the sensor node and base station codes there is a main programme and a sub-programme (subroutine), which is normally outside the flow of the main programme with extra functionality and called by the main program to achieve a specific task. When the embedded system runs, the main programs in both nodes will enter an infinite loop waiting for an interrupt or required parameters to be provided in order to execute their subroutines. When the subroutines finish executing, they return the result back to the main programme, which then carry out the rest of the queued tasks.

In the base station main programme (Fig. 9.3) two parameters are needed to be set before it starts executing. These are **operation mode** and **awake node**; the operation mode is holding which stage of the CM the embedded system will implement first, and helps in changing to fault diagnosis stage once a fault is detected. If the operation mode is set equal to 1, the embedded system will implement the fault detection stage, otherwise, 2 the fault diagnosis stage is performed. It can be set via the user interface, where the base station is communicating with the host computer via the serial channel B on the DSP board, using the letter F for first stage and S for second stage. The awake node variable is used to put the wireless node in an idle mode to wake it up when required, which helps in reducing the power consumption in the sensor node. This variable can be set manually or automatically, based on specific time period, from the user interface. In the default condition the sensor node is awake, however, if it receives the character D (for Disable) from the base station, the sensor node will idle; it can be enable again if character E (for Enable) is received.

When the system is initialised the wireless node sends a message to the base station telling it that it is awake and waiting to receive the desired sampling frequency (F_S) for the first stage signal capture, as shown in Fig. 9.5 which presents the flowchart of the sensor node programme. This stage is configurable to accommodate different robot tasks not just the pick and place task (described in Chap. 6). Figure 9.2 is a screenshot from the 'hyper terminal' used to communicate with the base station for presenting the analysis data from the embedded system. It was essential to setup the ADC to allow the sampling rate to be easily changed, using the on-chip Timer 3 module. Timers are one of the most important peripherals on embedded processors, e.g. as a counter that is incremented or decremented at the fixed time intervals.

On the base station side, for the first stage of operation the user must enter the sampling frequency to the wireless node. The second stage fault diagnosis is performed at a fixed speed for the joint therefore the sampling frequency is pre-set. If the embedded system is put in the first mode, the received trigger signal from the robot is translated to the ASCII character A and to B if the second mode is selected; these are then transferred to the sensor node. In the first mode the sensor node executes the first stage of the monitoring system, so the signals from the three axis accelerometers (X, Y and Z) will be captured by calling the subroutine code in the sensor node programme, as shown in Fig. 9.5. The captured signals are saved in three buffers (1, 2 and 3 respectively). In the main programme of the sensor node (Fig. 9.5) the STD value for each signal and the resultant STD value are computed; the latter is then pipelined into a vector array of 10 elements size. When the array is full, the control chart algorithm is implemented, and after a successive number of counts if they are (collectively) normally distributed within the control limits the robot is considered healthy; otherwise, there is an indication of a fault. The analysis result is then sent to the base station. If the sensor node indicates development of a fault in the robot, the

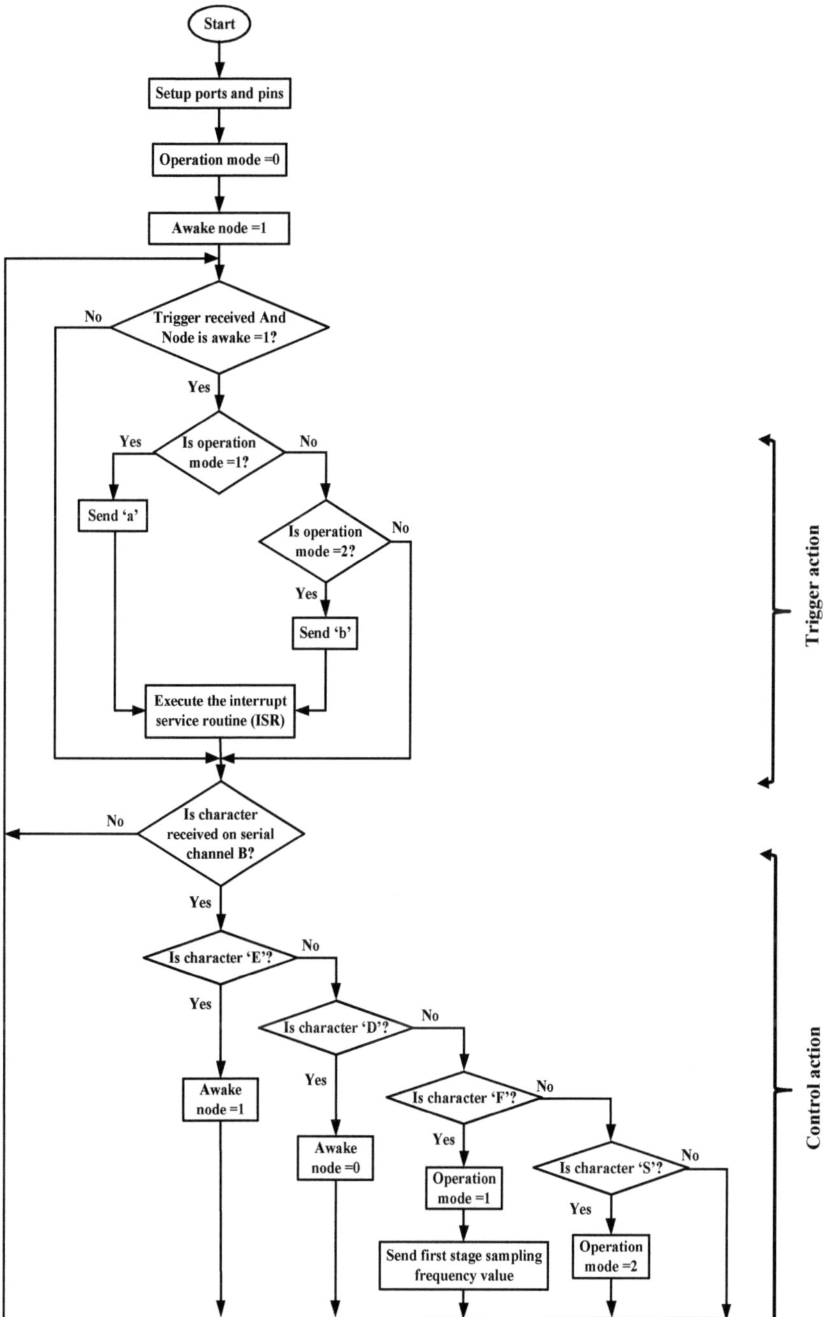

Fig. 9.3 Flow diagram of the main programme running on the base station

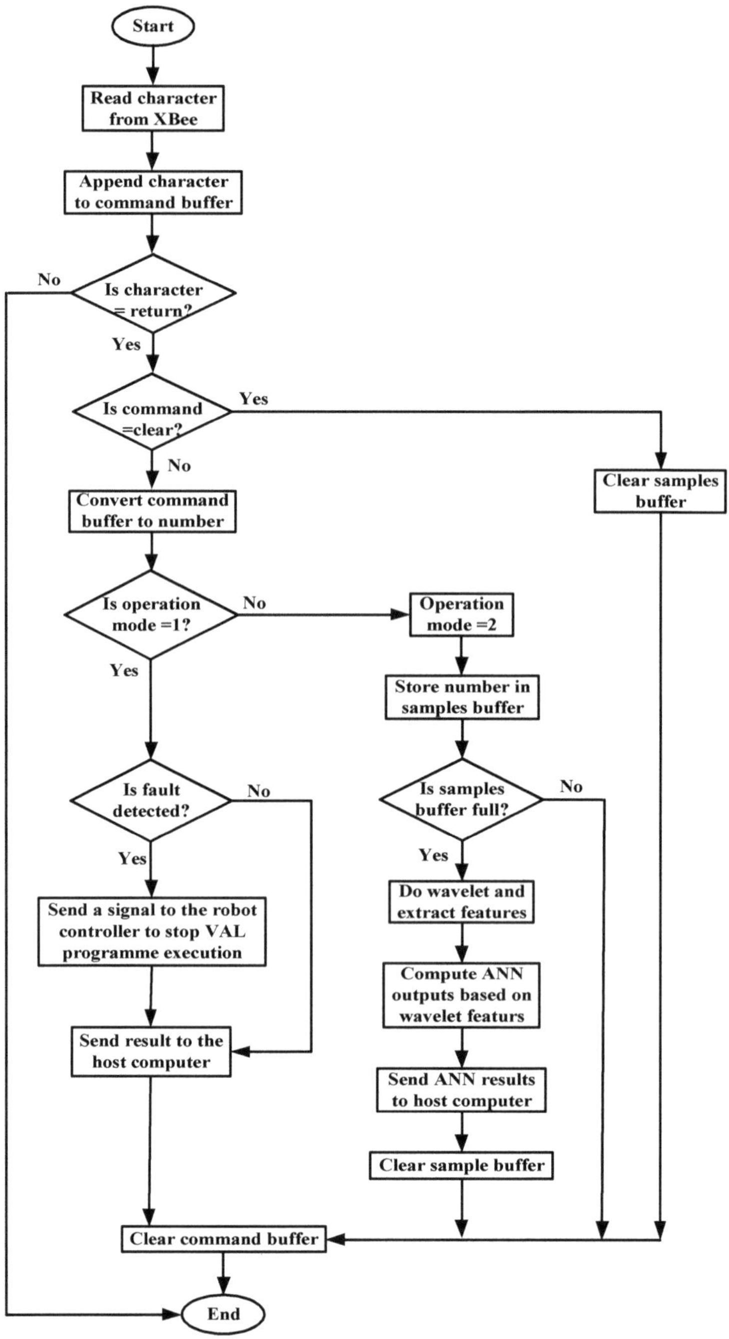

Fig. 9.4 Flow diagram of the subroutine programme running on the base station

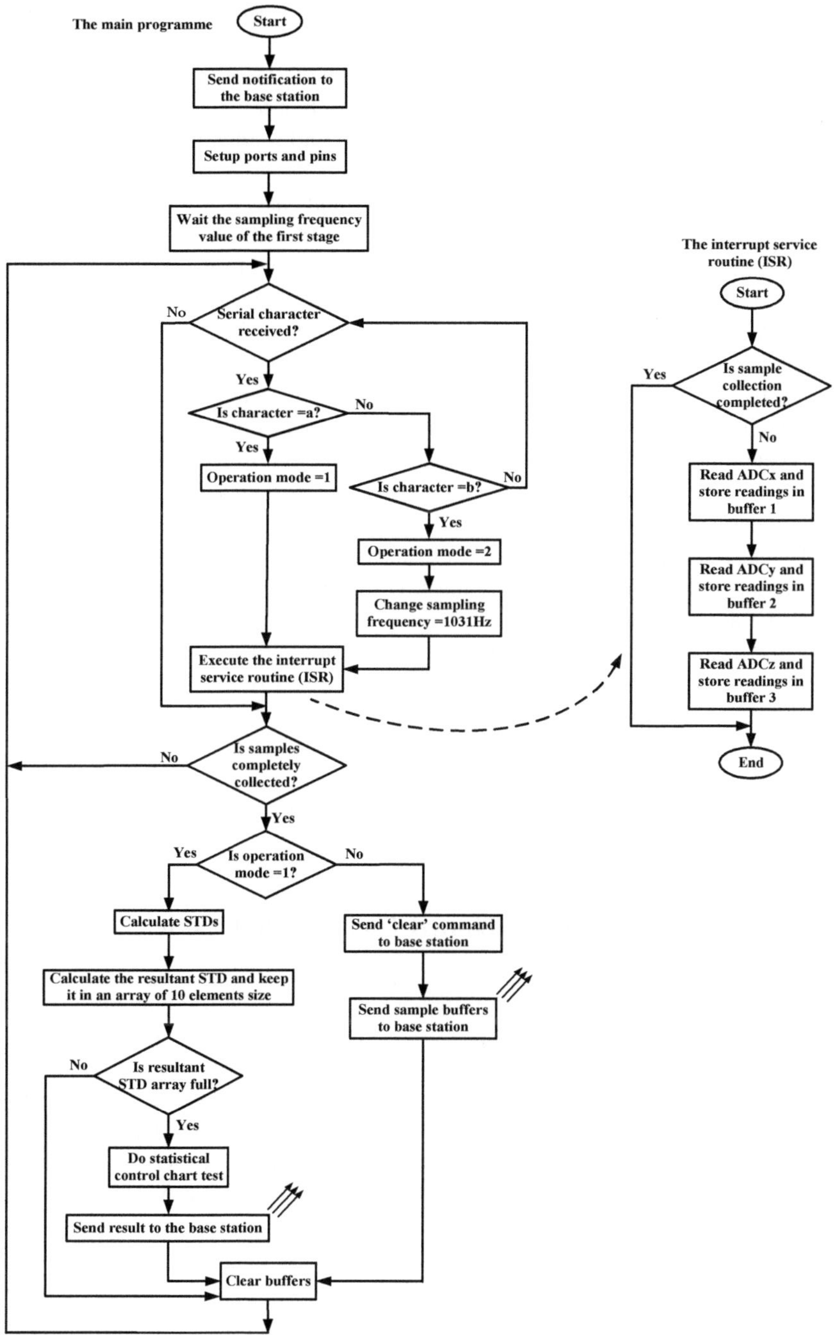

Fig. 9.5 Flow diagrams of the main and subroutine programmes running on the sensor node

base station transmits a signal to the robot controller to terminate task execution (see Fig. 9.4). This is achieved via the VAL code for the pick and place task, described in Chap. 6, using a subroutine named *check* (Appendix C.1) responsible for checking if the first line input on the input/output module is high or low (1 or 0). When the input signal is high, the programme will stop the robot and send a message to inform the operator that a fault has been detected.

Reception of the letter B on the sensor node indicates that the second stage has begun executing. At this level the sensor node will also capture the three accelerometer signals, but does not process them, instead forwards them to the base station, on which an interrupt service routine (see flowchart in Fig. 9.4) will be called. When the system is switched from the fault detection to diagnosis mode the sensor node sends a 'clear' command before transmitting the sample buffers, to inform the subroutine to delete any remaining samples if the fault detection cycle is incomplete, to prevent misdiagnosis of a fault. After receiving the requisite number of accelerometers samples, the wavelet transform is implemented, features are extracted and then passed to the neural network for classification; and the result sent to the host computer for visualization.

The base station programme is significantly more computationally intensive than the sensor node programme, since the DWT and ANN are algorithms processed, requiring a great amount of data, that has raised dynamic memory issues with the DSP board, due to the significant memory size required to store the DWT filter coefficients and the weights and biases of the designed ANN. To solve this issue, it was necessary to change several memory parameters using DSP/BIOS configuration tools. As a result, in the memory section manager (MEM), the IRAM was changed to SDRAM, which allows the use of additional memory to deal with large DWT decomposition levels and appropriate ANN architecture for accurate fault diagnosis.

9.4 Overview of the Developed Base Station Code

The integration and management of files for generating and running an executable file in the CCS code development process begins with the creation of what is called a 'project', which involves multiple source files, libraries, memory maps, and special command files. However, before executing any code on the base station, the C6713 DSK has to be initialized. The version of CCS supplied with the C6713 DSK has been configured to automatically load the general extension language (GEL) file '`DSK6713.gel`', which defines the memory map, sets some CPLD (complex programmable logic device) registers to configure components on the DSK board, and initializes the EMIF (external memory interface) for the memory on the board, while the developed program do the remaining initialization. Different header files, such as the chip support library (CSL) which contains C functions and macros for configuring and interfacing with all the C6713 on-chip peripherals and CPU interrupt controller and the DSK_COMM4 file that can be used to configure the interface of the DSK_COMM4 daughter card to the DSK board, are added to the project, at the beginning, using the '`#include`' directive, which is normally utilized to insert the contents of another file into the developed project. Also, to keep the prepared code as small as possible and easy to follow, two functions named `DWT.c` and `ANN.c` were developed as separate files and included into the main code by setting the search path, as shown.

```
#include "dsk_comm4_testcfg.h"                //DSK_COOM4 configuration file

#include "csl.h"                              //Chip support library file

#include "C:\Documents and Settings\DWT.c"

#include "C:\Documents and Settings\ANN.c"
```

In the main code, a function was developed to send a character via the RS232 to the user interface, which uses the second UART channel on the C6713 DSK. It takes a single ASCII character as an input and writes it to the UART output register; the UART then starts transmitting this character. For correct transmission without losing information a `Wait()` function is used to insert a small delay after each character.

```
void sendChar(char x)
{UART_THR2 = x; Wait () ;}
```

A similar function called void `sendCharXBEE(char x)` was employed to send a character via the XBee to the wireless node, using the third UART channel. To send string, which is defined as a sequence of characters, through the RS232 or the wireless module two functions were developed based on `sendChar` and `sendCharXBEE` functions. They loop through an array of characters (i.e. a string), sending one character at a time until they hit the "null terminator" ('\0'), which is a special character that marks the end of a string where the null terminator itself is not sent, as shown below.

```
void sendString(char *x)
{int i = 0; while (x[i] ! = '\0') sendChar(x[i ++]);}
```

Another two functions were developed to process the received information through the RS232 and XBee module. The first one is used for switching between the stages of the CM algorithm (fault detection or diagnosis) via the user interface whereas the second is employed to process the received data from the wireless node. After initialization and selection the required stage, the code execution waits within an infinite while loop until an interrupt occurs. The C6713 DSK has twelve interrupts (INT4–INT15), with INT4 having the highest and INT15 the lowest priority, and are employed for monitoring the occurrence of different events. INT4 was used for monitoring the trigger signal from the robot controller. The base station listens to the wireless node, after sending the triggering signal, until it receives the entire three accelerometer data packet and then executes the `DWT.c` and `ANN.c` functions, as follow:

```
DWT (samples, features);
ANN (features, results);
```

where 'samples' are the vector array of the received data packet and represents the input to the DWT function, 'features' are the extracted features from the wavelet analysis and 'results' are the output of the ANN. Lastly, the execution returns from interrupt to the while loop and then waits for the next interrupt to occur.

9.5 Overview of the Developed Sensor Node Code

Most Arduino code is based on using libraries to provide auxiliary functions that can be utilized in the main *sketch*. The computational algorithms of the developed code for the wireless node were written using functions, which allows their insertion into the main script by just naming the function and referencing the data used in the calculation. The three accelerometer outputs are connected directly to Arduino analog inputs A0, A2, and A4 pins, which are declared with their types at the start of the code as shown in the sample code below. The control chart limits (upper, lower and centre line) are then declared, along with three buffers of sizes 4096, one for each axis of the accelerometer. The buffers are used to store the captured data for the fault detection algorithm and are also transmitted to the base station for the fault diagnosis stage.

```
const int xpin = A0;                    // X-axis accelerometer
const int ypin = A2;                    // Y-axis accelerometer
const int zpin = A4;                    // Z-axis accelerometer
const double UCL = 0.745;               // Upper control limit
const double CL = 0.733;                // Centre line
const double LCL = 0.72;                // Lower control limit
unsigned short bufferX[4096];           // for x-axis signal
unsigned short bufferY[4096];           // for y-axis signal
unsigned short bufferZ[4096];           // for Z-axis signal
void setup()
{
Serial.begin(19600);                 // initialize serial communication
Serial.print("Sensor node AWAKE.\n\r");
Serial.print("Select OPMODE?\n\r");
analogReadResolution (12);
si = 0;      // An index counter used with the signals capturing process
changeSamplingFrequency(FREQ1_Hz);
}
```

As mentioned earlier the `void setup()` call only runs one time immediately the Arduino board is switch on or initialised. The data communication baud-rate between the Arduino board and the XBee is specified using `Serial.begin()` function, and messages are then transmitted to the base station using the `Serial.print()` function, which informs the user that the node is awake and the operation mode (fault detection or diagnosis) needs to be selected. The Arduino reads the analog voltages from the accelerometers and then digitalizes them using the internal ADC. The resolution of the ADC in the Arduino DUE comes preconfigured to only output 10 bit, however, it can be changed to 12 bit using the `analogReadResolution(bits)` function.

The `changeSamplingFrequency(FREQ1_Hz)` is a function that was developed to change the sampling frequency depending of the selected mode. A timer interrupt, using the *timer 3* module on Arduino DUE, was utilized for reading the signals from the accelerometers, using the `analogRead()` function as shown in the following code.

```
void TC3_Handler()

{

  TC_GetStatus(TC1, 0);

  if (si<4096)

  { bufferX[si] = analogRead(xpin);

    bufferY[si] = analogRead(ypin);

    bufferZ[si] = analogRead(zpin);

    si++; }

}
```

The void `loop()` section contains code that runs forever as long as the Arduino is on. In this section (see below), the control chart algorithm is performed using the function `testResult`. The operation mode in the wireless node can be changed depending on the received character from the base station. If `(c == 'a')`, the fault detection stage `(opmode = 1)` is selected while the fault diagnosis stage `(opmode = 1)` is selected if `(c == 'b')`. The function `Serial.available()` returns the number of received characters (or bytes) that have been buffered in the Arduino's incoming serial port; whenever it is greater than zero, the board will read the characters. The `Serial.read()` reads and returns the next character that is available in the serial port. A function was also developed to compute the medians of the captured data signals which was then subtracted to null the DC offset, as discussed in the developed data acquisition system (see Chap. 5). The function works by taking the samples of each signal, sorting them in ascending order, returning the central value, and then subtracts it.

```
void loop()
{
 boolean testResult;
 int c;
 while (Serial.available() > 0)
 {c = Serial.read();
  if (c == 'a')
  {opmode = 1; si = 0;
  changeSamplingFrequency(FREQ1_Hz);}
 else if (c == 'b')
  {si = 0; opmode = 2;
  changeSamplingFrequency(FREQ2_Hz);}
}
```

9.6 Embedded System Testing and Result

Prior to deploying the intelligent embedded condition monitoring system software, the major components were tested and validated, these include the captured vibration signals and the fault detection implemented on the sensor node, and the DWT and ANN realized on the base station. The testing was performed by setting up the system for real-time robot vibration analysis. The established results from the NI data acquisition system (Chap. 5), which were used for constructing both stages of the proposed CM system, are re-called here for comparison with their peer results generated by the embedded system, to establish the accuracy of the prepared C codes, as discussed in the following subsections.

9.6.1 Wireless Signal Capturing

The NI 6009 DAQ device used for building the preliminary data acquisition and analysis system has 14-bit resolution; the extracted robot vibration features using this system were utilized in Chap. 7 for designing the intelligent condition based monitoring (iCBM) system. Whereas the Arduino DUE board used in the sensor node has only 12-bit ADC resolution, means that 2-bit less than the used NI DAQ. Thus, it was important, as the first step of testing the designed embedded system, to establish if the produced accuracy of the signal capture using the DUE board, with respect to NI DAQ, is reasonable. Significant differences between signals are not preferred, as it could lead to increase rate of false fault detection and diagnosis.

Generally, ADCs transform analogue signal, after sampling at a constant time interval, produces a discrete signal in both time and amplitude; known as the *quantization* process. Conveniently, an n-bit ADC will assign a finite number of amplitude levels corresponding to discrete values of input signals between the range of 0 and the full-scale value of the sensor output. Mathematically, if an ADC has a range of 12-bit resolution, 4096 different values (where $2^{12} = 4096$) of the input voltage can be represented. Thus, a 12-bit ADC with a maximum input range of 3.3 V, which is the case here, the resolution would be (approximately) 3.3/4096 = 0.8 mV. Similarly, for the same voltage range, a 14-bit ADC resolution is $3.3/2^{14} = 0.2$ mV. Concluding that the higher the resolution, the larger the number of divisions the range is broken into and the smaller the detectable voltage change, while with low resolution ADC some information will be lost.

The result of the healthy signals comparisons for each of the three axes of measurements is presented in Fig. 9.6. Signals from both NI DAQ and Arduino DUE are compared with the robot executing the pick and place task, using the same sampling frequency (383 Hz). The time-domain signals are associated with their probability distributions, to find out if the signal distortion, due to the reduced resolution in Arduino, has skewed the distribution of the signals. It can be

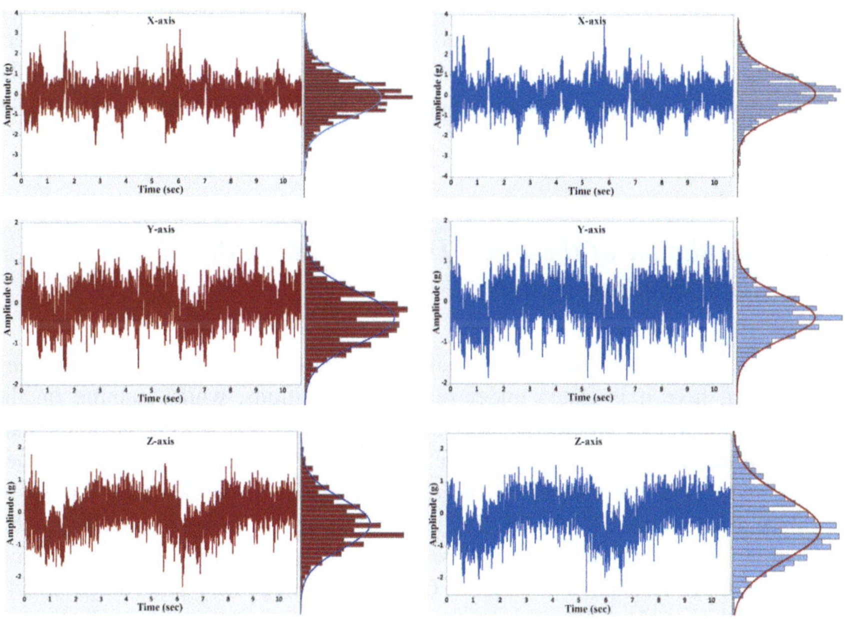

(a) Signals from theNI DAQ **(b) Signals from Arduino DUE**

Fig. 9.6 Time-domain signals from the X, Y and Z accelerometers captured using NI 6009 DAQ and Arduino DUE board

Table 9.1 Statistical features extracted from time-domain signals captured using NI 6009 DAQ and Arduino DUE board

Axis of measurement	Used ADC	STD (g)	RMS (g)	Maximum (g)	Minimum (g)
X	NI DAQ	0.62613	0.626	3.18387	−2.50729
	Arduino DUE	0.6636	0.664	3.5530	−2.5378
Y	NI DAQ	0.45126	0.448	1.39296	−1.67155
	Arduino DUE	0.50545	0.506	1.63550	−1.91748
Z	NI DAQ	0.52054	0.521	1.79090	−2.30827
	Arduino DUE	0.57752	0.578	1.52271	−2.42505

seen that the embedded system signals are identical with NI DAQ signals, and normally distributed. Also, the lost information from the signals does not affect their extracted statistical features significantly, as shown in Table 9.1. It can be observed that the signals from the Arduino are having slightly higher amplitudes, also the other statistical features, i.e. the standard deviation (STD) and root mean square (RMS). This could be attributed to attaching the sensor node to the robot structure, which cannot be avoided, making the system capturing extra vibration from the robot, while this is not the case when the NI DAQ has been used. Nevertheless, it was concluded that the signals captured using the Arduino DUE board have very reasonable accuracy, and thus it is expected that by employing the DUE board the desired credibility level of the designed robot health monitoring system can be achieved.

9.6.2 Validation of the Fault Detection Stage Based on Embedded System

In Chap. 7 the resultant standard deviation control chart was tested offline using unseen data, here it is tested under real-time conditions, whilst running on the wireless sensor node. The testing was done when the robot is in the healthy state and with different backlash levels, while it executes its repetitive task (as described in Chap. 6). As in the offline test the robot was stopped periodically in order to introduce different backlash levels. The result of the test is shown in Fig. 9.7. In this test extra limits are added, these shall be called the upper and lower warning limits (*UWL* and *LWL*) respectively, equal to $CL \pm 2\sigma$ (Montgomery and Runger 2014), where σ is the standard deviation of the samples used for constructing the control chart, and can be used to give an early indication about the robot health deterioration.

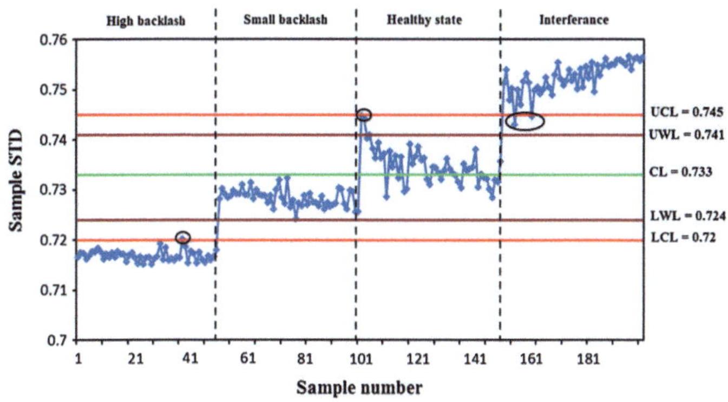

Fig. 9.7 Fault detection result from the designed embedded system

From the figure it can be seen that the extracted features for the low and interference backlash cases are outside the control limit; this is a very obvious indication of a fault in the robot. In the small backlash level it is observed that all the STD features are inside the warning limits, on the same side below the centre line. This is also an indication of a fault, based on the fourth rule of the Western Electric Company rules (Montgomery and Runger 2014), which states that a run of eight (or more) consecutive points on one side of the centre line (Chap. 7). Samples of the healthy case are randomly distributed inside the warning limits, indicating a healthy condition. Occasional some non-consecutive points fall outside the control limits or one of successive faulty points located within the control limits, these are marked in black ellipses in the figure and is due to the intrinsic variation within the calculated features (Montgomery and Runger 2014). In addition to the rule of eight consecutive points (mentioned above), if there are three or more successive points exceeding the warning limits or six points in a row steadily increasing or decreasing, a fault development in the robot is considered. These three rules are applied in the embedded system for the early robot fault detection.

To simulate faults similar to the natural fault development standard, by which a machine deteriorates gradually from its healthy condition, it would have been better if the backlash fault could be developed progressively while the robot is executing the pick and place task, which currently cannot be achieved. One possible idea is to fix a controlled device to the backlash adjustment grub-screws on the robot and programme it to increase or decrease the backlash level concurrently during the robot task execution. However, this is out of the scope of this research, but could be applied in future work. Thus, in this work to present something similar to gradual deterioration fault, first from the robot healthy condition the backlash was changed (increased or decreased) and ten measurement points were taken; then the power was disconnected from the robot arm and also the backlash was changed and another ten points were captured. Each time the grub-screw was turned by a quarter turn, where each turn is equivalent to change the backlash by 0.072 mm (Chap. 5). Figure 9.8 shows the captured result when the robot was healthy (backlash = 0.68 mm) and the

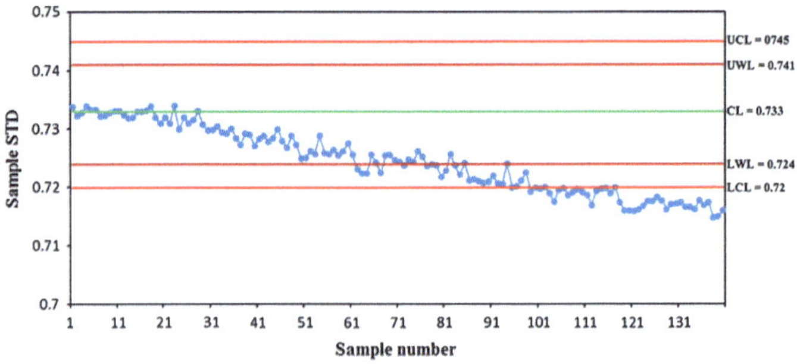

Fig. 9.8 Fault detection result from the designed embedded system when the backlash was gradually increased

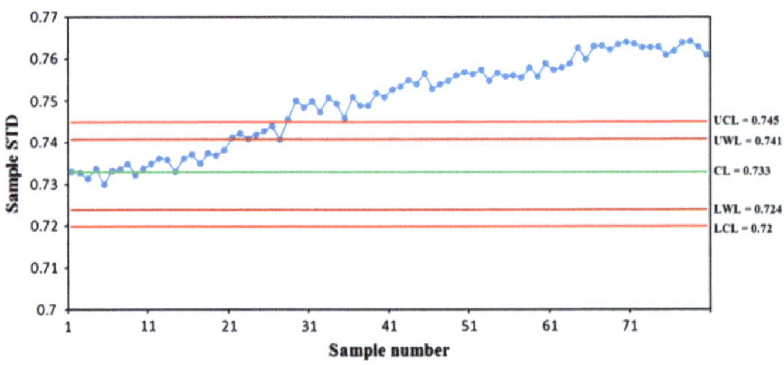

Fig. 9.9 Fault detection result from the designed embedded system when the backlash was gradually decreased

backlash progressively increased (each time by 0.018 mm) until the excessive backlash is reached (backlash = 0.92 mm), while Fig. 9.9 presents the result of decreasing the backlash (also each time by 0.018 mm) starting from the robot healthy condition until gears interference was introduced (backlash = 0.56 mm). A total of 140 points were captured in the case of backlash increasing, whereas 80 points were captured in the case of backlash decreasing (Figs. 9.8 and 9.9), as a higher number of steps are required to obtain excessive backlash than obtaining the gear interference.

It can be clearly seen from these figures that the resultant STD values are gradually decreased, as the backlash is increased, and increased, as the backlash is decreased. Obviously, the resultant STD values indicated by the samples 1–30 in Fig. 9.8 and to 20 in Fig. 9.9 are randomly distributed around the centre line (CL), signalling a healthy robot condition. This is, however, at the healthy backlash range, which is a little higher or lower 0.68 mm (Chap. 5). In both cases when a fault was detected, which was at sample 38 in Fig. 9.8 (as eight consecutive points

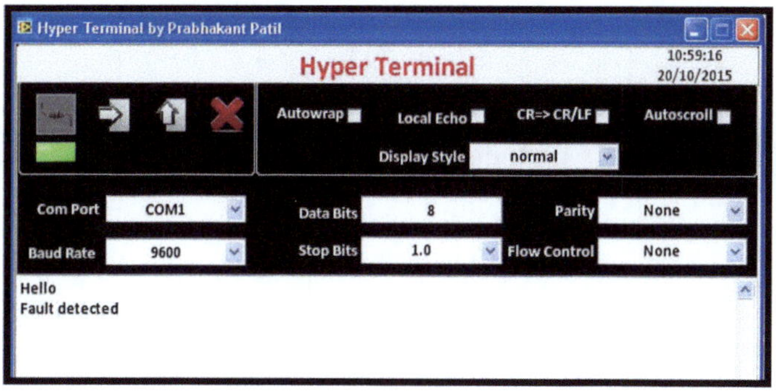

Fig. 9.10 A screenshot from the serial communication software indicating detection of a fault in the robot

under the *CL*) and sample 23 in Fig. 9.9 (as three consecutive points above the *UWL*), the embedded system showed a message flagging that a fault was detected, Fig. 9.10. It can be concluded that the designed control chart is sensitive enough to the small changes in the backlash level.

As stated previously the sampling frequency in the fault detection stage is 383 Hz, therefore the time required to capture three signals, each with 4096 samples for the three accelerometers is about 10.7 s. Based on the developed control chart, to obtain one data point 10 signals from each axis must be collected in order to calculate the resultant standard deviation; thus, the time required is 107 s ($\approx$1.8 min). However, to apply the stated fault detection rules at least 10 points should be computed, in order to apply the fault detection rules, and therefore the total time required for the entire fault detection process is approximately 18 min.

9.6.3 Validation of DWT Based on Embedded System

After ensuring that the captured and wirelessly received signals on the DSP board (or base station) are correct, it is important to validate the DWT signal analysis. The developed C code for DWT analysis was downloaded on the base station and the robot was programmed to perform the cyclic rotation task of joint 3 (Chap. 6). The received signals on the base station are analysed in a real-time manner, then, the break point and watch window debugging tools in CCS IDE were utilized to save the gathered signals to a text file for offline analysis. These signals were analysed using the Matlab wavelet toolbox; the result of DWT from DSP and Matlab for X-axis signal is depicted in Fig. 9.11.

The sub-band signals produce by Matlab and DSP board are highly correlated with each other. This is readily apparent in D1, D2, D3 sub-bands and even D4, due

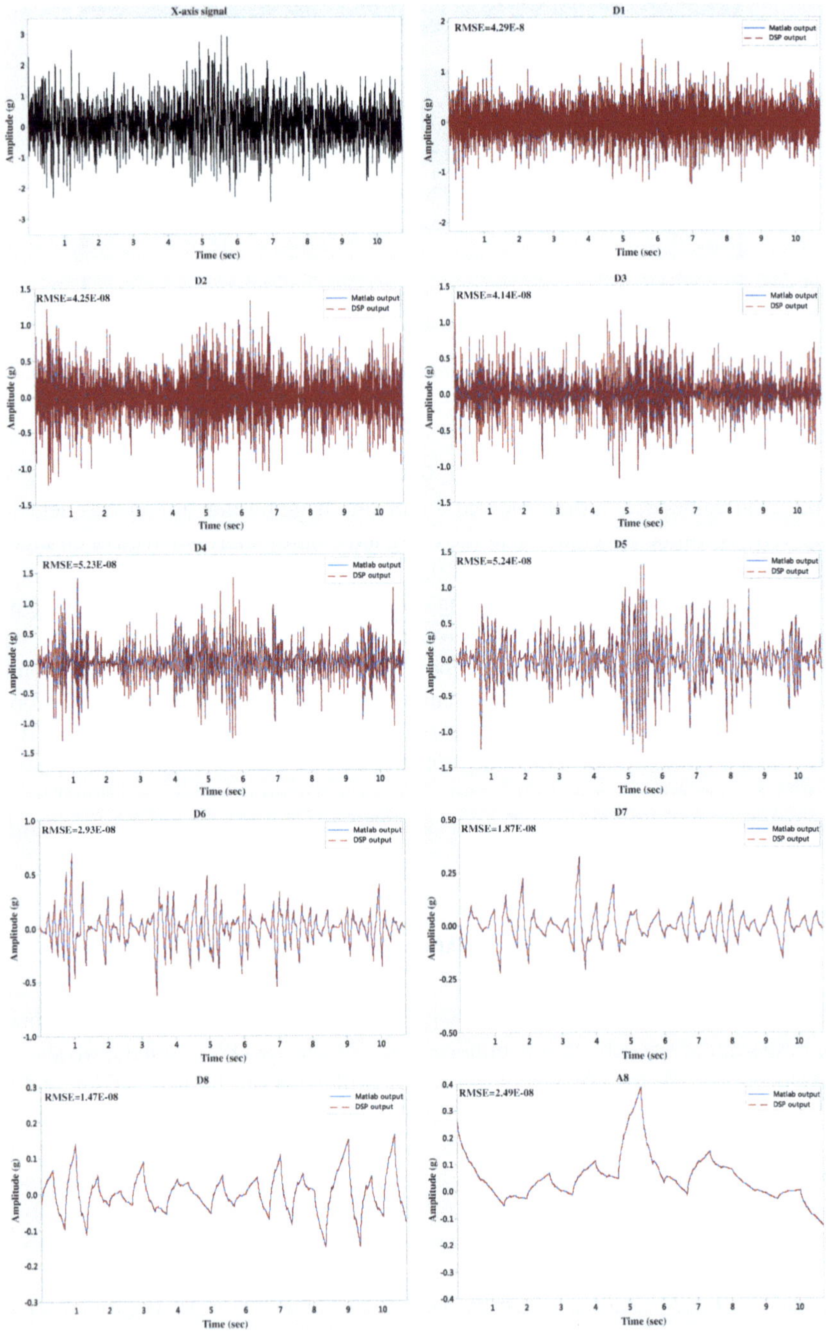

Fig. 9.11 DWT analysis result using Matlab and designed embedded system

to the significant frequency contents at the higher levels of analysis. Matlab uses higher precision and accuracy; however, the DWT analysis based DSP exhibits certain errors with respect to the DWT based Matlab. The difference between each level signals is evaluated using the root-mean-square error (RMSE), which is also indicated in Fig. 9.11. However, the difference appears negligible and can be ignored, indicating a correct and accurate DWT implementation on the base station.

9.6.4 Validation of the Designed ANN Based on Embedded System

As mentioned earlier in the thesis, once the wavelet analysis is done, the standard deviation feature for each decomposition level is calculated, and a vector of 24 element (8 from each axis signal) is passed to the fault diagnosis stage. The designed ANN using Matlab software for fault classification (discussed in Chap. 7) is realized in the embedded system. The established network structure (number of layers and neurons in each layer) and weight matrices are adapted for developing the C code for ANN to be operating in the based station. The ANN code was successfully built and downloaded, along with DWT code, on the DSP board; and has to be tested before use. Actually, testing the ANN means testing the entire fault diagnosis stage, as this is based on DWT analysis and ANN fault classification.

The wireless node was attached to the robot; and the robot 'healthy' condition and the backlash faults (high, small and interference) were tested, as in the fault detection stage. To verify the system capability in classifying the gear and bearing faults the vibration signals related to these faults, which were captured using the developed LabVIEW data acquisition software (Chap. 6), are re-called here. These signals were sent wirelessly from the sensor node to the base station in order to be classified. It was stated when the ANN was designed (Chap. 7) that the output layer contains 11 neurons, and that each one indicates a specific robot fault type, except the first one, which is related to the robot healthy condition. In order for the signal classification consider to be correct the output of just one node must be 1 (or close to 1) and the remaining nodes are zero (or close to zero) depending on the robot health condition. The result of this validation is presented in Fig. 9.12, where the X-axis is the neuron's order and Y-axis its output value. It can be seen that when the robot is healthy the output of the first neuron is 1 and the others are close to zero, and the same is true with other faults. For each fault condition, the embedded system was tested many times and the classification success rate was found close to 100 %.

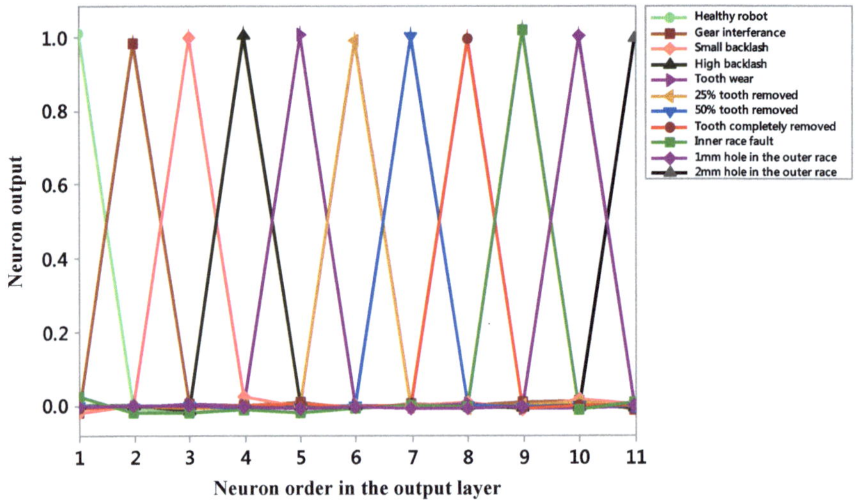

Fig. 9.12 Embedded system fault classification result

9.7 Sensor Node Power Consumption Analysis

Rechargeable batteries represent a finite energy source, and they need to be charged time to time or replaced as their functioning life deteriorates. These disadvantages will increase the maintain cost of a wireless node, and also the maintenance problem becomes more complex if sensor nodes are used in remote locations. In general, the consumed power in wireless sensor nodes is determined by the hardware platform features and the programme that runs within the node. In this work the designed embedded system will be executing the fault detection stage most of the time and if a fault is detected the diagnosis stage is executed one time, therefore, the sensor node power consumption analysis was done at just the fault detection stage.

To monitor the consumed current on the sensor node, a multi-meter device has been placed in series between the power source (the battery) and Arduino power jack. A simple LabVIEW code was prepared to record the drawn current that required for calculating one resultant STD. It was found that the average demanded current is 116.6 mA, meaning that the node will run continuously for up to 8.5 h; this has come from dividing the used battery current capacity, which is 1000 mAh, on the established average consumed current. However, this is not a very long time for prospective industrial wireless system.

To overcome the issue of powering wireless systems, some researchers have developed energy harvesting system to scavenge energy from the available ambient sources, such as solar and vibration energy, to either power the sensor node or as an auxiliary powering system to charge the battery when it gets discharged (Li et al. 2013; Zhang et al. 2014). Investing the solar energy for power supply to the node in this work is not possible, as the required solar panel would be too big size and thus cannot be fixed on the robot. However, for future work, robot movement energy could be invested in building a small energy harvesting system for the wireless sensor node. On the other hand, since the embedded system could be manufactured and sold as a part from the robot system, for future development some arrangements might be considered in the design of the robotic system in order to power the node from the robot controller.

9.8 Summary

In this chapter the available software development tools for the proposed embedded system hardware were deliberated. The designed software was divided into two parts, as the proposed condition monitoring algorithm; the first part was implemented on the sensor node and mainly responsible for robot health assessment and fault detection based on the designed control chart (Chap. 7). The second part of the code involves the fault diagnosis stage, which relies on the proposed wavelet family for signal analysis (Chap. 3) and the developed ANN for extracted feature classification (Chap. 7). Arduino IDE and CCS studio software, which are based on C language, were utilized for programming the sensor node and base station, respectively.

In order to demonstrate the validation of the developed wireless embedded system, captured vibration data were compared to their peers in Chap. 6. The three backlash levels were then introduced in the robot and the embedded system showed successful robot health change detection. The DWT analysis of the wirelessly received signals on the base station was examined to investigate the developed code correctness and accuracy; the results showed very high correlation with Matlab results. In terms of fault diagnosis, the system has proved high capability in classifying the seeded faults. However, it would be better if a wider range of real defects were considered; due to the time required to develop such defects the investigation of the full range of possible defects which might occur would have been beyond the scope of this research.

Optimistically, the established wireless embedded system, with a little modification on the code if required, can be envisaged for fault detection and diagnostic for any machine that its monitoring system is realized on vibration signal analysis. Also, with some modification on the system hardware, especially the sensor node hardware, or the software different condition monitoring techniques, such as acoustic emission or motor current signature analysis, can be applied.

References

Faludi, R. (2011). *Building wireless sensor networks*. Sebastopol: O'Reilly Media, Inc.
Kehtarnavaz, N. (2008). *Digital signal processing system design: LabVIEW-based hybrid programming*. Amsterdam: Elsevier.
Li, Y., Li, X., & Wang, P. (2013). A module harvesting wind and solar energy for wireless sensor node. In *Communications in computer and information science*. Berlin: Springer.
Montgomery, D. C., & Runger, G. C. (2014). *Applied statistics and probability for engineers*. New York: Wiley.
Zhang, X., Fang, J., Meng, F., & Wei, X. (2014). A novel self-powered wireless sensor node based on energy harvesting for mechanical vibration monitoring. *Mathematical Problems in Engineering, 2014*.

Chapter 10
Conclusions and Future Work

Abstract This work has been carried out to design an online, intelligent health monitoring system for an industrial robot based on embedded electronic technology. Throughout the journey of this research a large amount of effort and time was employed to achieve this aim using a systematic and logical approach. In this chapter, the research work that has been described in this thesis is concluded and the limitations of the study are discussed. A summary of the main scientific contributions and achievements in the thesis are presented, followed by a discussion of the potential directions for further research in terms of both extending and development of the present work to make it more suitable for practical applications.

10.1 Conclusions

This research has been based on a mechatronics approach, by which different fields are integrated into an application. Throughout this thesis, a preliminary summary to conclude the isolated findings was integrated at the end of each chapter; however, this section is devoted to summarising the entire thesis, discussing the important conclusions and the links between the chapters.

In the first chapter the necessary background related to industrial robots and their fault categories and the available maintenance strategies, with focus on condition based maintenance, was provided. Also, the main challenges that were expected to be faced during the development process of the robot condition monitoring (CM) system along with the study aims and objectives were discussed. A thorough review of previous and contemporary research in the state of the art was conducted in Chap. 2 to investigate and evaluate the exiting condition monitoring approaches and techniques. A model-free (or data-driven) approach using vibration monitoring techniques was selected for robot health monitoring, due to the difficulty of constructing an accurate analytical model for the robot and also its capability in developing high performance CM systems. Furthermore, the statistical and artificial intelligence tools that have been commonly utilized in developing model-free CM systems were reviewed in this chapter.

A.A. Jaber, *Design of an Intelligent Embedded System for Condition Monitoring of an Industrial Robot*, Springer Theses,
DOI 10.1007/978-3-319-44932-6_10

Due to the complexity of the robot dynamic behaviour, as it is completely different from the behaviour of rotating machines for which most of CM systems were developed, and to fulfil the aim of creating an intelligent robot health monitoring system that is able to detect progression of a fault and then diagnose it, a two-stage CM algorithm was proposed (Chap. 3), composed of fault detection and fault diagnosis stages. The first stage is responsible for detection of the fault, regardless of its location and type, when the robot is executing a repetitive task. The diagnosis stage is responsible for locating where the fault is and its type; it is performed when the robot runs one joint at a time in a cyclic motion. To accomplish this, different signal processing techniques, which usually have an important role in analysing the captured signals and then extracting features in any CM applications, were evaluated (Chap. 3). The decision was, taking into account the implementation of the selected techniques in the designed embedded system in Chap. 8, to use the time-domain signal analysis in conjunction with statistical control chart (SCC) in the fault detection stage and the time-frequency domain analysis, based on discrete wavelet transform (DWT), along with artificial neural network (ANN) in the diagnosis stage. The time-domain signal analysis and SCC were selected for the first stage because they do not required a lot of computational effort and also the fault type is not known in this stage, so just signalling the progression of a fault was important. The DWT and ANN were used in the second stage due to the capability of DWT in detecting the transitory nature in vibration signals, which is very common in the presence of a fault, and the high performance of the ANN in solving various classification problems.

The background information on PUMA 560 robot mechanical construction, methods of control and programming was provided in Chap. 4, along with an experimental modal analysis (EMA) carried out to find the robot natural frequencies. The theory behind the EMA and the used hardware and software to accomplish it has been presented. The experiments showed that the robot arm configurations significantly affect the emergence of these frequencies. The established frequencies were compared with the results of other research and acceptable correlation was found. The usage of these frequencies in preliminary robot condition monitoring was evaluated, as presented in Chap. 6, and that introducing of a fault in the robot will excite some of the robot natural frequencies, therefore they could be utilized for robot health monitoring.

In general, several transmission fault types that could be progressed in gears and bearings were investigated along with their characteristic frequencies in Chap. 5. In the literature concerning industrial robot fault diagnosis it was established that the majority of them are focusing on gear backlash faults diagnosis, implying there are essential shortcomings in the diagnosis of other types of transmission faults, such as gear tooth wear and breakage and inner and outer race bearing faults. Thus, it was decided to fill part of this gap by simulating different levels of gear and bearing faults; however, due to the time limitation only one joint (the elbow) of the robot was selected to seed the fault.

To study the robot vibration signals when different faults are presented a preliminary data acquisition system was developed using the capabilities of LabVIEW and Matlab in signal capturing and analysis (Chap. 5). The selected signal processing techniques (time-domain and DWT) were implemented in the developed system. Initially, different levels of backlash (high, small and interference) were simulated; after that dismantling and reassembling the selected joint was practiced and the bevel pinion and one of the bearings used to support the input shaft were used to seed the gear and bearing faults, which are (as shown in Chap. 5) tooth wear, 25 % tooth removed, 50 % tooth removed, tooth completely removed, a rectangular slot with 1 mm width in the inner race, 1 mm hole in the outer race and 2 mm in the outer race.

Three single axis, MEMS type accelerometers were utilized to measure the robot vibration in three axes (X, Y and Z). The optimum location that can be used to affix the accelerometers on the robot was justified and found to be close to the robot wrist. Sensor mounting methods were deliberated and an aluminium adapter was designed to hold the accelerometers in an orthogonal configuration; attached on the identified location using cyanoacrylate adhesive. The vibration analysis was conducted with the robot executing a simple and repetitive pick and place task, which mimics one of its common robot operation, and then when the robot programmed to cyclically rotate just joint 3. Programming of the robot was achieved using its dedicated control language named VAL II; the developed VAL II codes are presented in Appendix C. The measured vibration and extracted features that are related to different health conditions were discussed in Chap. 6. It was established that as the backlash increased the robot vibration level decreased, attributed to the damping effect of the grease lubricant, and vice versa. On the other hand, as the gear tooth or bearing fault severity increased the vibration level is increased as a result.

Using the established robot natural frequencies and the estimated gears and bearing fault frequencies an approach based on combining the capabilities of the DWT and the Fast Fourier transform (FFT) was investigated for the robot fault detection and diagnosis, as discussed in Chap. 6. Although this approach has shown good ability in detecting the changes in frequencies related to the natural frequencies and gears and bearings it was not applied in the designed embedded system, as it is more suitable for offline condition monitoring. The basis and theory behind the SCC and ANN were presented in Chap. 7; the Shewhart average level chart and multilayer perceptron neural network (MLPN) were selected. Then, based on the extracted features from the robot vibration when it is healthy and with different seeded faults were used to design the SCC and ANN; a test procedure was accomplished and the results were discussed and found that very reasonable performance was shown in fault detection and classification.

A wireless embedded system was developed for online implementation of the proposed intelligent robot health monitoring algorithm. It comprised two parts: the sensor node for vibration signal acquisition and conducting the fault detection analysis, and the base station for controlling the embedded system operation and implementing the diagnosis stage. As presented in Chap. 8, several commercially available proprietary electronic components employed in embedded system development were evaluated, in order to identify and select the main embedded processors. A low-cost Arduino DUE microcontroller was selected as the core of the sensor node, whereas the DSK C6713 digital signal processor board from Texas Instruments was utilized in the base station. To power the sensor node a lithium-ion polymer (Li-Po) rechargeable battery was used. Other auxiliary electronics, such as a wireless shield for Arduino and DSK_COMM4 board for the DSK C6713, were used to integrate the wireless capability in the developed system. Evaluation of different wireless technologies, include Wi-Fi, Bluetooth and ZigBee, was also undertaken; it was established that ZigBee is the best candidate, as it has a feasible communication range, cost-effective and low-power technology. Embedded code based on the C programming language was designed to apply the developed CM algorithm using the developed wireless embedded system; a discussion and the flow charts that explain the main steps and working concept of the code were realized in Chap. 9. Also, in the same chapter the results of system testing and validation was deliberated and very good performance was realized. In this respect it is felt that the main aims and objectives of the thesis have been achieved.

On the other hand, as previously stated the fault detection stage was designed to be able to detect the fault in any of the robot parts regardless of location, which was made possible using the SCC approach, however, the ANN was trained to diagnosis only a limited number of faults in only joint 3; this represents one of the developed system limitations. Other issue was faced is the limited power supply source that was used with the designed wireless node, as the battery powering method was found lasting for limited period of time (approximately 8.5 h). Also, integrating such a long antenna (RPSMA) on the sensor node could lead to interaction between the sensor node and the robot working environment in the practical field. The issue of limited memory size of the DSK C6713 board was an obstacle in developing a big size ANN, which could bound the design of (at least) a semi-comprehensive robot condition monitoring system.

10.2 Contribution to Knowledge

The research described in this thesis has made the following main contributions towards the state of the art in condition monitoring systems:

1. An intelligent condition monitoring algorithm based on vibration analysis of an industrial robot for fault detection and diagnosis has been proposed, and which combines the use of a statistical control chart with time-domain signal analysis for detecting a fault, and the use of an artificial neural network with the discrete wavelet transform for diagnosing the specific fault type. The algorithm can be

implemented when the robot is in an operational state and undertaking any given task, and not for a specific task, as in previous research.

2. The design and development of an embedded micro-processor system, based on wireless communication, for online implementation of the intelligent condition monitoring algorithm, described above. A wireless sensor node, with integral accelerometers, was developed and attached to the robot arm for data acquisition and fault detection. A base station was also developed, which can be interfaced to a PC, for presenting the results of the analysis, and for fault diagnosis.
3. The implementation of a discrete wavelet transform with statistical feature extraction for the robot fault diagnosis is also considered to be novel. In this method, the vibration signals are first decomposed into eight levels of wavelet coefficients. Then different statistical parameters, such as the standard deviation and root mean square for each level are calculated and analysed in order to select the one sensitive to specific features. The standard deviation parameter was found to be the most fault sensitive and thus selected as a quantitative feature to be extracted from the wavelet analysis and subsequently used to recognize the onset of faults in the robot joint transmissions.
4. The simulation of different faults types in the robot joints was also considered to be a novel development. In addition to variances in different levels of backlash, which is the most common type of fault studied by previous researchers, gear tooth and bearing faults with several levels of severity were successfully introduced and detected in the robot transmission. The simulation of gear tooth and bearing faults in industrial robot transmissions is considered to be rather challenging which is why these have not been reported before in the previous research. A considerable amount of time was devoted to practicing the dismantling/assembly of the elbow joint of an industrial robot to permit the simulation of these faults.
5. The adverse effects of backlash faults were highlighted in the robot. Generally, in rotating machinery the increase of backlash in pairs of mating gears will increase the vibration level, however, it was found that in the robot used in thus study an increase in backlash was consistent with decreasing vibration amplitudes. This was attributed to the copious amounts of grease lubricant entering between the mating teeth at increased backlash levels leading to damping of the vibration, whereas with reduced backlash levels the lubricant is effectively squeezed out of the mating teeth giving rise to a much stiffer dynamic system.

10.3 Summary of Achievements

The following provides a list of the summary of achievements attained during the execution of this study:

1. An extensive literature review was undertaken to establish the recent trends in the condition monitoring algorithms that are applied to a wide range of rotating machinery, with an emphasis on industrial robots.
2. The different signal processing techniques that can be applied in condition monitoring systems were extensively assessed, and their differences and suitability for the application in this study were evaluated. A time-domain signal analysis approach was adopted in the detection stage and the time-frequency analysis based on discrete wavelet transform used in the diagnosis stage.
3. A frequency response function analysis, using experimental modal analysis, was carried out to establish the robot natural frequencies.
4. A data acquisition system for online vibration signal capturing, analysis and feature extraction was developed using National Instrument hardware (NI 6009) and software (LabVIEW) in conjunction with Matlab software.
5. Preliminary vibration analysis, when the robot is healthy and with different faults introduced, was conducted, and important features were extracted. Preliminary robot fault detection and diagnosis based on the extracted features and robot natural frequencies was carried out.
6. The design, training and testing of the proposed control chart and neural network for fault detection and diagnosis were successfully achieved. In this regard, the extracted features using the developed data acquisition system were utilized.
7. The design of a sophisticated embedded software suite, using C programming language, was developed, upon which both stages of the proposed condition monitoring algorithm were implemented on the designed embedded system.
8. An extensive series of tests were conducted on a 'healthy' robot, and subsequently when a fault was introduced to evaluate the performance and effectiveness of the designed embedded condition monitoring system.

10.4 Recommendations for Future Work

Having achieved the main objectives, there are still some potential challenges remaining that needed to be overcome in order to implement the developed monitoring system into widespread industrial application and thus the total success of the research is achieved. For future development of this work the following recommendations should be considered:

Recommendations for future experimental work

1. The investigated faults were mainly related to the elbow joint transmission (gears and bearings). Other categories of faults related to this joint motor, such as motor bearings, stator and rotor faults, or faults in the other joints of the robot form a logical direction to further extend the capabilities of the designed

embedded-based health monitoring system. Also, the examined faults scenarios were simple, as only one fault is seeded at a time; however complex scenarios could be studied by combining multiple fault types at the same time.

2. In the present study, due to the limitation of time accelerated fault types were seeded in the joint gearbox. For future research, the algorithm could be extended to monitor the progressive development of faults in the robot transmission. This can be achieved by allowing the robot to operate for a considerable time period and the vibration level is monitored. Otherwise, a control device that can be connected to the backlash adjustment grub-screws and allow the progressive increase or decrease of the gear backlash while the robot is running could be developed.
3. Use different type of condition monitoring techniques, such as acquiring the acoustic emission signal, motor current signal, noise signal… etc., or combining two (or more) techniques together, such as vibration with online motor current analysis, for robot health monitoring, which could be conducted using hall-effect sensors attached to the motor power cables.
4. The proposed condition monitoring approach was developed to monitor a rather old, but operational, PUMA 560 robot, and thus it follows that more experimental work is needed to further evaluate its reliability in fault detection and diagnosis with the robot executing a variety of different tasks. Also, it should be tested on another PUMA 560 robot in order to verify that it can confidently be utilized with this model of robots. Of course it would be much more appropriate if the designed system is generalised for any industrial robot model, however, a considerable effort and time are needed to investigate the required scientific analysis methods, and the appropriate approaches for hardware and software development.
5. Only one robot was monitored throughout this study, therefore, the hardware and software could be developed to accommodate monitoring of more than a robot. This can be achievable by building and using more than one wireless sensor node and updating the designed embedded software to be able to communicate with more than one sensor node.

Recommendations for improving the condition monitoring algorithm

6. The designed control chart in this thesis, which is of a univariate type, Shewhart average level chart, has a weakness that is relatively insensitive to small shift in population mean, since a large number of plotted values are required to detect the change (Montgomery and Runger 2014). To overcome this problem, more advanced univariate control charts, such as the cumulative sum (CUSUM) and the exponentially weighted moving average (EWMA) could be designed. These charts have better performance for detecting small shifts than Shewhart chart, as they allow each plotted point depends not only on the most recent subgroup average but also on some of the other subgroup averages.
7. Another significant problem with control chart is that it assumes that the monitored system behaviour is steady-state (or static) and ignores the influence of the dynamic effect, meaning that the intended or unintended shift in the

process mean or variance when the system is still healthy. A possible solution for this problem would be to explore other approaches that could help in developing control charts with adaptive limits that can follow system dynamics in the sense that big variation from central line, which is consequence of system dynamics and not system fault, is treated as a normal operation mode (Kisić et al. 2013).

8. The control chart was utilized to monitor only one quality character, which is the resultant standard deviation of the captured vibration signals, to detect the progression of a fault. As a future work, however, multivariate control charts, which provide simultaneous monitoring of several correlated variables and superior to univariate charts monitoring (El-Din et al. 2006), are needed for developing more robust fault detection stage.
9. In this study, the implemented neural network for fault diagnosis uses supervised-learning algorithm for training the network, meaning that the network has to be given the historical information corresponding to input (extracted features related to different fault types) and output (identities of faults). However, the major drawback of this learning method is that only faults that are included in the training data can be diagnosed. Thus, to build a comprehensive fault diagnosis stage, the historical data of all possible robot fault types must be available for training, which in practice is extremely difficult. As an extension to the present work, unsupervised-learning can be used to handle this issue.
10. Although the designed ANN has shown high classification capability, it seems that there is still a significant scope for research in AI techniques, aiming to simplify the complex nonlinear systems and realizing cost effective hardware. Beside others, support vector machines, fuzzy logic and practical swarm optimization should be investigated and their suitability to be applied on an embedded system with an expanded number of the classified fault categorizes established.
11. To reduce the effect of the progressed fault in the robotic system performance and in order to allow the robot to operate as long as possible before a catastrophic failure is taken place a fault tolerant control (FTC) system should be developed; which is defined as a control system that should be performed immediately after a successful detection of a fault and it possess the ability to maintain the desirable performance and stability properties while accommodating a safe component failures (Ding 2014)..
12. The main problem that was faced during the deployment of the designed intelligent condition monitoring system on the embedded system was the limited memory size of the base station digital signal processor. This was mostly due to the size of the designed neural network for fault classification, which is strongly connected with number of the neurons in the input, hidden and output layers which in turn associated with the input features and output fault identities. However, the neurons number in the output layer cannot be changed, as it depends on the number of classified faults, instead statistical techniques could be utilized to reduce the size of the input vector. As an example, the principle component analysis (PCA) could be applied to reduce

the input data dimensionality so that only the more essential information is retained.

13. Other time-frequency signal analysis techniques should be tried and the sensitivity of extracted features to the seeded fault analysed. These techniques include, but not limited to, discrete wavelet packet transform (DWPT), which is a continuation to DWT by which the detail information of the signal is further decomposed, and empirical mode decomposition (EMD), which is a numerical approximation algorithm designed to extract the intrinsic oscillatory components of a signal (Yan 2015). The efficiency of these methods in terms of sensitivity to fault, processing speed and required computational effort could be compared with the used DWT and the best is selected.
14. A realistic mathematical model that allows simulation of different fault severities and considers the different sources of uncertainties present in practice, such as joint flexibility, torque ripple, temperature and speed and load variations, could be developed for the joint gearbox. This will help in understanding the induced vibration signals under different healthy and faulty conditions.

Recommendations for embedded system development

15. The use of a restricted power supply and the consequent requirement of periodic battery replacement represent the main issue with the developed wireless sensor node. To defeat this limitation, energy harvesting approaches show promising technology for powering wireless systems, which makes them more independent. Several possible solutions could be integrated with developed sensor node, for example, electrical energy can be produced by harvesting the ambient vibrations using magnetostrictive material (Chalasani and Conrad 2008), or using ambient RF energy to remotely power the wireless unit (Bouzid 2013).
16. The used Arduino and DSK C6713 boards have superfluous peripherals. Thus, to make the proposed embedded system suitable for industrial applications the wireless sensor node and the base station are required to be miniaturised, for ease of development and installation. This can be achieved by integrating the antenna, processor, power supply and the other required components in a single board; this would contribute in lowering the cost of the complete system and reduce the weight, size and power consumption, especially for the sensor node.
17. It would also be much better if the embedded system is consolidated into a single design specific electronic board (attached to the robot) that has the capability to implement both stages of the condition monitoring algorithm and communicating wirelessly with a PC, for data transmitting. For example a small size embedded device with high hardware and computational specifications from National Instruments called myRIO (www.ni.com/myrio), or any other device with a higher specification, could be utilized to for this purpose.
18. In the fault detection stage the control chart was developed offline depending on the extracted features from healthy time-domain vibration signals that were captured while the robot performing the pick and place task. However, if the robot is programmed to achieve different task, or as it is recommended that the

control limits should be re-estimated between time to time even if the robot is healthy (Montgomery and Runger 2014), the upper and lower control limits of the chart must be re-computed again based on the feature extracted from the newly captured vibration signals. This process is time consuming, thus it is very important that the capability of online control chart design is developed on the designed wireless sensor node.

19. For the fault diagnosis stage in Chap. 7, the weights and the biases for the ANN were determined using offline training performed in Matlab; which were then downloaded on the embedded system. However, for real time application would be more practical if an online training strategy by which the ANN parameters are adjusted in the DSP board itself. This would make the training process easier and the adoption of other fault types on the embedded system is simpler.
20. Programming embedded systems using C language was found to be a non-trivial task especially for a researcher with very little experience in programming. As a result, a lot of challenges were encountered during the embedded software development, specifically with the development of the base station software using the CCS IDE. However, it would be preferable if instead of writing lines of code in text-based programming, the user simply uses graphical programming languages, as they offer an interactive and a more intuitive approach toward building embedded systems in a relatively short amount of time, by which blocks from different libraries are grabbed and placed into the workspace and connected with arrows. Both Arduino and DSK C6713 can be interfaced to Simulink and LabVIEW, therefore in future it is very crucial that this approach is investigated and adopted for designing embedded systems.
21. The developed embedded system should be interfaced to a user-friendly graphical user interface (GUI) that provides an intuitive environment to guide the user through the steps of operating the embedded system and has clear indicators that assist in easy interpretation of the vibration analysis result even for non-expert maintenance technicians. Furthermore, it could be integrated with the capability to remotely communicate with mobile platforms, such as mobile phones and PDAs, via sending messages/emails that enable the concern users to be aware of the robot working condition. The GUI can be designed based on LabVIEW or Matlab software, however, they are expensive and a licence is required. Instead, Visual Basic programming could be adopted, which is free and readily compatible with Microsoft Windows programs.

References

Bouzid, O. M. (2013). *In-situ health monitoring for wind turbine blade using acoustic wireless sensor networks at low sampling rates*. Ph.D. Thesis, Newcastle University.

Chalasani, S., & Conrad, J. M. (2008). A survey of energy harvesting sources for embedded systems. *Conference Proceedings—IEEE SOUTHEASTCON* (pp. 442–447).

Ding, S. X. (2014). *Data-driven design of fault diagnosis and fault-tolerant control systems*. Springer.

El-Din, M. A. S., Rashed, H. I., & El-Khabeery, M. M. (2006). Statistical process control charts applied to steelmaking quality improvement. *International Journal of Quality Technology and Quantitative Management, 3*, 473–491.

Kisić, E., Petrović, V., Jakovljević, M., & Đurović, Ž. (2013). Fault detection in electric power systems based on control charts. *Serbian Journal of Electrical Engineering, 10*, 73–90.

Montgomery, D. C., & Runger, G. C. (2014). *Applied statistics and probability for engineers*. Wiley.

Appendix A
Datasheets of the Used Equipment in EMA

A.1 ICP Impact Hammer Model 086c0

Performance	**ENGLISH**	**SI**
Sensitivity(± 15 %)	10 mV/lbf	2.25 mV/N
Measurement Range	± 500 lbf pk	± 2224 N pk
Resonant Frequency	≥ 22 kHz	≥ 22 kHz
Non-Linearity	≤ 1 %	≤ 1 %
Electrical		
Excitation Voltage	20 to 30 VDC	20 to 30 VDC
Constant Current Excitation	2 to 20 mA	2 to 20 mA
Output Impedance	<100 ohm	<100 ohm
Output Bias Voltage	8 to 14 VDC	8 to 14 VDC
Discharge Time Constant	≥ 2000 sec	≥ 2000 sec
Physical		
Sensing Element	Quartz	Quartz
Sealing	Epoxy	Epoxy
Hammer Mass	0.34 lb	0.16 kg
Head Diameter	0.62 in	1.57 cm
Tip Diameter	0.25 in	0.63 cm
Hammer Length	8.5 in	21.6 cm
Electrical Connection Position	Bottom of Handle	Bottom of Handle
Extender Mass Weight	2.6 oz	75 gm
Electrical Connector	BNC Jack	BNC Jack

CE [2]

All specifications are at room temperature unless otherwise specified.
In the interest of constant product improvement, we reserve the right to change specifications without notice.

ICP® is a registered trademark of PCB Group, Inc.

A.A. Jaber, *Design of an Intelligent Embedded System for Condition Monitoring of an Industrial Robot*, Springer Theses,
DOI 10.1007/978-3-319-44932-6

A.2 ICP Accelerometer Model 352c68

Performance	ENGLISH	SI	
Sensitivity (±10 %)	100 mV/g	10.2 mV/(m/s^2)	
Measurement Range	±50 g pk	±491 m/s^2 pk	
Frequency Range (±5 %)	0.5 to 10000 Hz	0.5 to 10000 Hz	
Frequency Range (±10 %)	0.3 to 12000 Hz	0.3 to 12000 Hz	
Frequency Range (±3 dB)	0.2 to 20000 Hz	0.2 to 20000 Hz	
Resonant Frequency	≥35 kHz	≥35 kHz	
Phase Response (±5 °) (at 70&#176F [21&#176C])	2 to 6000 Hz	2 to 6000 Hz	
Broadband Resolution (1 to 10000 Hz)	0.00016 g rms	0.0015 m/s^2 rms	[1]
Non-Linearity	≤1 %	≤1 %	[3]
Transverse Sensitivity	≤5 %	≤5 %	[4]
Environmental			
Overload Limit (Shock)	±5000 g pk	±49050 m/s^2 pk	
Temperature Range (Operating)	-65 to +200 °F	-53 to +93 °C	[2]
Base Strain Sensitivity	≤0.005 g/με	≤0.05 (m/s^2)/με	[1]
Electrical			
Excitation Voltage	18 to 30 VDC	18 to 30 VDC	
Constant Current Excitation	2 to 20 mA	2 to 20 mA	
Output Impedance	≤300 Ohm	≤300 Ohm	
Output Bias Voltage	8 to 12 VDC	8 to 12 VDC	
Discharge Time Constant	0.8 to 2.4 sec	0.8 to 2.4 sec	
Settling Time (within 10% of bias)	<10 sec	<10 sec	
Spectral Noise (1 Hz)	60 μg/√Hz	588 (μm/sec^2/√Hz	[1]
Spectral Noise (10 Hz)	16 μg/√Hz	157 (μm/sec^2/√Hz	[1]
Spectral Noise (100 Hz)	5 μg/√Hz	49 (μm/sec^2/√Hz	[1]
Spectral Noise (1 kHz)	1.5 μg/√Hz	14.7 (μm/sec^2/√Hz	[1]
Physical			
Size (Height)	0.73 in	18.5 mm	
Weight	0.070 oz	2.0 gm	[1]
Sensing Element	Ceramic	Ceramic	
Size (Hex)	9/32 in	9/32 in	
Sensing Geometry	Shear	Shear	
Housing Material	Titanium	Titanium	
Sealing	Welded Hermetic	Welded Hermetic	
Electrical Connector	10-32 Coaxial Jack	10-32 Coaxial Jack	
Electrical Connection Position	Top	Top	
Mounting Thread	5-40 Male	5-40 Male	
Mounting Torque	8 to 12 in-lb	90 to 135 N-cm	

CE [5]

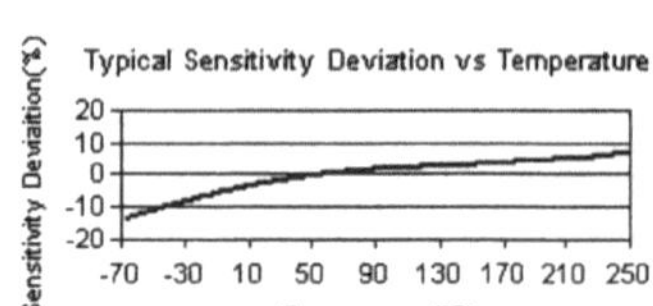

All specifications are at room temperature unless otherwise specified.
In the interest of constant product improvement, we reserve the right to change specifications without notice.

A.3 Data Physics Quattro Analyser

HARDWARE SPECIFICATIONS

INPUT

2 to 4 channels
ADC Resolution (Analog AAF): 24-bits
Sample Resolution (Digital AAF): 32-bit floating point
Coupling: AC/DC, DIFF/SE. ICP, TEDS
Anti-alias Filters: 100 dB protection, all ranges
Dynamic Range: 120 dB
Input Ranges: 0.1, 1, 10 V Full Scale (20 V optional)
Input Impedance:
100 kΩ symmetric for Diff.;
100 kΩ with 50 Ω shield to GND for SE
Max Input Voltage: 80 vPeak, 2.5 vRmsShield (SE)
CMRR: 60 dB (typical), f < 40 kHz
Amplitude Accuracy:
+/- 0.020 dB (0.2% FS) at 1 kHz for 15° C < T < 40° C
Amplitude Ripple: (Digital AAF) - 0.005 dB for 0 < f < fS / 2.56
Amplitude Droop:
(Analog AAF) - 0.005 dB at 5 kHz; 0.010 dB at 25 kHz;
0.050 dB at 49 kHz
Residual Offset: +/- 0.1% FS AND not larger than 3 mvDC
Phase Accuracy: 0.05° to 0.5° for DC to 40 kHz
Crosstalk between Inputs: < -100 dB
Crosstalk between inputs and source: < -90 dB
THD: -100 dB @ 1 kHz
Minimum SampleRate: < 1 Sps
Maximum SampleRate: 102.4 kHz standard (204.8 kHz optional)
Maximum useful Frequency: 40 kHz standard (94 kHz optional)
Frequency Accuracy: 25 ppm
Time Accuracy: 25 ppm

TACHOMETER INPUT

1 channel
Maximum Frequency: 200 kHz
Input Range: +/- 24 V FS
Adjustable threshold, holdoff, prescaling

OUTPUT

1 to 2 channels
Dynamic Range: > 100 dB
Resolution: 24-bit
Voltage Range: 10 V FS
Output Current: 1 mA min., continuous short
THD: -100 dB @ 1 kHz
Output Waveforms: 65536 max blocksize for arbitrary; unlimited for recorded (optional)

CHASSIS DETAIL

Dimensions: 5.6" x 4.0" x 0.9"
Weight: 1.2 lbs
Operating Temperature: 0 to 55° C
Power: USB 2.0 Bus Powered
Lights: Input OK (4), Output Active (2), Trigger Signal (1), DSP Active (1), USB Active (1)

NOTE: Continued product improvement necessitates that Data Physics reserve the right to modify these specifications without notice.

Data Physics Corporation
Discover more at www.dataphysics.com

DATA PHYSICS CORPORATION
1741 Technology Drive, Suite 260, San Jose, CA 95110
TEL: (408) 437-0100 FAX: (408) 437-0509

Appendix B
Setting up the Preliminary Data Acquisition System

B.1 RMS of Time-Domain Signals for Selection the Best Accelerometer Location

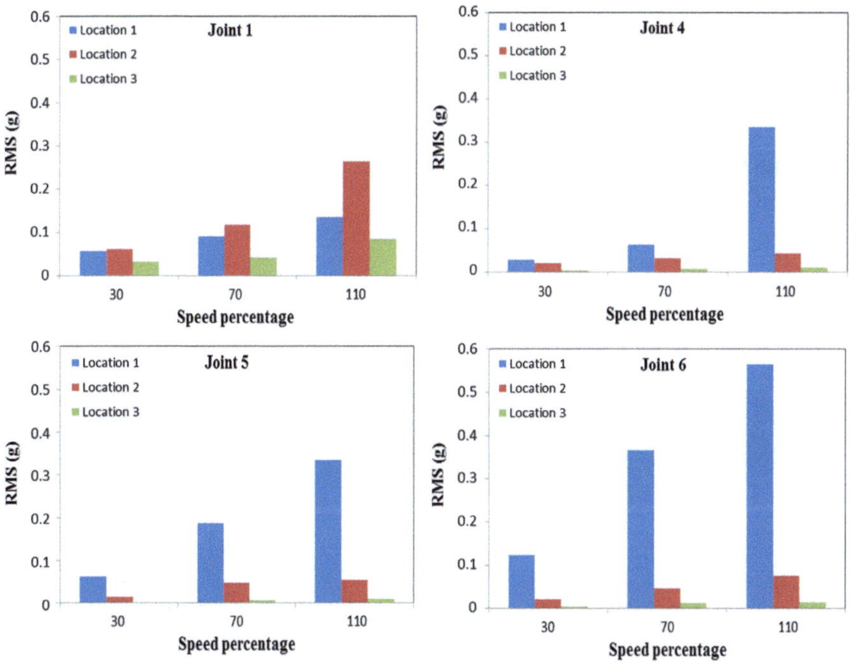

RMS values when the accelerometer on different locations and the robot moving different joints at different speeds

A.A. Jaber, *Design of an Intelligent Embedded System for Condition Monitoring of an Industrial Robot*, Springer Theses,
DOI 10.1007/978-3-319-44932-6

B.2 Calibration Result of the MEMS Accelerometer Against the Piezoelectric One

Time-domain signals

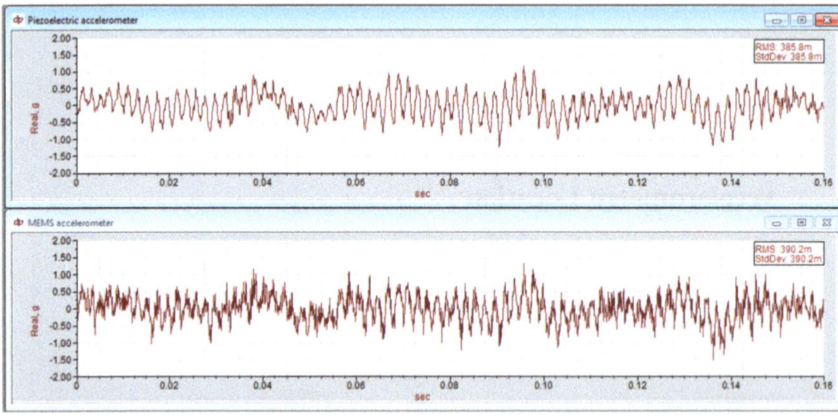

Frequency-domain signals

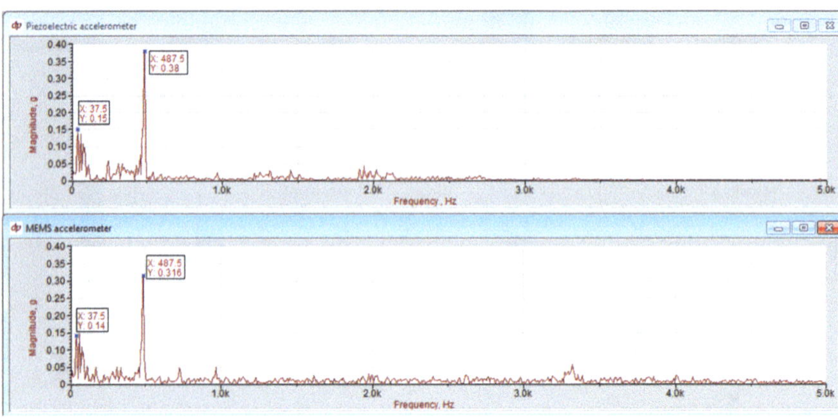

B.3 NI USB-6009 DAQ Card Pins Layout and Description

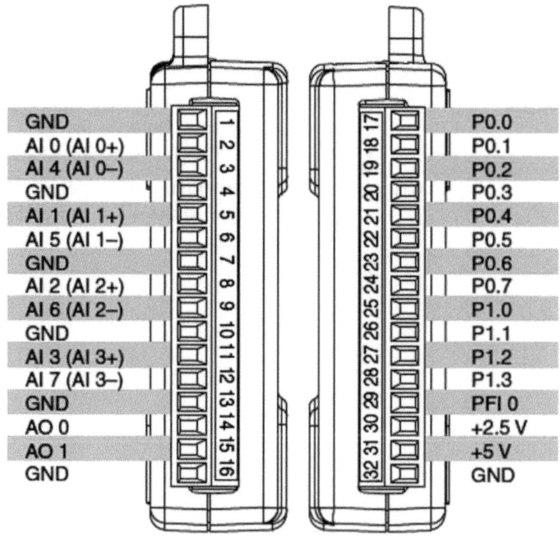

Signal Name	Reference	Direction	Description
GND	–	–	**Ground**—The reference point for the single-ended analog input measurements, analog output voltages, digital signals, +5 VDC supply, and +2.5 VDC at the I/O connector, and the bias current return point for differential mode measurements
AI⟨0–7⟩	Varies	Input	**Analog Input Channels 0 to 7**—For single-ended measurements, each signal is an analog input voltage channel. For differential measurements, AI 0 and AI 4 are the positive and negative inputs of differential analog input channel 0. The following signal pairs also form differential input channels: AI⟨1, 5⟩, AI⟨2, 6⟩, and AI⟨3, 7⟩. Refer to the *Analog Input* section for more information
AO⟨0, 1⟩	GND	Output	**Analog Output Channels 0 and 1**—Supplies the voltage output of AO channel 0 or AO channel 1. Refer to the *Analog Output* section for more information
P0⟨0–7⟩	GND	Input or Output	**Port 0 Digital I/O Channels 0 to 7**—You can individually configure each signal as an input or output. Refer to the *Digital I/O* section for more information
P1⟨0–3⟩	GND	Input or Output	**Port 1 Digital I/O Channels 0 to 3**—You can individually configure each signal as an input or output. Refer to the *Digital I/O* section for more information

(continued)

(continued)

Signal Name	Reference	Direction	Description
PFI 0	GND	Input	**PFI 0**—This pin is configurable as either a digital trigger or an event counter input. Refer to the *PFI 0* section for more information
+2.5 V	GND	Output	**+2.5 V External Reference**—Provides a reference for wrap-back testing. Refer to the *+2.5 V External Reference* section for more information
+5 V	GND	Output	**+5 V Power Source**—Provides +5 V power up to 200 mA. Refer to the *+5 V Power Source* section for more information

B.4 Used Schmitt Trigger Datasheet

74HC14

Hex Schmitt-Trigger Inverter

High–Performance Silicon–Gate CMOS

The 74HC14 is identical in pinout to the LS14, LS04 and the HC04. The device inputs are compatible with Standard CMOS outputs; with pullup resistors, they are compatible with LSTTL outputs.

The HC14 is useful to "square up" slow input rise and fall times. Due to hysteresis voltage of the Schmitt trigger, the HC14 finds applications in noisy environments.

Features

- Output Drive Capability: 10 LSTTL Loads
- Outputs Directly Interface to CMOS, NMOS and TTL
- Operating Voltage Range: 2.0 to 6.0 V
- Low Input Current: 1.0 μA
- High Noise Immunity Characteristic of CMOS Devices
- In Compliance With the JEDEC Standard No. 7A Requirements
- ESD Performance: HBM > 2000 V; Machine Model > 200 V
- Chip Complexity: 60 FETs or 15 Equivalent Gates
- These are Pb–Free Devices

ON Semiconductor®

http://onsemi.com

MARKING DIAGRAMS

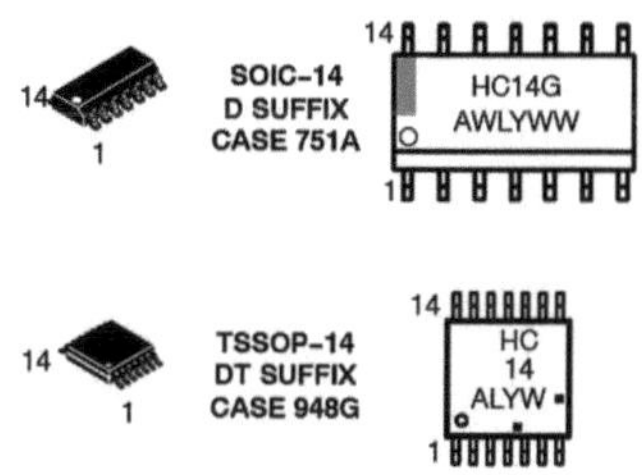

HC14 = Device Code
A = Assembly Location
L, WL = Wafer Lot
Y = Year
W, WW = Work Week
G or ▪ = Pb–Free Package
(Note: Microdot may be in either location)

Pinout: 14–Lead Packages (Top View)

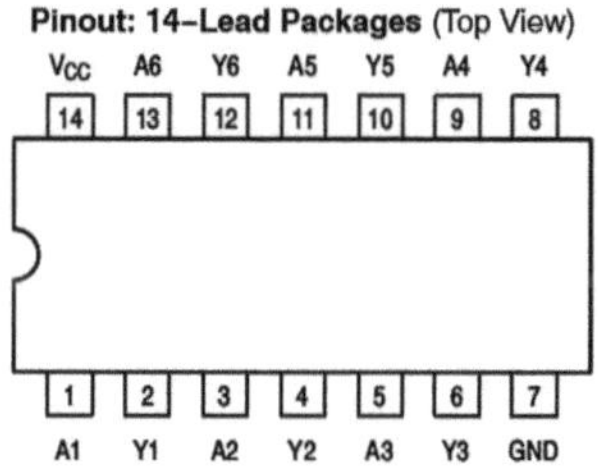

FUNCTION TABLE

Inputs	Outputs
A	Y
L	H
H	L

B.5 Develop LabVIEW Data Acquisition and Analysis Software

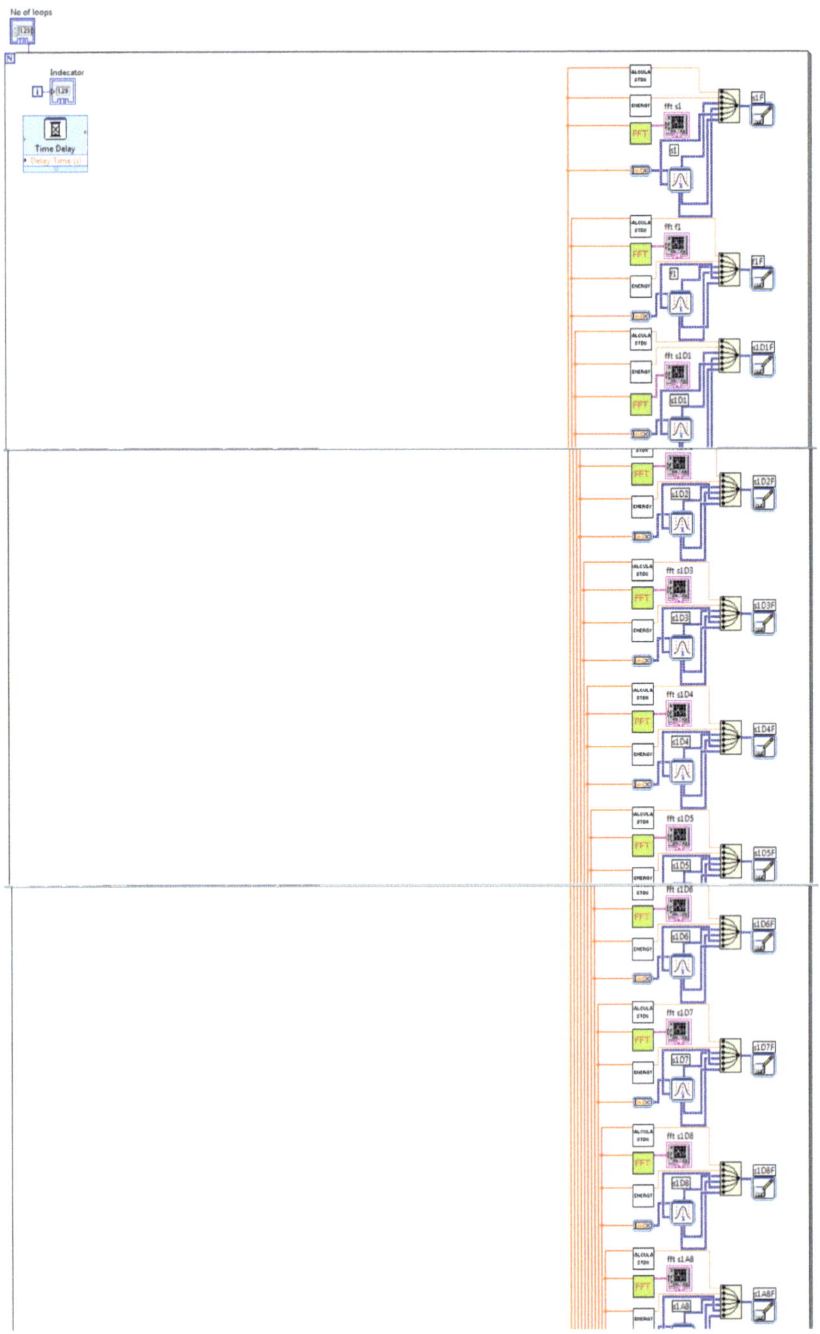

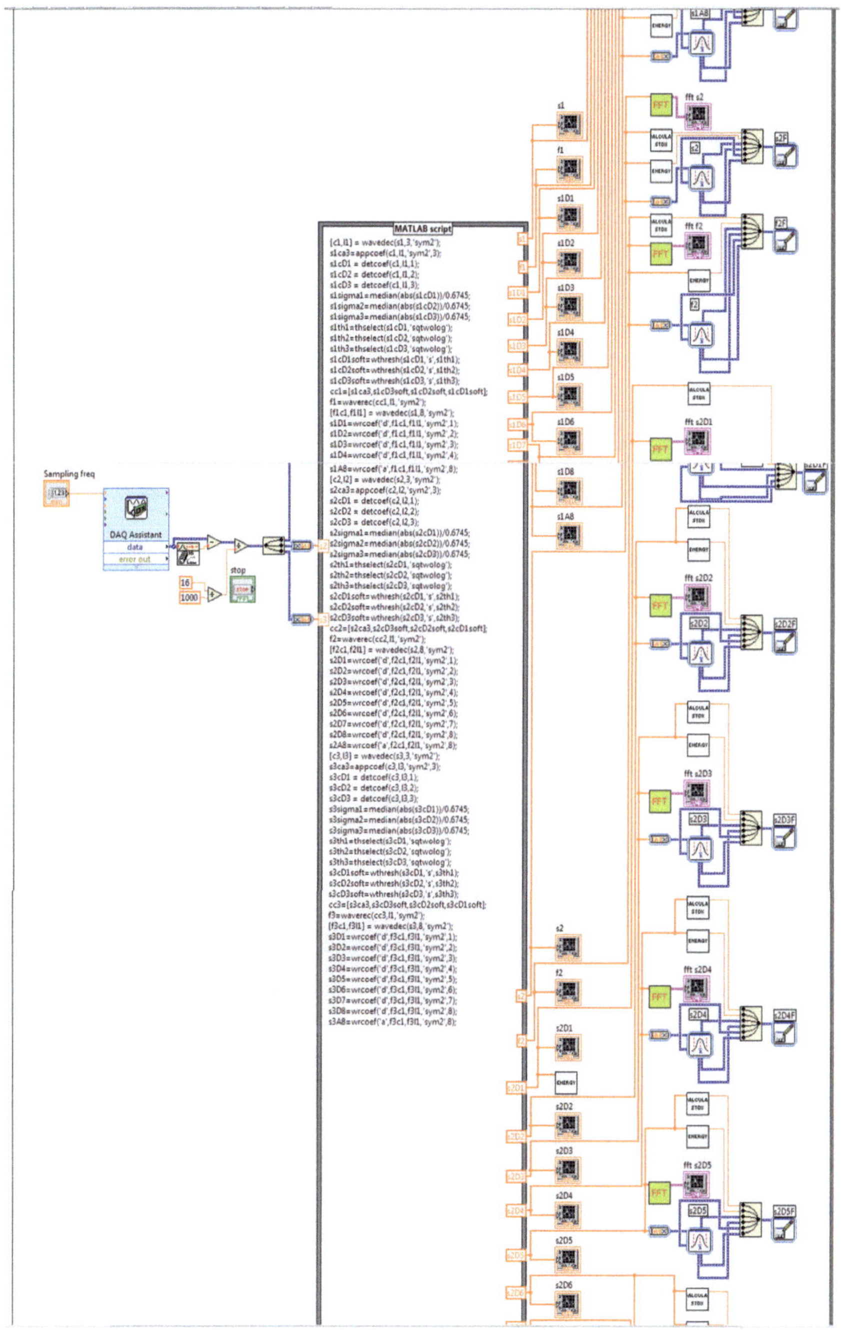
MATLAB script
[c1,l1] = wavedec(s1,3,'sym2');
s1ca3=appcoef(c1,l1,'sym2',3);
s1cD1 = detcoef(c1,l1,1);
s1cD2 = detcoef(c1,l1,2);
s1cD3 = detcoef(c1,l1,3);
s1sigma1=median(abs(s1cD1))/0.6745;
s1sigma2=median(abs(s1cD2))/0.6745;
s1sigma3=median(abs(s1cD3))/0.6745;
s1th1=thselect(s1cD1,'sqtwolog');
s1th2=thselect(s1cD2,'sqtwolog');
s1th3=thselect(s1cD3,'sqtwolog');
s1cD1soft=wthresh(s1cD1,'s',s1th1);
s1cD2soft=wthresh(s1cD2,'s',s1th2);
s1cD3soft=wthresh(s1cD3,'s',s1th3);
cc1=[s1ca3,s1cD3soft,s1cD2soft,s1cD1soft];
f1=waverec(cc1,l1,'sym2');
[f1c1,f1l1] = wavedec(s1,8,'sym2');
s1D1=wrcoef('d',f1c1,f1l1,'sym2',1);
s1D2=wrcoef('d',f1c1,f1l1,'sym2',2);
s1D3=wrcoef('d',f1c1,f1l1,'sym2',3);
s1D4=wrcoef('d',f1c1,f1l1,'sym2',4);
s1A8=wrcoef('a',f1c1,f1l1,'sym2',8);
[c2,l2] = wavedec(s2,3,'sym2');
s2ca3=appcoef(c2,l2,'sym2',3);
s2cD1 = detcoef(c2,l2,1);
s2cD2 = detcoef(c2,l2,2);
s2cD3 = detcoef(c2,l2,3);
s2sigma1=median(abs(s2cD1))/0.6745;
s2sigma2=median(abs(s2cD2))/0.6745;
s2sigma3=median(abs(s2cD3))/0.6745;
s2th1=thselect(s2cD1,'sqtwolog');
s2th2=thselect(s2cD2,'sqtwolog');
s2th3=thselect(s2cD3,'sqtwolog');
s2cD1soft=wthresh(s2cD1,'s',s2th1);
s2cD2soft=wthresh(s2cD2,'s',s2th2);
s2cD3soft=wthresh(s2cD3,'s',s2th3);
cc2=[s2ca3,s2cD3soft,s2cD2soft,s2cD1soft];
f2=waverec(cc2,l1,'sym2');
[f2c1,f2l1] = wavedec(s2,8,'sym2');
s2D1=wrcoef('d',f2c1,f2l1,'sym2',1);
s2D2=wrcoef('d',f2c1,f2l1,'sym2',2);
s2D3=wrcoef('d',f2c1,f2l1,'sym2',3);
s2D4=wrcoef('d',f2c1,f2l1,'sym2',4);
s2D5=wrcoef('d',f2c1,f2l1,'sym2',5);
s2D6=wrcoef('d',f2c1,f2l1,'sym2',6);
s2D7=wrcoef('d',f2c1,f2l1,'sym2',7);
s2D8=wrcoef('d',f2c1,f2l1,'sym2',8);
s2A8=wrcoef('a',f2c1,f2l1,'sym2',8);
[c3,l3] = wavedec(s3,3,'sym2');
s3ca3=appcoef(c3,l3,'sym2',3);
s3cD1 = detcoef(c3,l3,1);
s3cD2 = detcoef(c3,l3,2);
s3cD3 = detcoef(c3,l3,3);
s3sigma1=median(abs(s3cD1))/0.6745;
s3sigma2=median(abs(s3cD2))/0.6745;
s3sigma3=median(abs(s3cD3))/0.6745;
s3th1=thselect(s3cD1,'sqtwolog');
s3th2=thselect(s3cD2,'sqtwolog');
s3th3=thselect(s3cD3,'sqtwolog');
s3cD1soft=wthresh(s3cD1,'s',s3th1);
s3cD2soft=wthresh(s3cD2,'s',s3th2);
s3cD3soft=wthresh(s3cD3,'s',s3th3);
cc3=[s3ca3,s3cD3soft,s3cD2soft,s3cD1soft];
f3=waverec(cc3,l1,'sym2');
[f3c1,f3l1] = wavedec(s3,8,'sym2');
s3D1=wrcoef('d',f3c1,f3l1,'sym2',1);
s3D2=wrcoef('d',f3c1,f3l1,'sym2',2);
s3D3=wrcoef('d',f3c1,f3l1,'sym2',3);
s3D4=wrcoef('d',f3c1,f3l1,'sym2',4);
s3D5=wrcoef('d',f3c1,f3l1,'sym2',5);
s3D6=wrcoef('d',f3c1,f3l1,'sym2',6);
s3D7=wrcoef('d',f3c1,f3l1,'sym2',7);
s3D8=wrcoef('d',f3c1,f3l1,'sym2',8);
s3A8=wrcoef('a',f3c1,f3l1,'sym2',8);
Sampling freq
DAQ Assistant
data
error out
16
1000
stop
s1
f1
s1D1
s1D2
s1D3
s1D4
s1D5
s1D6
s1D8
s1A8
s2
f2
s2D1
s2D2
s2D3
s2D4
s2D5
s2D6
fft s2
fft f2
fft s2D1
fft s2D2
fft s2D3
fft s2D4
fft s2D5
FFT

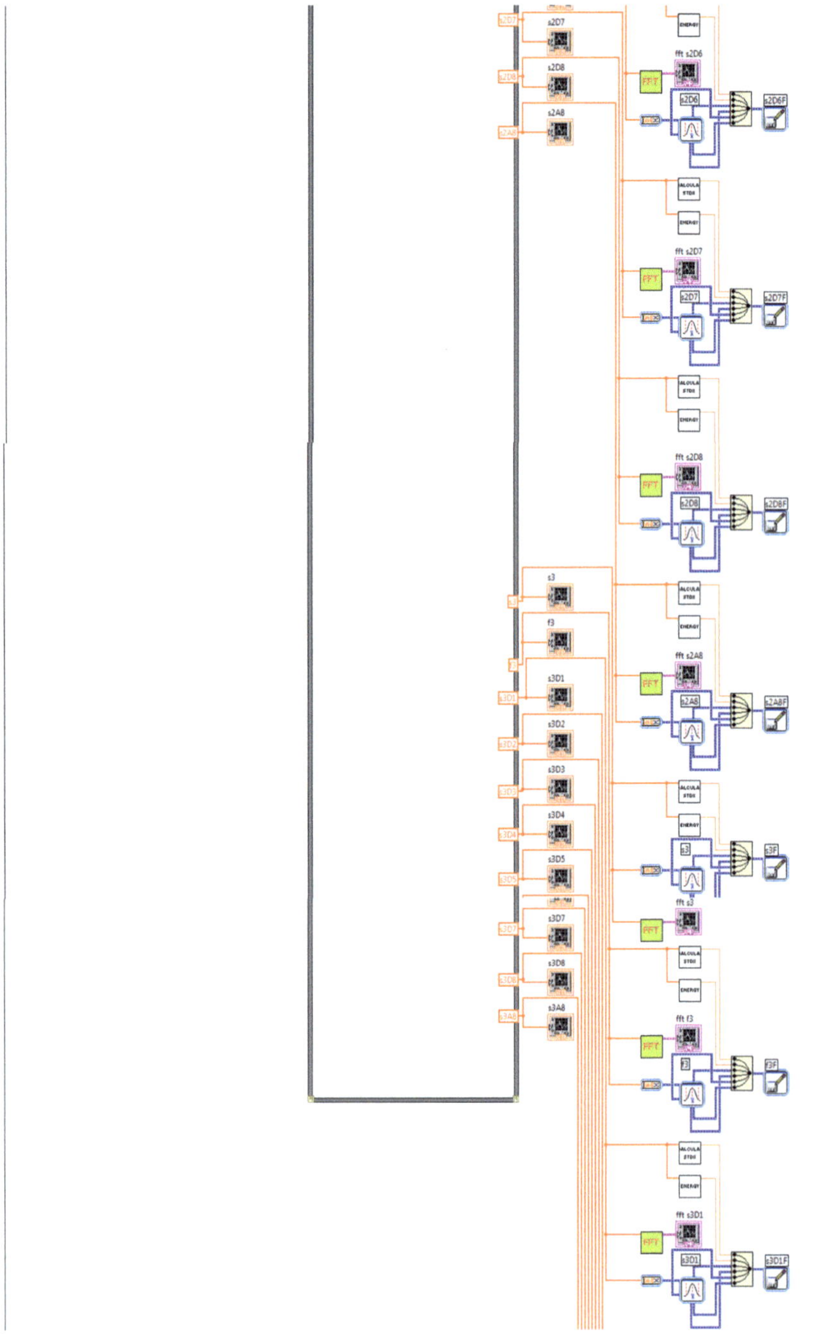
s2D7
s2D8
s2A8
s3
f3
s3D1
s3D2
s3D3
s3D4
s3D5
s3D7
s3D8
s3A8
fft s2D6
s2D6F
fft s2D7
s2D7F
fft s2D8
s2D8F
fft s2A8
s2A8F
s3F
fft s3
fft f3
f3F
fft s3D1
s3D1F

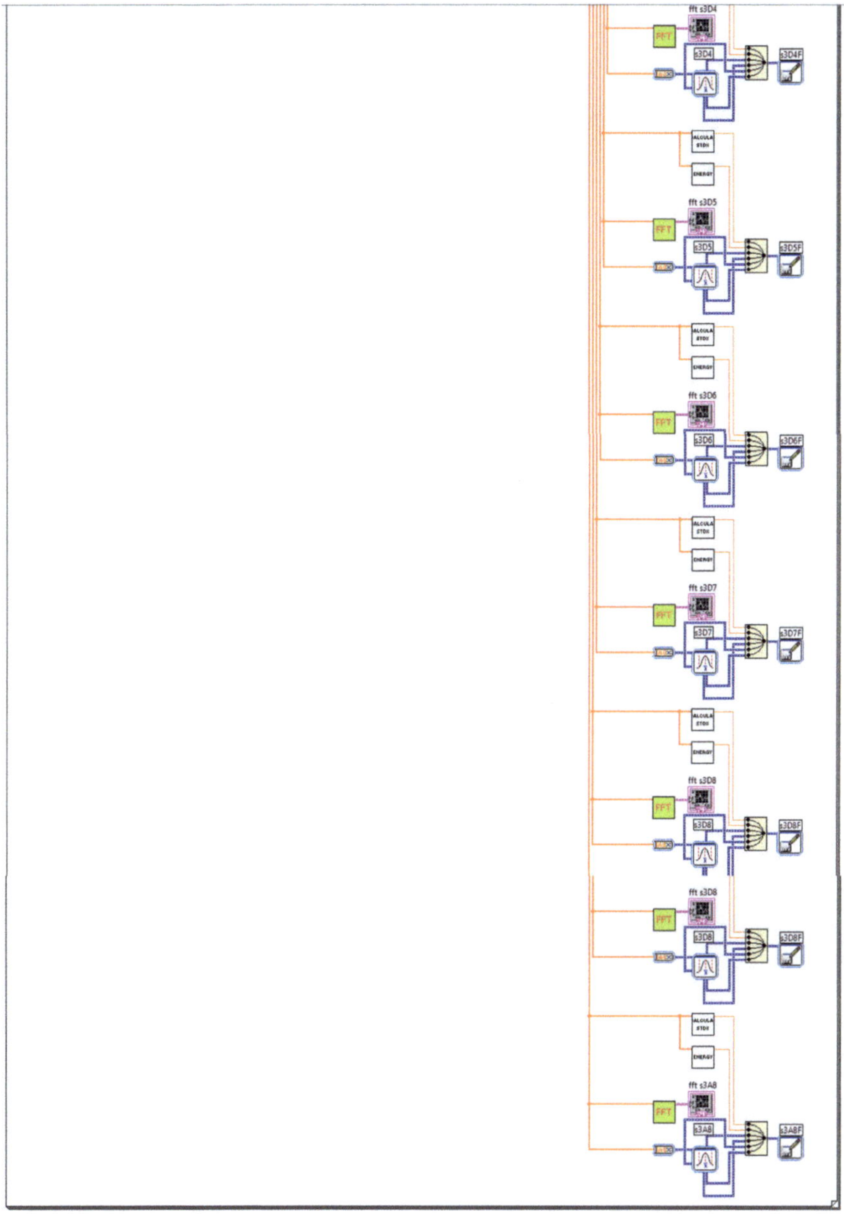
fft s3D4
FFT
s3D4
s3D4F
ENERGY
fft s3D5
FFT
s3D5
s3D5F
ENERGY
fft s3D6
FFT
s3D6
s3D6F
ENERGY
fft s3D7
FFT
s3D7
s3D7F
ENERGY
fft s3D8
FFT
s3D8
s3D8F
fft s3D8
FFT
s3D8
s3D8F
ENERGY
fft s3A8
FFT
s3A8
s3A8F

File Dialog
selected path
Excel
Sheet 1 Data
Sheet 2 Data
Sheet 3 Data
Sheet 4 Data
Sheet 6 Data
Sheet 7 Data
Sheet 8 Data
Sheet 9 Data
Sheet 10 Data

Appendix C
VAL II Programme and Result of the DWT and FFT

C.1 VAL II Program for Pick and Place Task

Below is the pick and place VAL II program which was used in this thesis. The operation was to swap the positions of two snooker balls. The programme contain a subroutine that is executed to check if the external input signal is on, in order to halt any more execution of the programme.

The main programme:

1 `SIGNAL 1`	{Turn the output signal on to trigger the embedded system}
2 `APPRO p1, 100`	{Moves to 100 mm above position 1}
3 `MOVE p1`	{Moves robot so gripper positioned to pick up the red ball}
4 `DELAY 0.1`	{Delay of 100 ms}
5 `CLOSEI`	{Closes gripper}
6 `DELAY 0.1`	{Delay of 100 ms}
7 `DEPART 100`	{Moves to 100 mm above position 1}
8 `APPRO p3, 100`	{Moves to 100 mm above position 3}:
9 `MOVE p3`	{Moves robot so gripper positioned to release the red ball}
10 `DELAY 0.1`	{Delay of 100 ms}
11 `OPENI`	{Opens gripper}
12 `DELAY 0.1`	{Delay of 100 ms}
13 `DEPART 100`	{Moves to 100 mm above position 3}
14 `APPRO p2, 100`	{Moves to 100 mm above position 2}
15 `MOVE p2`	{Moves robot so gripper positioned to pick up the yellow ball}
16 `DELAY 0.1`	{Delay of 100 ms}
17 `CLOSEI`	{Closes gripper}
18 `DELAY 0.1`	{Delay of 100 ms}
19 `DEPART 100`	{Moves to 100 mm above position 2}
20 `APPRO p1, 100`	{Moves to 100 mm above position 1}
21 `MOVE p1`	{Moves robot so gripper positioned to release the yellow ball}

A.A. Jaber, *Design of an Intelligent Embedded System for Condition Monitoring of an Industrial Robot*, Springer Theses,
DOI 10.1007/978-3-319-44932-6

```
DELAY 0.1               {Delay of 100 ms}
OPENI                   {Opens gripper}
DELAY 0.1               {Delay of 100 ms}
DEPART 100              {Moves to 100 mm above position 1}
APPRO p3, 100           {Moves to 100 mm above position 3}
MOVE p3                 {Moves robot so gripper positioned to pick up the red ball}
DELAY 0.1               {Delay of 100 ms}
CLOSEI                  {Closes gripper}
DELAY 0.1               {Delay of 100 ms}
DEPART 100              {Moves to 100 mm above position 3}
APPRO p2, 100           {Moves to 100 mm above position 2
MOVE p2                 {Moves robot so gripper positioned to release the red ball}
DELAY 0.1               {Delay of 100 ms}
OPENI                   {Opens gripper}
DELAY 0.1               {Delay of 100 ms}
DEPART 100              {Moves to 100 mm above position 2}
SIGNAL -1               {Turn the output signal off}
DELAY 0.1               {Delay of 100  ms}
CALL Check              {Starts Check subroutine checking for external input
                        signal}
END                     {The programme finished}
```

External signal check subroutine:

```
IF SIG(1001)==1 THEN    {Check if the external signal on the input 1 is
                        high}
TYPE fault detected     {Print fault-detected on the operator}
Halt                    {Stop the programme execution}
END                     {End of the subroutine programme}
RETURN                  {Returns to the main programme}
```

C.2 Result of the DWT analysis for different fault types

See Figs. C.1, C.2, C.3, C.4, C.5, C.6, C.7, C.8, C.9, C.10, C.11, C.12, C.13, C.14 and C.15.

(a) Healthy robot **(b) High backlash**

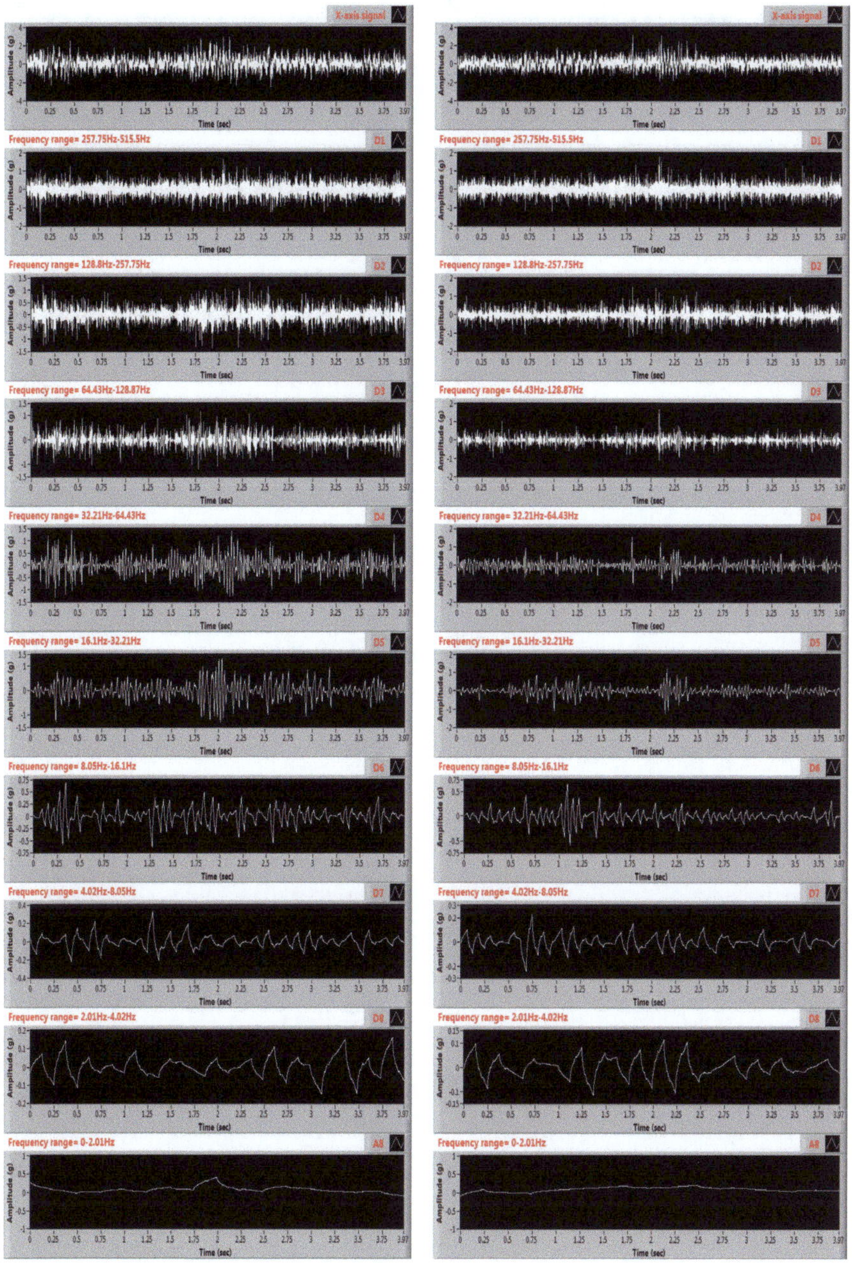

Fig. C.1 DWT analysis of the X-axis vibration signals from the robot arm when it is healthy and high backlash fault simulated (auto scaled)

(a) Healthy robot **(b) High backlash**

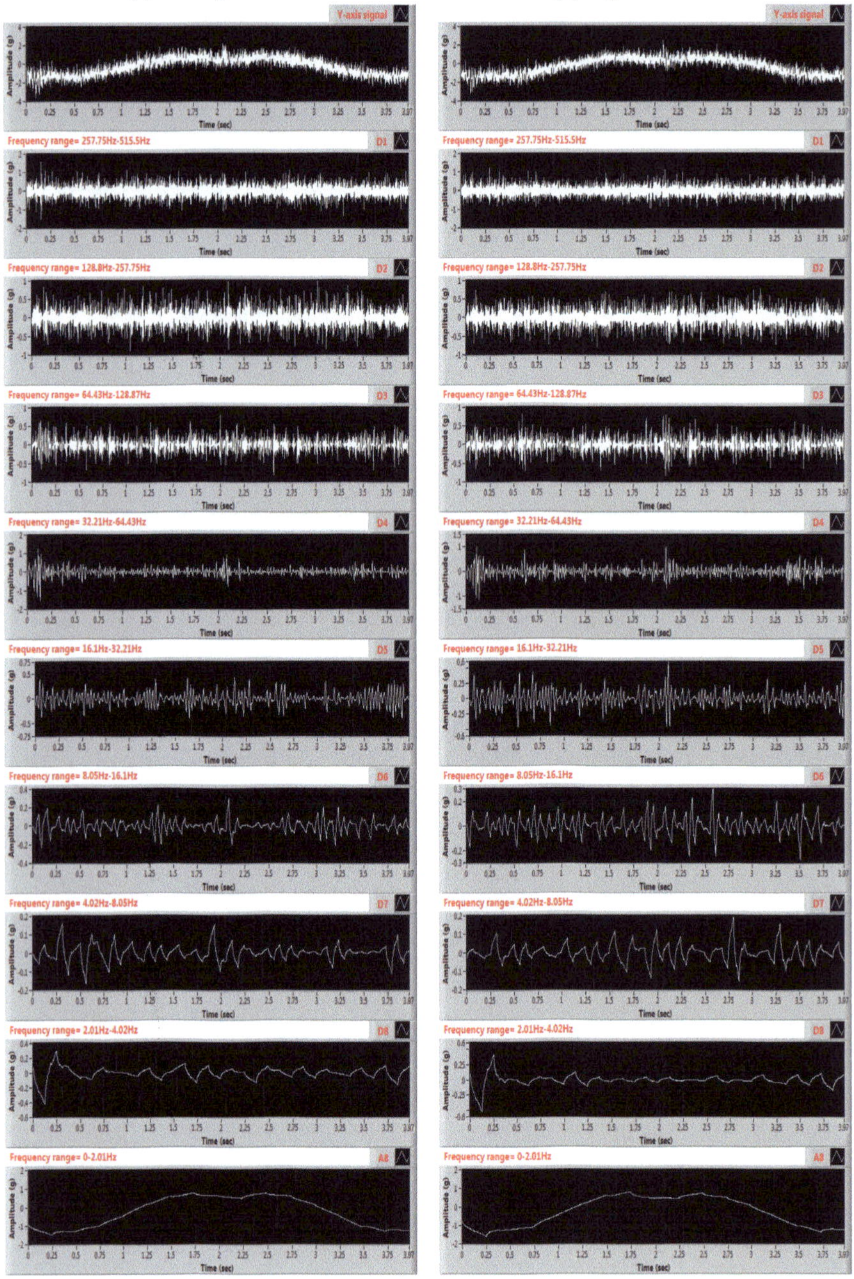

Fig. C.2 DWT analysis of the Y-axis vibration signals from the robot arm when it is healthy and high backlash fault simulated (auto scaled)

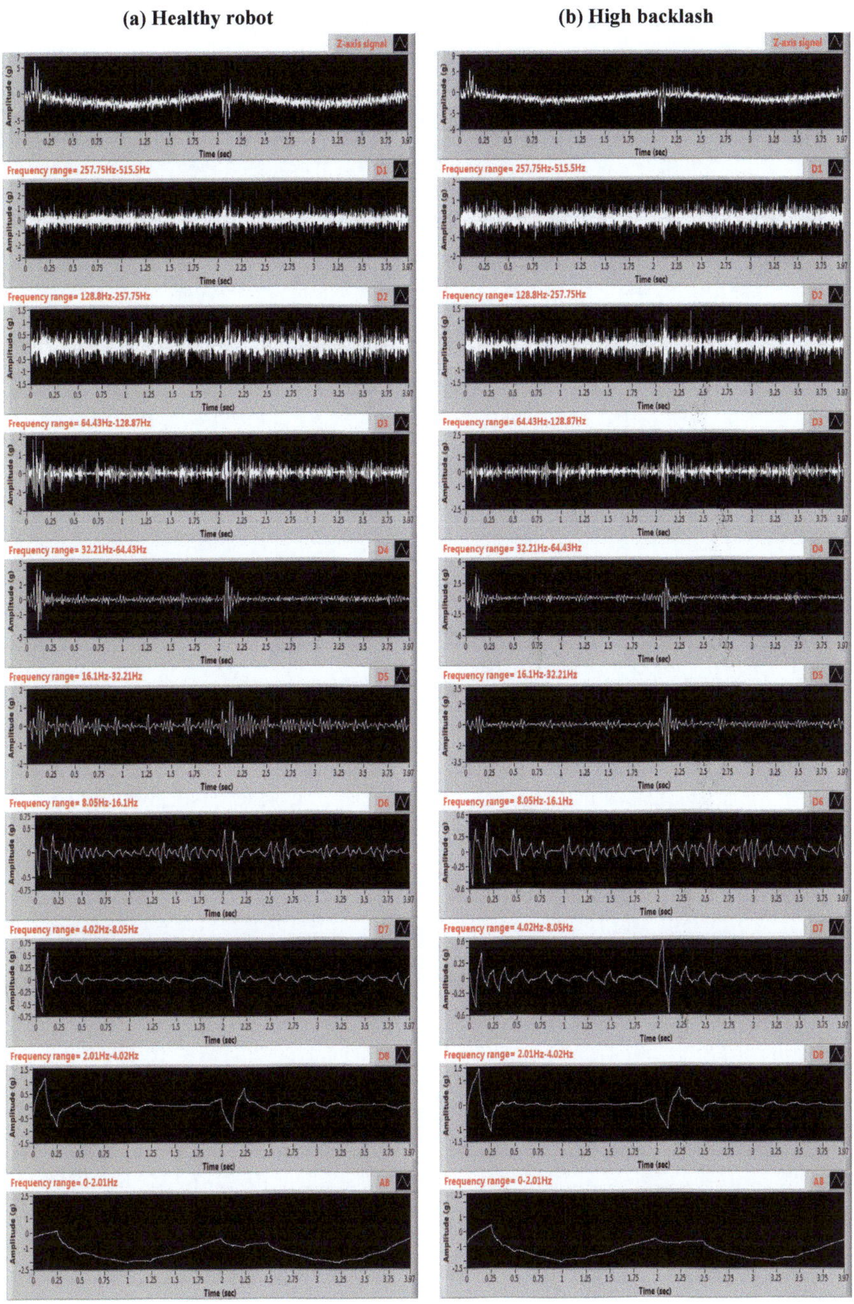

Fig. C.3 DWT analysis of the Z-axis vibration signals from the robot arm when it is healthy and high backlash fault simulated (auto scaled)

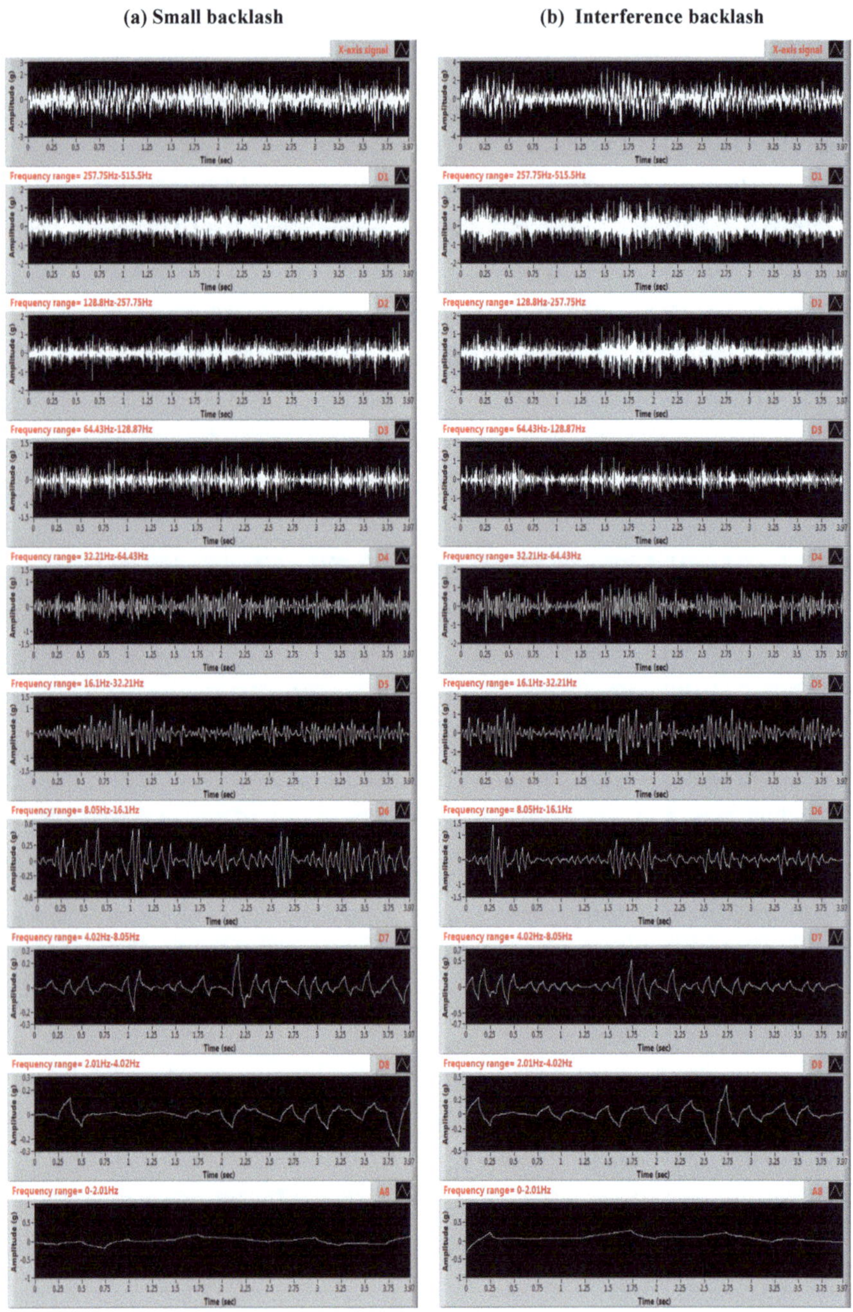

Fig. C.4 DWT analysis of the X-axis vibration signals from the robot arm when small and interference backlash faults simulated (auto scaled)

(a) Small backlash **(b) Interference backlash**

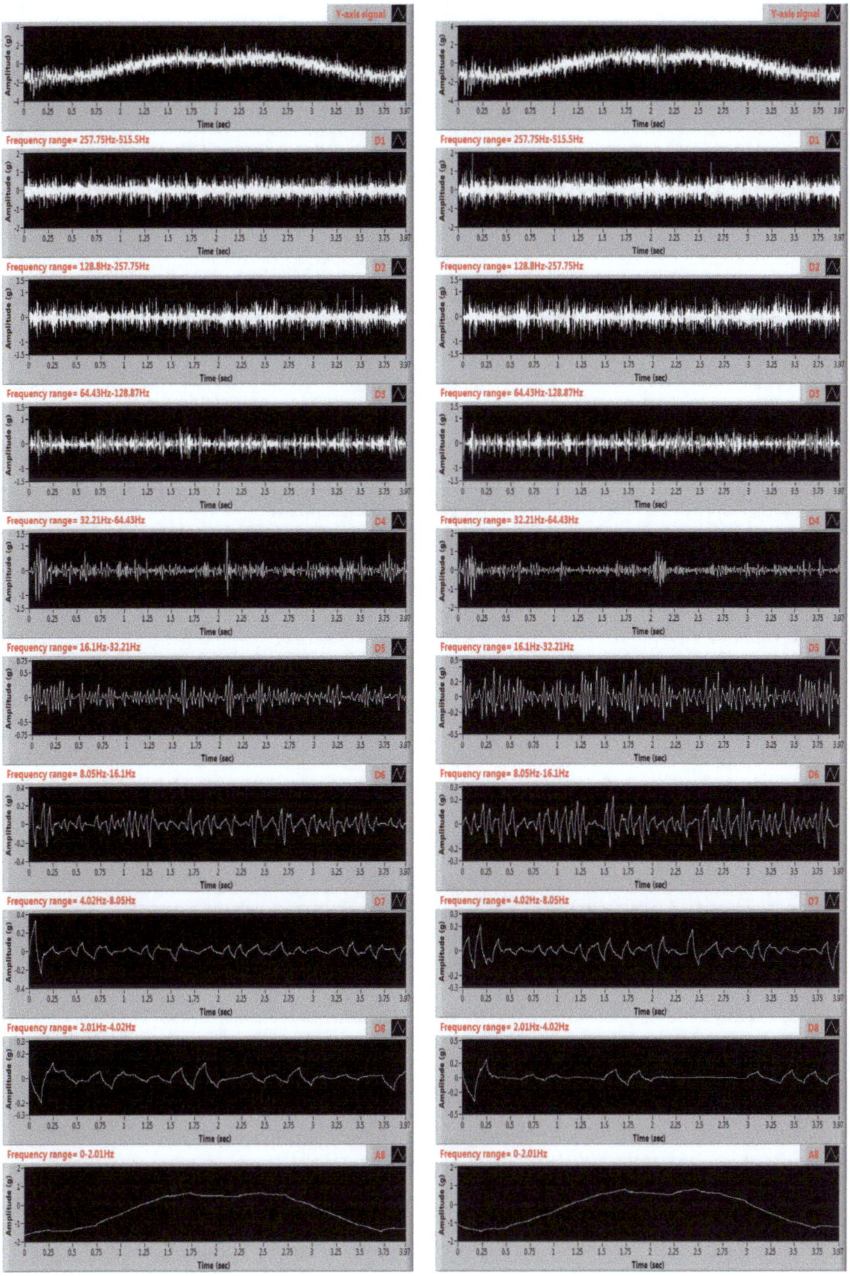

Fig. C.5 DWT analysis of the Y-axis vibration signals from the robot arm when small and interference backlash faults simulated (auto scaled)

(a) Small backlash (b) Interference backlash

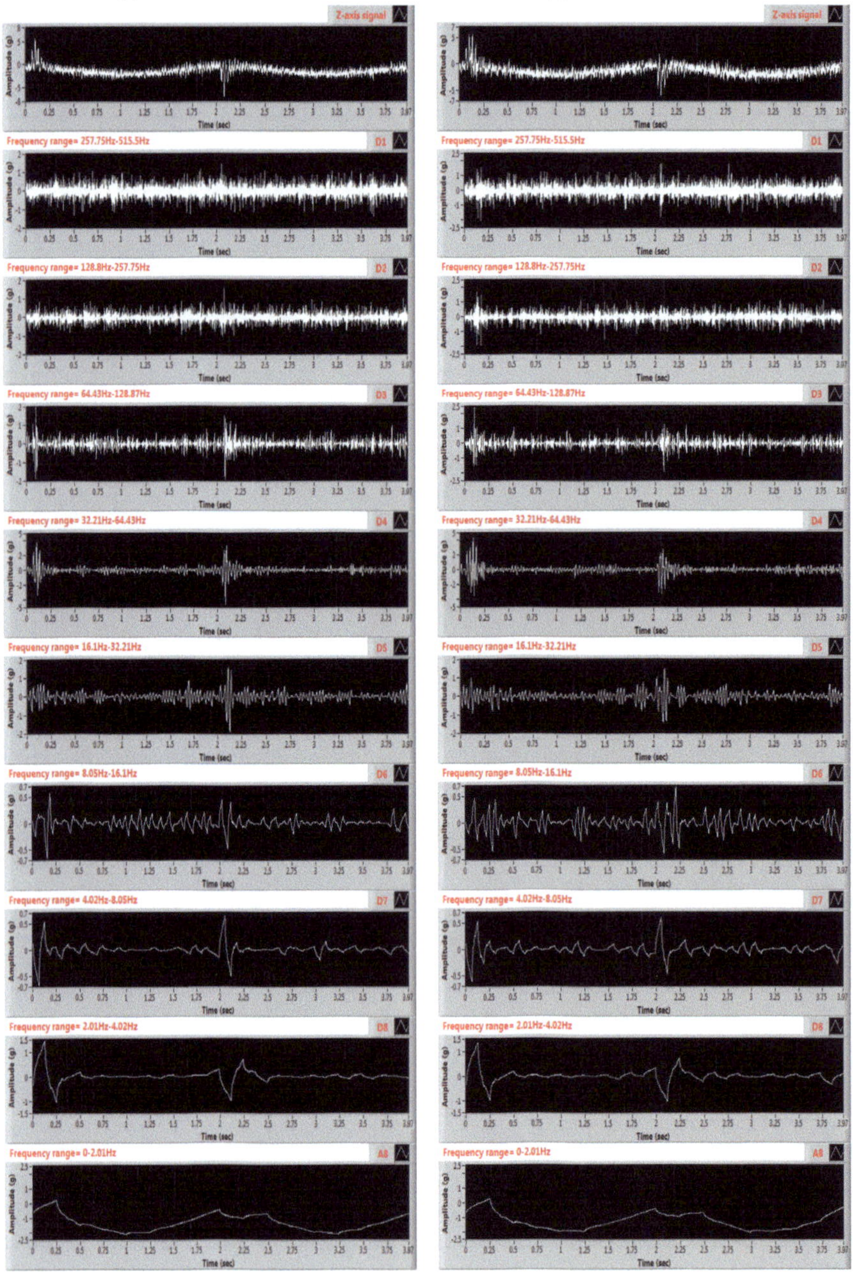

Fig. C.6 DWT analysis of the Z-axis vibration signals from the robot arm when small and interference backlash faults simulated (auto scaled)

(a) Tooth wear **(b) 25% tooth removed**

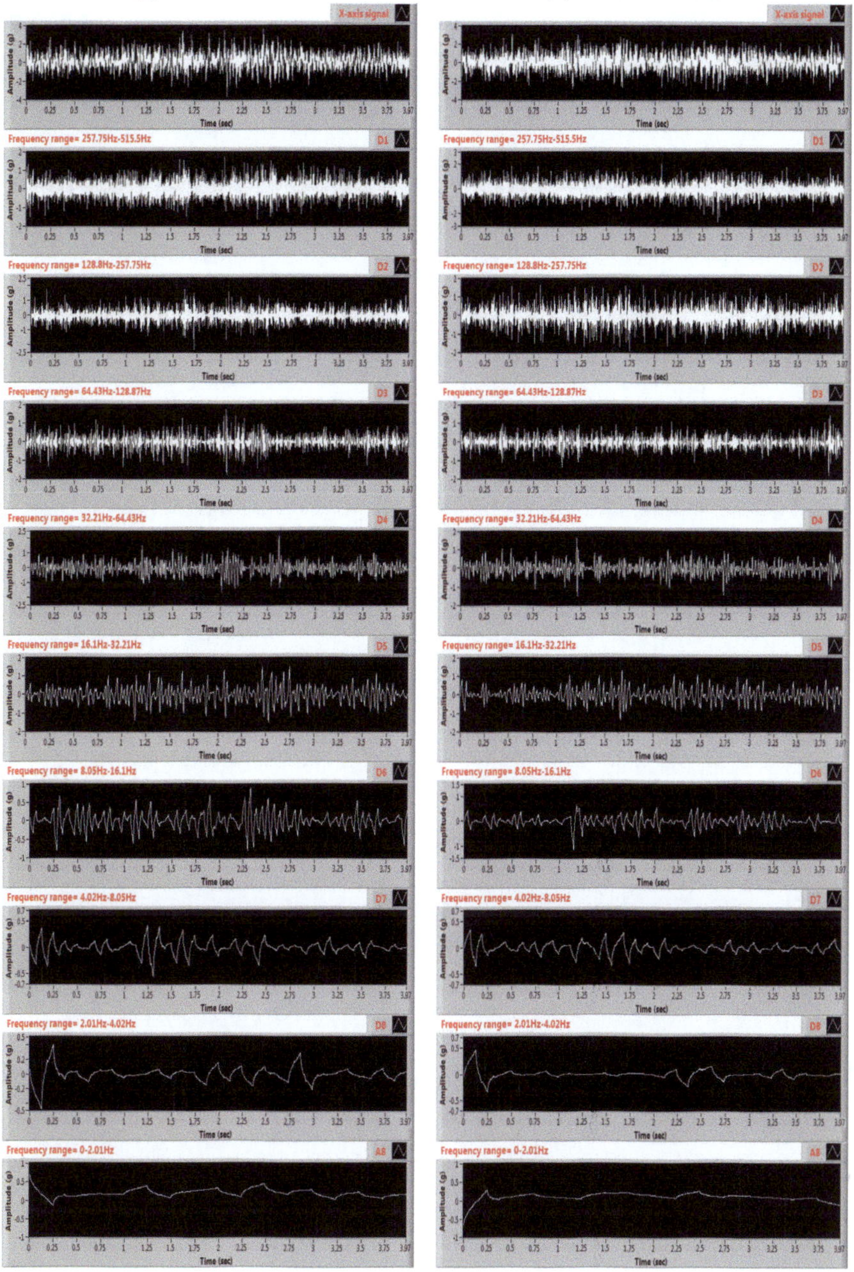

Fig. C.7 DWT analysis of the X-axis vibration signals from the robot arm when gear tooth wear and 25 % removed faults simulated (auto scaled)

(a) Tooth wear **(b) 25% tooth cut**

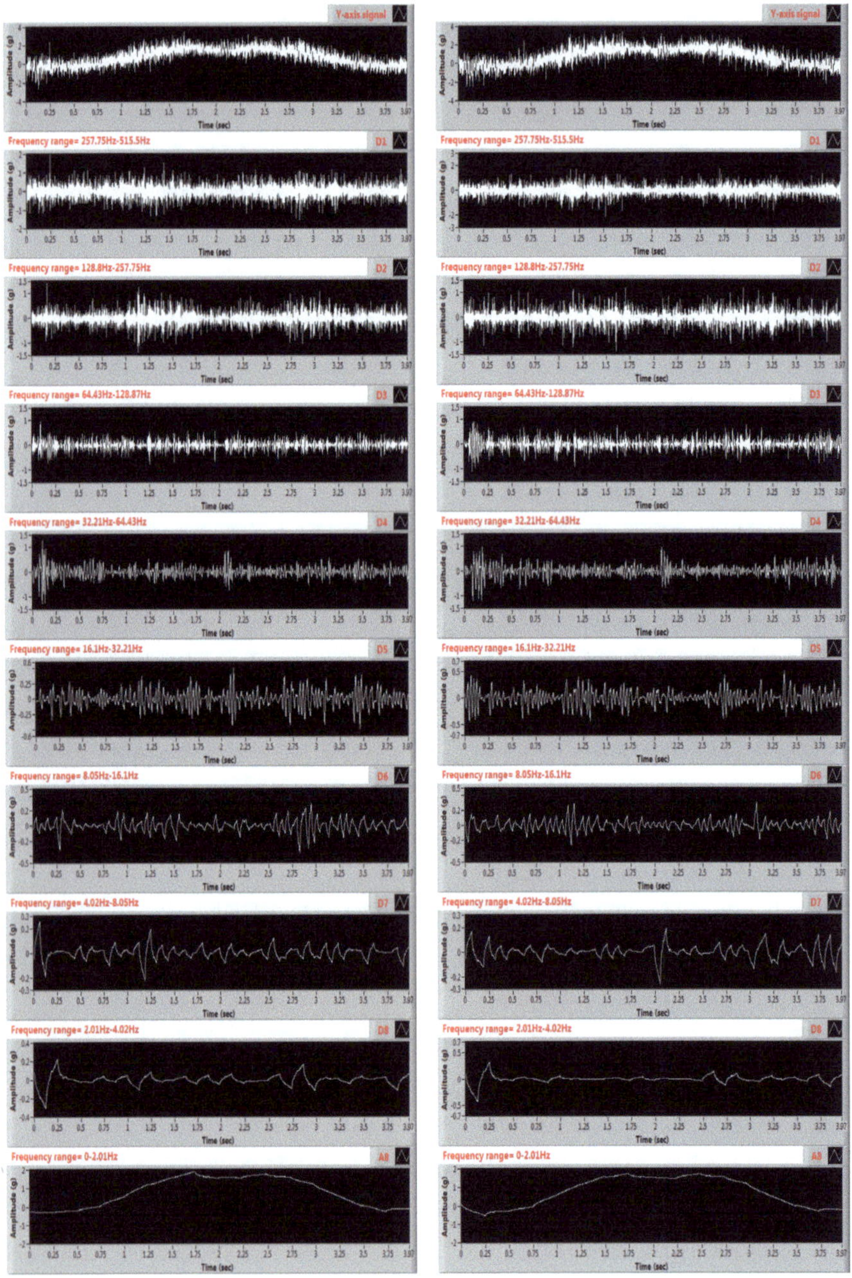

Fig. C.8 DWT analysis of the Y-axis vibration signals from the robot arm when gear tooth wear and 25 % removed faults simulated (auto scaled)

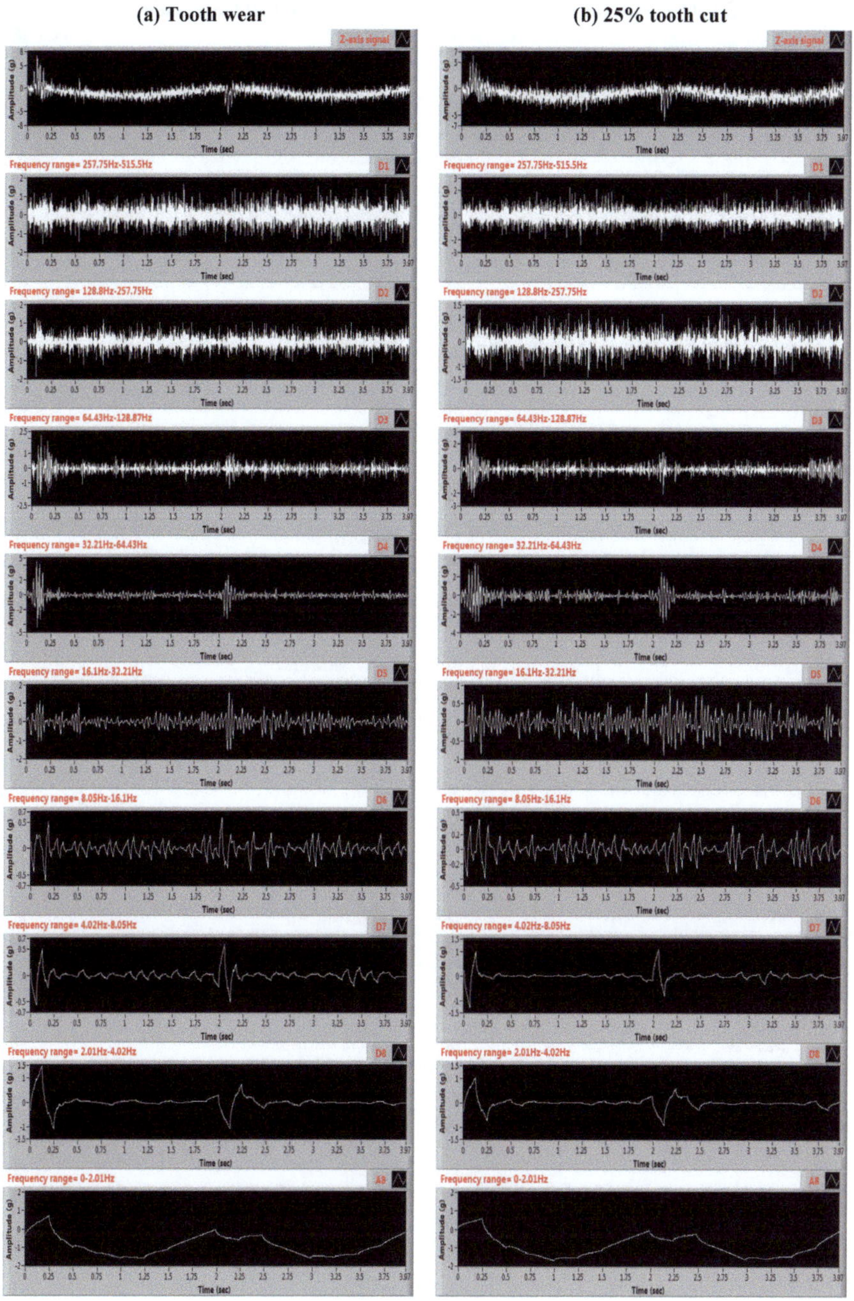

Fig. C.9 DWT analysis of the Z-axis vibration signals from the robot arm when gear tooth wear and 25 % removed faults simulated (auto scaled)

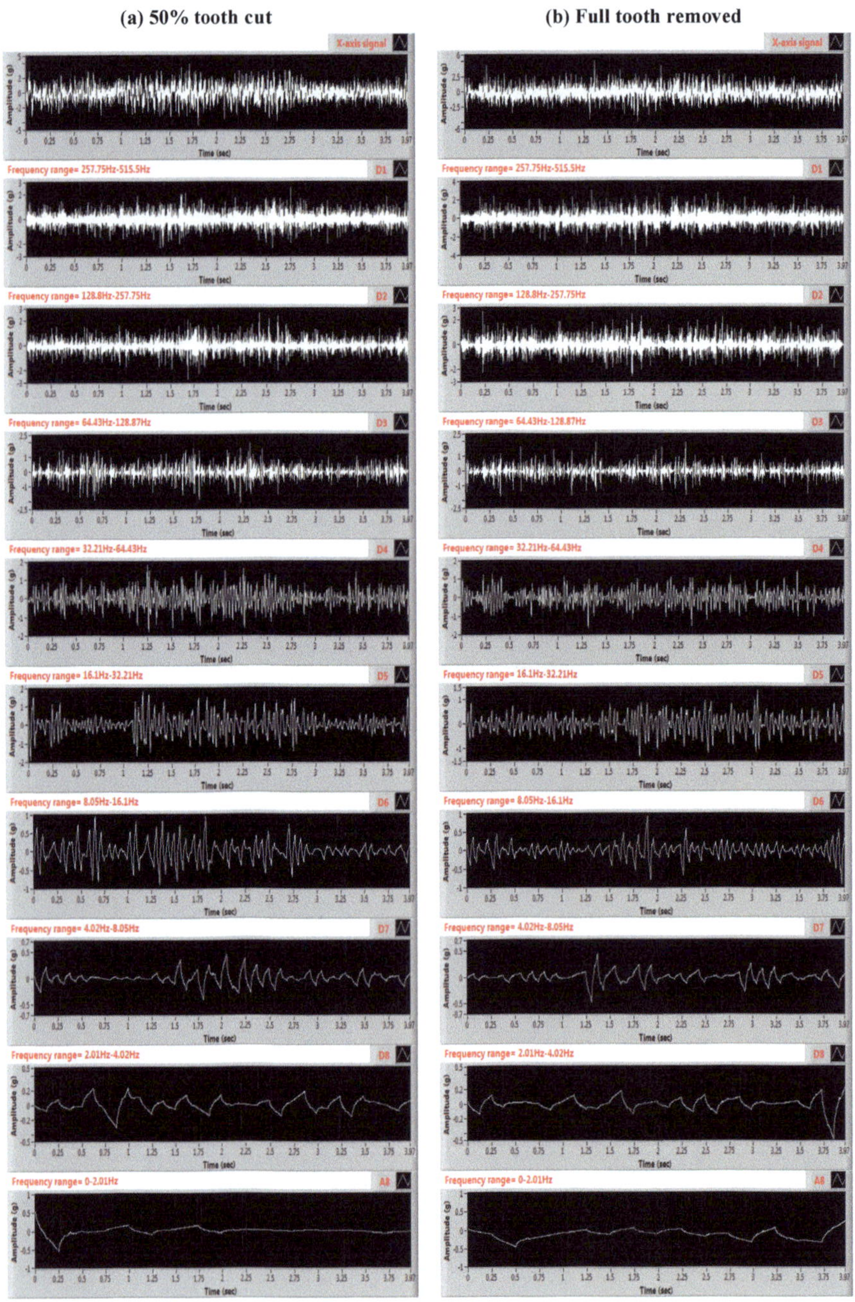

Fig. C.10 DWT analysis of the X-axis vibration signals from the robot arm when 50 % and full gear tooth removed faults simulated (auto scaled)

(a) 50% tooth cut **(b) Full tooth removed**

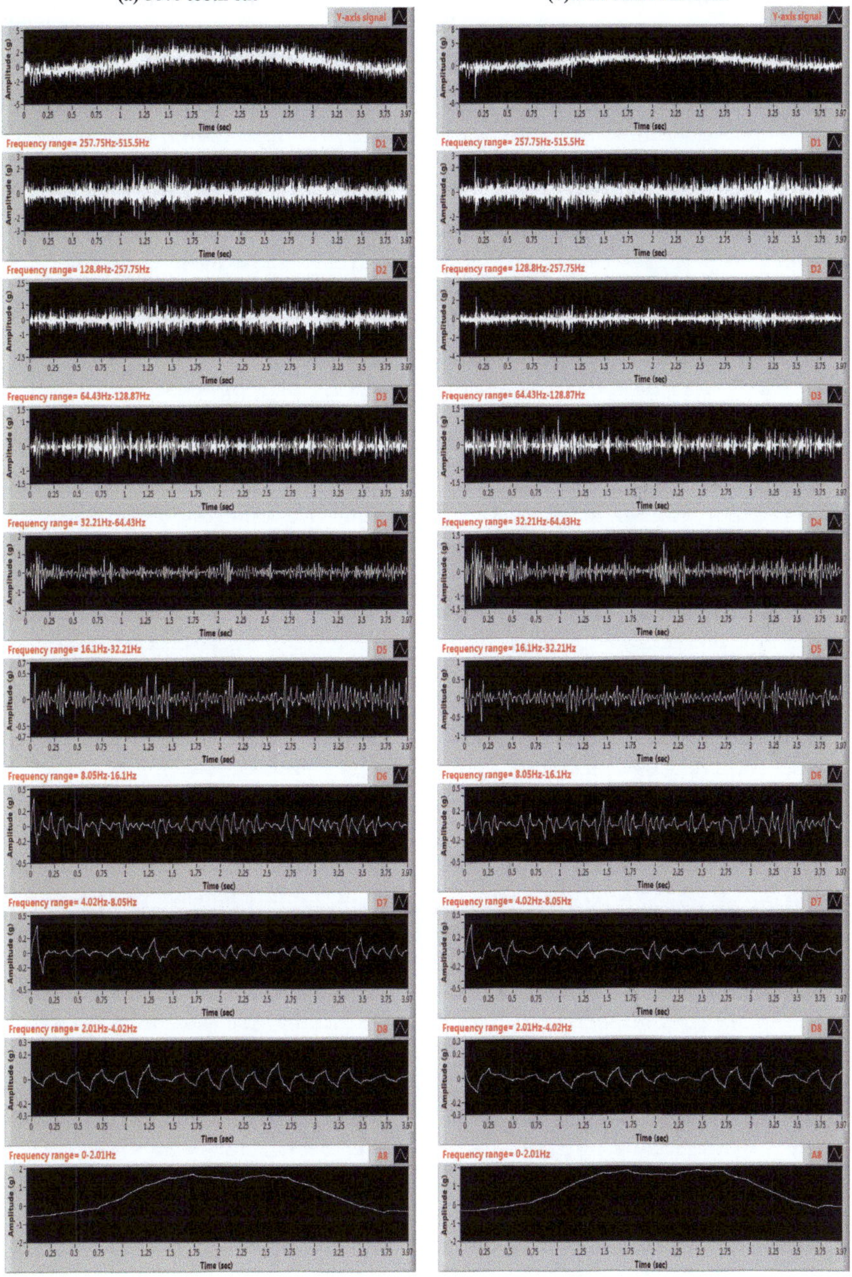

Fig. C.11 DWT analysis of the Y-axis vibration signals from the robot arm when 50 % and full gear tooth removed faults simulated (auto scaled)

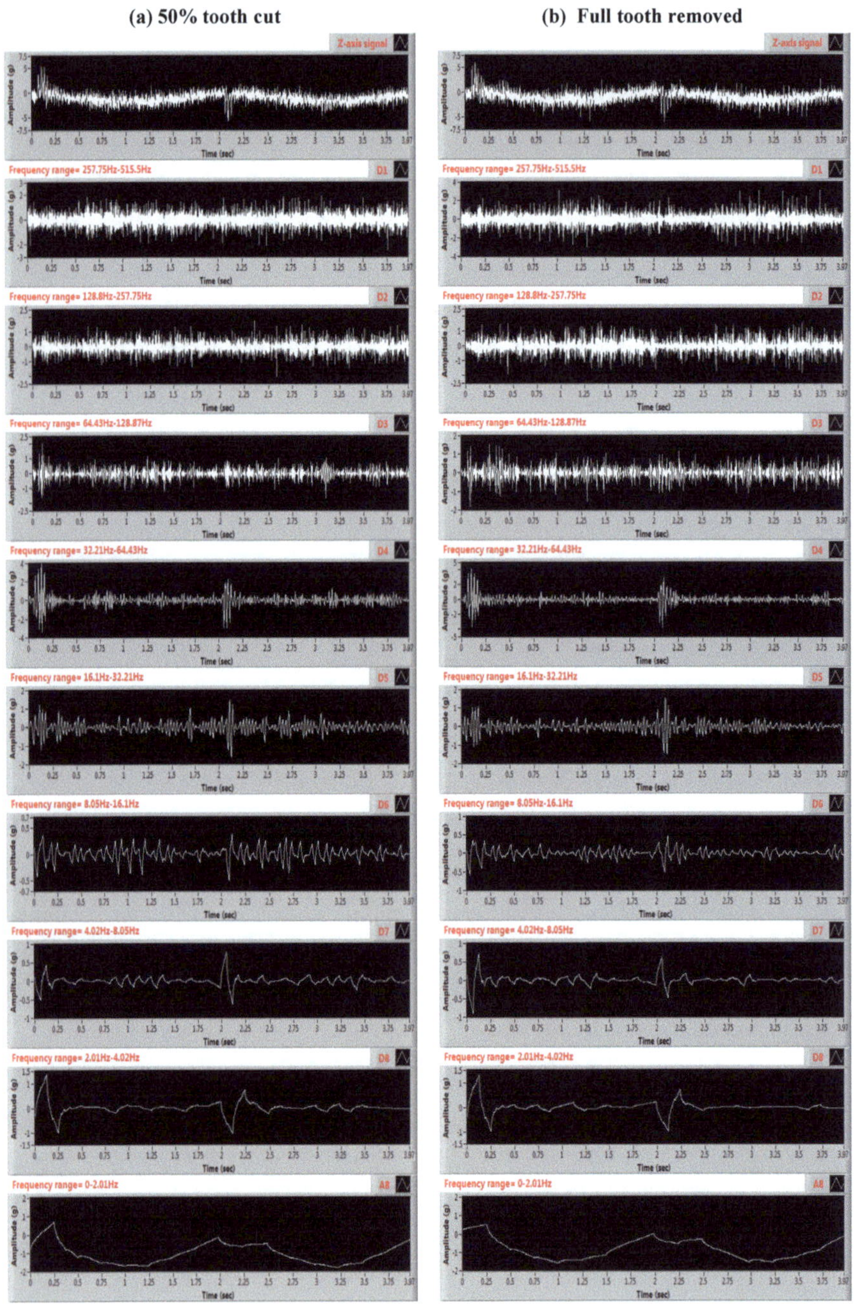

Fig. C.12 DWT analysis of the Z-axis vibration signals from the robot arm when 50 % and full gear tooth removed faults simulated (auto scaled)

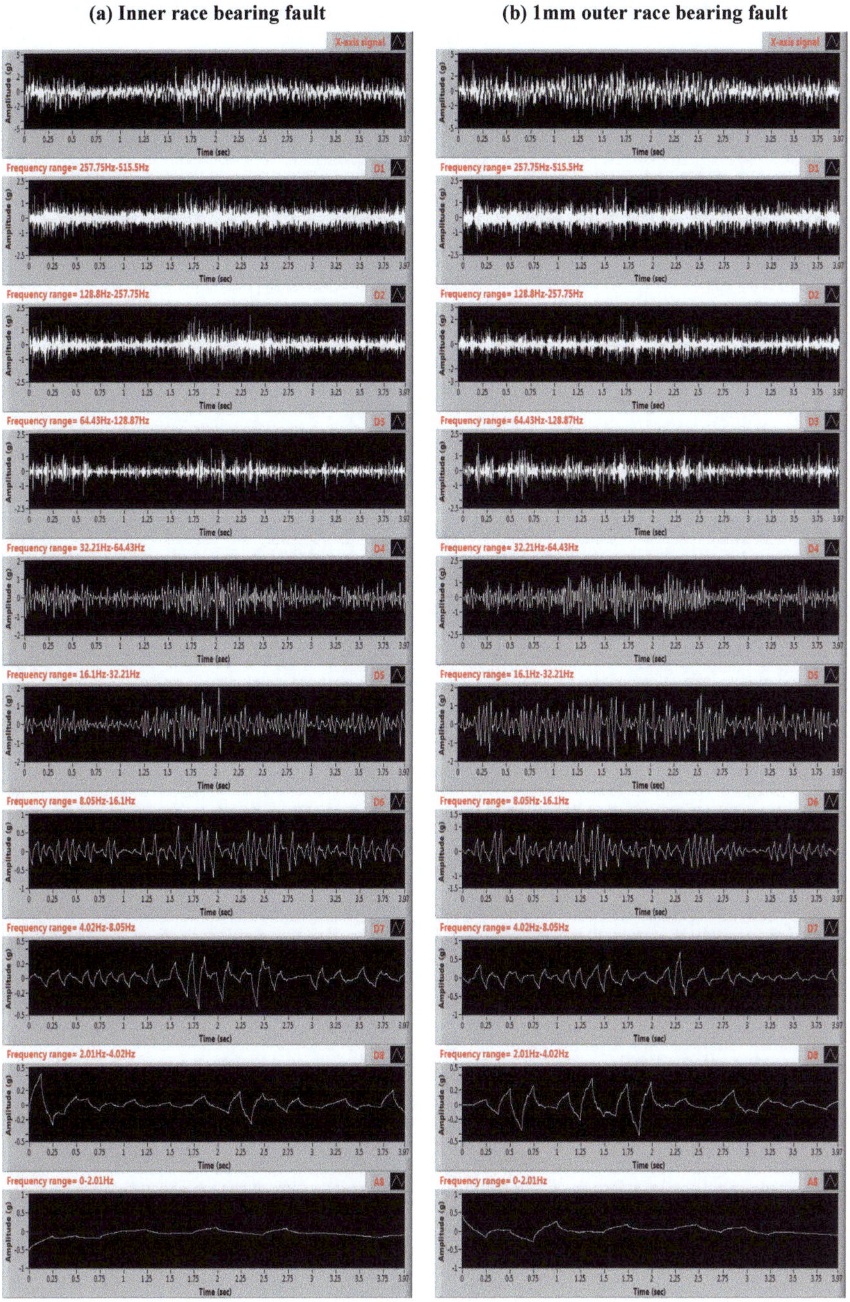

Fig. C.13 DWT analysis of the X-axis vibration signals from the robot arm when inner race and 1mm hole in the outer race bearing faults simulated (auto scaled)

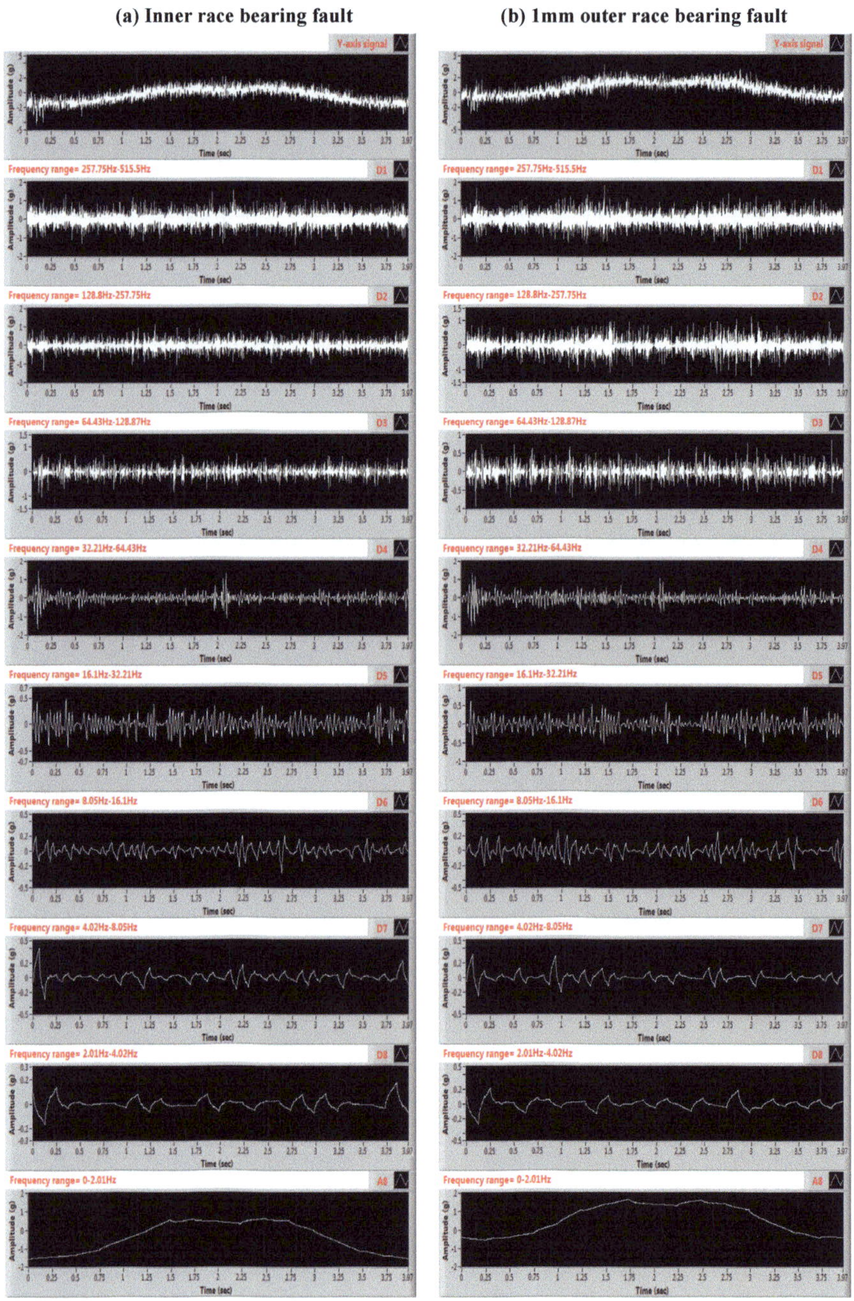

Fig. C.14 DWT analysis of the Y-axis vibration signals from the robot arm when inner race and 1mm hole in the outer race bearing faults simulated (auto scaled)

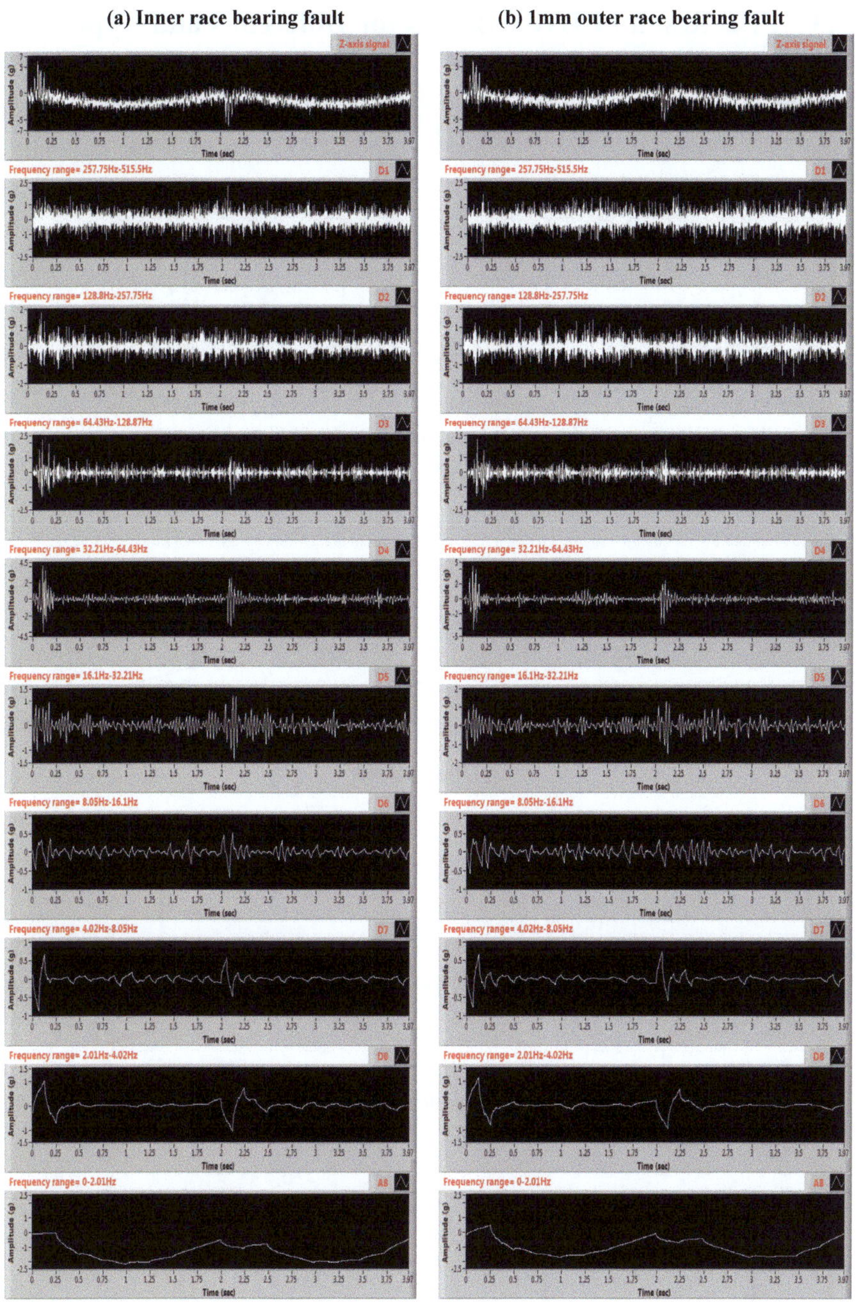

Fig. C.15 DWT analysis of the Z-axis vibration signals from the robot arm when inner race and 1 mm hole in the outer race bearing faults simulated (auto scaled)

C.3 FFT Result of DWT Details for Different Fault Types

See Figs. C.16 and C.17.

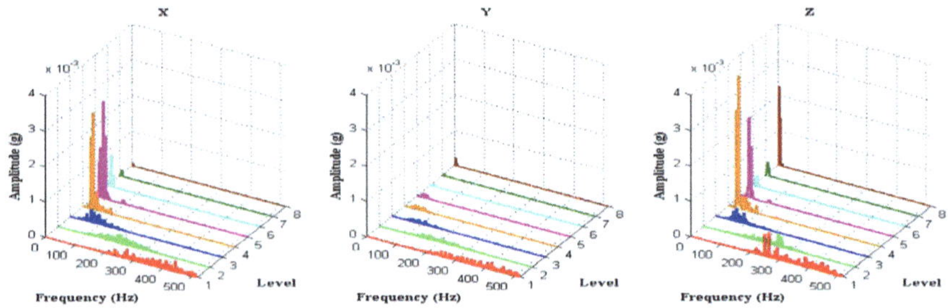

(a) Healthy robot

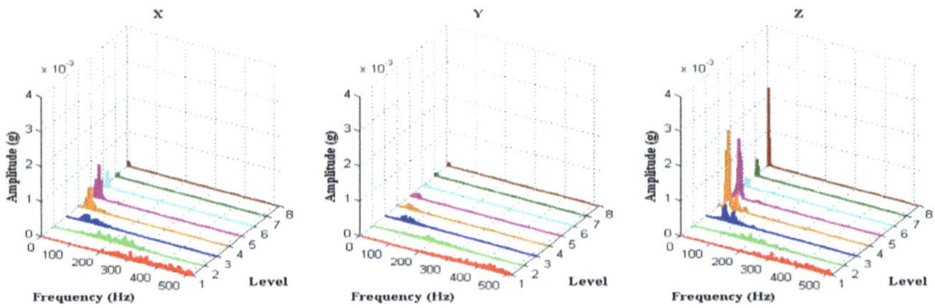

(b) High backlash

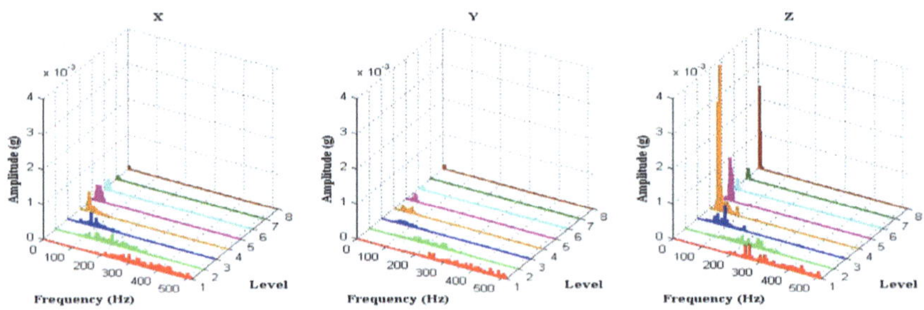

(c) Small backlash

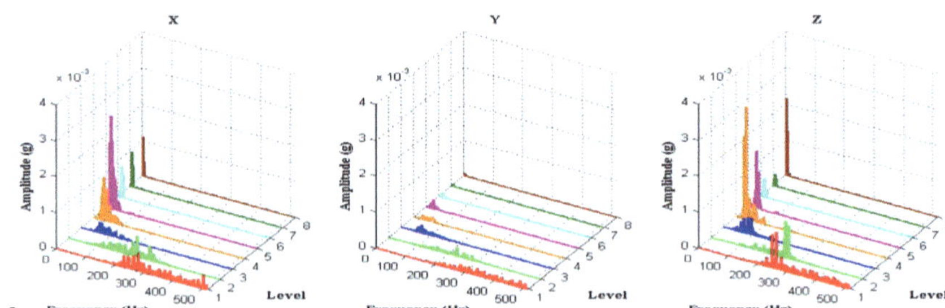

(d) Interference backlash

Fig. C.16 Frequency spectrum of each detail signal in the X, Y and Z axes in healthy and backlash fault cases

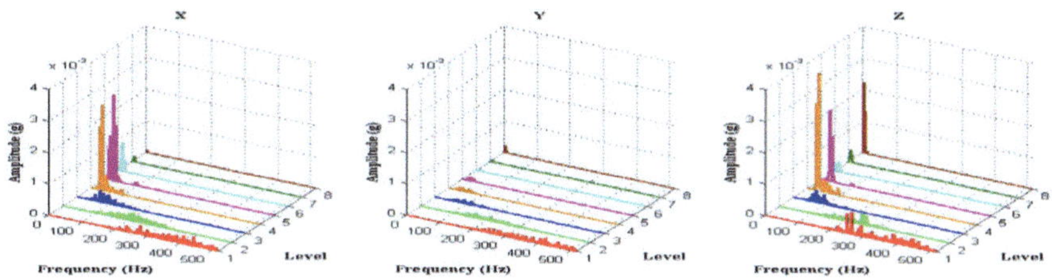

(a) Healthy robot

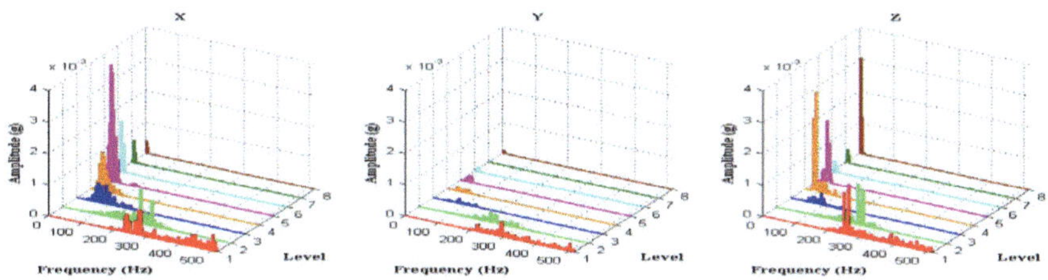

(b) Gear tooth wear

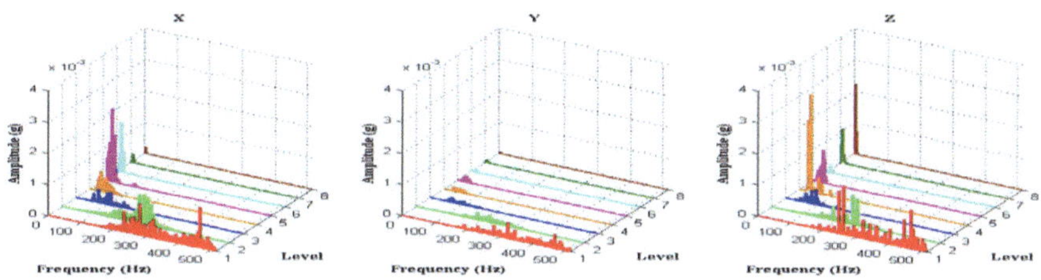

(c) 25% gear tooth removed

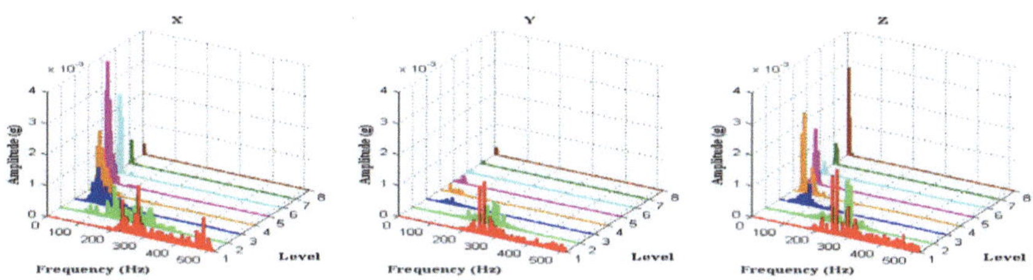

(d) 50% gear tooth removed

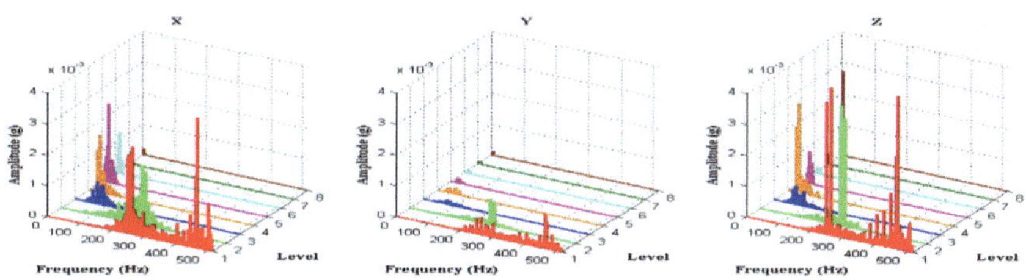

(e) Full gear tooth removed

Fig. C.17 Frequency spectrum of each detail signal in the X, Y and Z axes in healthy and gear tooth fault cases

Appendix D
Performance Testing Result of the Designed ANN

See Fig. D.1.

A.A. Jaber, *Design of an Intelligent Embedded System for Condition Monitoring of an Industrial Robot*, Springer Theses,
DOI 10.1007/978-3-319-44932-6

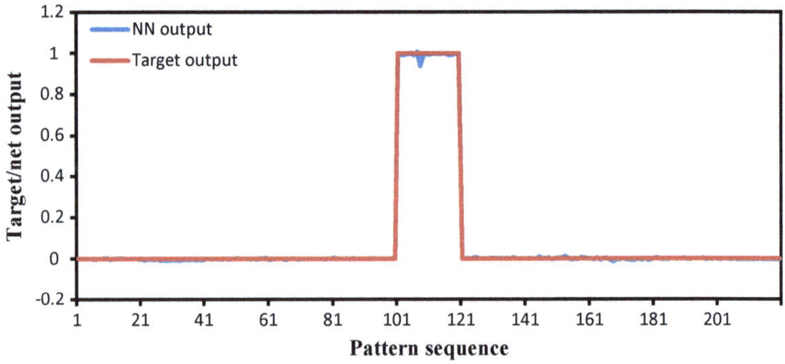

(a) 25% gear tooth removed

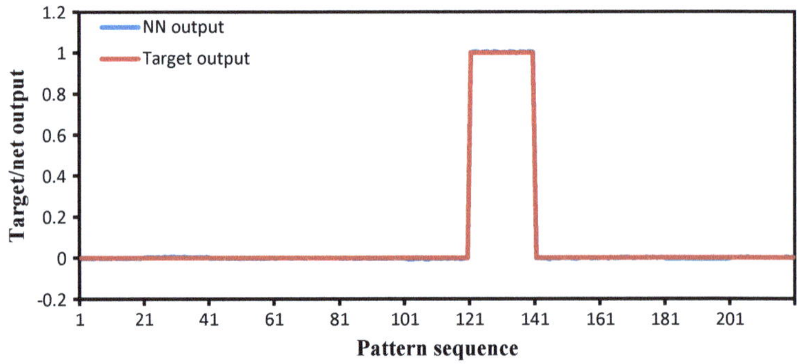

(b) 50% gear tooth removed

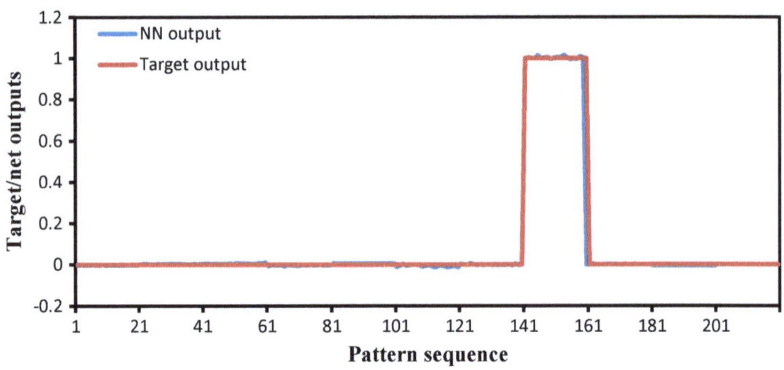

(c) Gear tooth completely removed

Fig. D.1 The designed ANN network performance testing with unseen healthy and faulty data

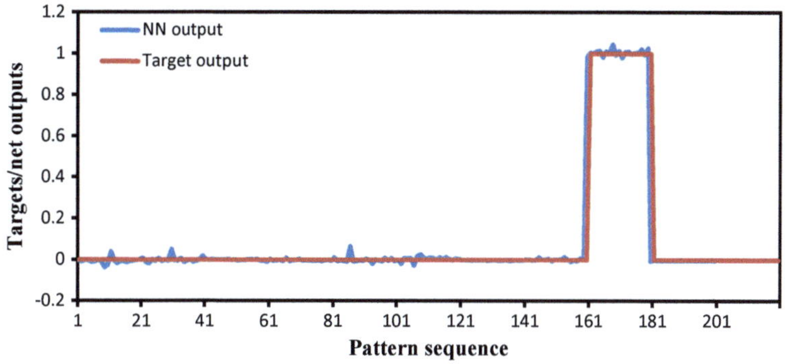

(d) Inner race bearing fault

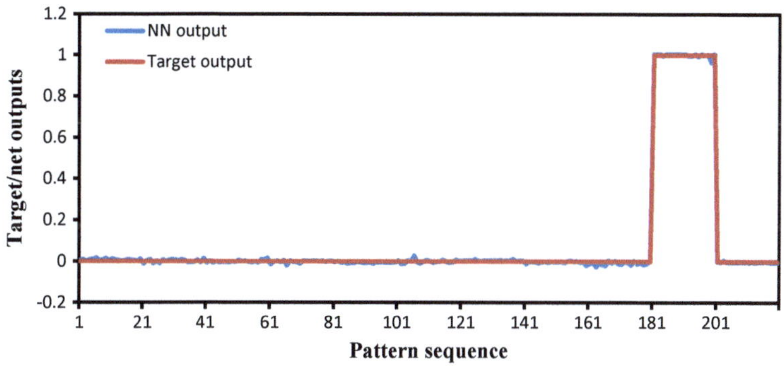

(e) 1 mm hole in the bearing outer race

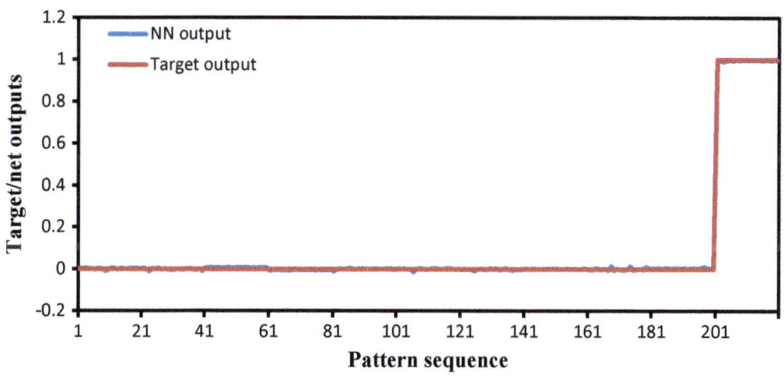

(f) 2 mm hole in the bearing outer race

Fig. D.1 (continued)

MIX
Papier aus verantwortungsvollen Quellen
Paper from responsible sources
FSC® C105338

If you have any concerns about our products,
you can contact us on
ProductSafety@springernature.com

In case Publisher is established outside the EU,
the EU authorized representative is:
Springer Nature Customer Service Center GmbH
Europaplatz 3, 69115 Heidelberg, Germany

Printed by Libri Plureos GmbH
in Hamburg, Germany